Essential Circuit Analysis using NI Multisim™ and MATLAB®

T0234115

Farzin Asadi

Essential Circuit Analysis using NI Multisim™ and MATLAB®

 Springer

Farzin Asadi
Department of Electrical and Electronics Engineering
Maltepe University
Istanbul, Turkey

ISBN 978-3-030-89852-6   ISBN 978-3-030-89850-2   (eBook)
https://doi.org/10.1007/978-3-030-89850-2

© The Editor(s) (if applicable) and The Author(s), under exclusive license to Springer Nature Switzerland AG 2022

This work is subject to copyright. All rights are solely and exclusively licensed by the Publisher, whether the whole or part of the material is concerned, specifically the rights of translation, reprinting, reuse of illustrations, recitation, broadcasting, reproduction on microfilms or in any other physical way, and transmission or information storage and retrieval, electronic adaptation, computer software, or by similar or dissimilar methodology now known or hereafter developed.

The use of general descriptive names, registered names, trademarks, service marks, etc. in this publication does not imply, even in the absence of a specific statement, that such names are exempt from the relevant protective laws and regulations and therefore free for general use.

The publisher, the authors and the editors are safe to assume that the advice and information in this book are believed to be true and accurate at the date of publication. Neither the publisher nor the authors or the editors give a warranty, expressed or implied, with respect to the material contained herein or for any errors or omissions that may have been made. The publisher remains neutral with regard to jurisdictional claims in published maps and institutional affiliations.

This Springer imprint is published by the registered company Springer Nature Switzerland AG
The registered company address is: Gewerbestrasse 11, 6330 Cham, Switzerland

In loving memory of my father Moloud Asadi
and my mother Khorshid Tahmasebi,
always on my mind, forever in my heart.

Preface

A computer simulation is an attempt to model a real-life or hypothetical situation on a computer so that it can be studied to see how the system works. By changing variables in the simulation, predictions may be made about the behavior of the system. So, computer simulation is a tool to virtually investigate the behavior of the system under study.

Computer simulation has many applications in science, engineering, education, and even entertainment. For instance, pilots use computer simulations to practice what they learned without any danger and loss of life.

A circuit simulator is a computer program which permits us to see the circuit behavior, that is, circuit voltages and currents, without making it. Use of circuit simulator is a cheap, efficient, and safe way to study the behavior of circuits. A circuit simulator even saves your time and energy. It permits you to test your ideas before you go wasting all that time building it with a breadboard or hardware, just to find out it doesn't really work.

This book shows how a circuit can be simulated in NI Multisim™ environment. Multisim is a very powerful schematic capture and circuit simulation software for analog, digital, and power electronics in education and research. Multisim was originally created by a company named Electronics Workbench Group, which is now a division of National Instruments (NI). Multisim integrates industry-standard SPICE simulation with an interactive schematic environment to instantly visualize and analyze electronic circuit behavior.

This book contains 89 sample simulations. All the details of studied examples are shown, so you can follow and redo them easily. It is recommended to do some hand calculations for the given examples and compare it with the simulation result. Try to find the source of discrepancy if hand analysis and simulation results are not the same. This helps you to learn the concepts deeply. For instance, the forward voltage drop of diodes is neglected in hand analysis of diode rectifiers. So, there is a little bit difference between hand analysis result (which ignores the voltage drop of diodes) and simulation result (which considers the voltage drop of diodes).

The main audiences of this book are students of engineering (for instance, electrical, biomedical, mechatronics, and robotic, to name a few) and engineers who want to learn the art of circuit simulation with Multisim. By ignoring the theoretical discussions behind the circuits, anyone who is interested in circuits can use the book to learn how to analyze a circuit with the aid of computer.

This book is composed of five chapters and one appendix. A brief summary of book chapters and appendix is as follows:

Chapter 1 introduces MATLAB® and its common daily uses for electrical engineers. MATLAB is used widely in this book to do the required calculations.

Chapter 2 introduces Multisim and shows how it can be used to analyze electric circuits. Students who take/took electric circuits I/II course can use this chapter as a reference to learn how to solve an electric circuit problem with the aid of computer. This chapter has 44 sample simulations.

Chapter 3 focuses on the simulation of electronic circuits (i.e., circuits which contain diode, transistor, and ICs) with Multisim. Students who take/took electronic I/II/III course can use this chapter as a reference to learn how to analyze an electronic circuit with the aid of computer. This chapter has 27 sample simulations.

Chapter 4 focuses on the simulation of digital circuits with Multisim. Students who take/took digital design course can use this chapter as a reference to learn how to simulate a digital circuit on computer. This chapter has eight sample simulations.

Chapter 5 focuses on the simulation of power electronics circuits with Multisim. Students who take/took power electronics/industrial electronics course can use this chapter as a reference to learn how to simulate a power electronics circuit with the aid of computer. This chapter has 10 sample simulations.

Appendix reviews some of the important theoretical concepts used in the book.

I hope that this book will be useful to the readers, and I welcome comments on the book.

Istanbul, Turkey   Farzin Asadi

Contents

Chapter 1
Essential of MATLAB®

1.1 Introduction

MATLAB® (an abbreviation of MATrix LABoratory) is a programming platform designed specifically for engineers and scientists to analyze and design systems and products that transform our world. The heart of MATLAB is the MATLAB language, a matrix-based language allowing the most natural expression of computational mathematics. Keywords of MATLAB language are shown in Table 1.1.

Table 1.1 Keywords of MATLAB language

break	case	catch	classdef
continue	else	elseif	end
for	function	global	if
otherwise	parfor	persistent	return
spmd	switch	try	while

Millions of engineers and scientists worldwide use MATLAB for a range of applications, in industry and academia. MATLAB is both an analysis and design tool. This chapter introduces the MATLAB and its common daily uses for electrical engineers.

1.2 MATLAB Environment

The MATLAB environment is shown in Fig. 1.1. It is composed of four parts. First part is a collection of icons that are required frequently. The second section (current folder browser) enables you to interactively manage files and folders in MATLAB. Use the current folder browser to view, create, open, move, and rename files and folders in the current folder. Use the icon shown in Fig. 1.2 to open the desired

© The Author(s), under exclusive license to Springer Nature Switzerland AG 2022
F. Asadi, *Essential Circuit Analysis using NI Multisim™ and MATLAB®*,
https://doi.org/10.1007/978-3-030-89850-2_1

folder. The third section of MATLAB is command window. The MATLAB commands are entered here and their results are shown here. You need to press Enter key of your key board to run the commands written in command window. The fourth section of MATLAB (the workspace) contains variables that you create within or import into MATLAB from data files or other programs.

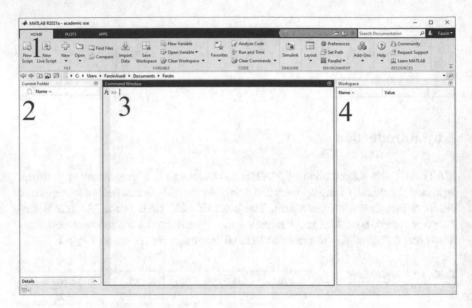

Fig. 1.1 MATLAB environment

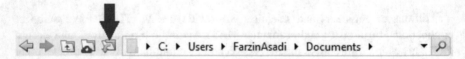

Fig. 1.2 Brows for folder icon

The layout of MATLAB can be changed using the layout icon (Fig. 1.3). This book uses the three column layout.

Fig. 1.3 Layout icon

1.3   Basic Operation with MATLAB

Let's define a complex variable a with value of 1 + 2i. This can be done with the aid of commands shown in Figs. 1.4, 1.5 or 1.6. In MATLAB, both of i and j can be used to enter the imaginary part of a complex number.

Fig. 1.4 Defining variable a

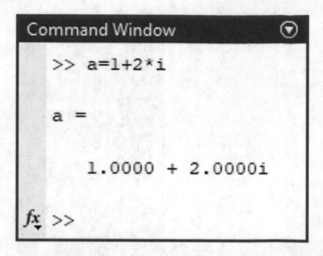

Fig. 1.5 Defining variable a

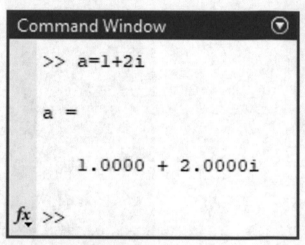

Fig. 1.6 Defining variable a

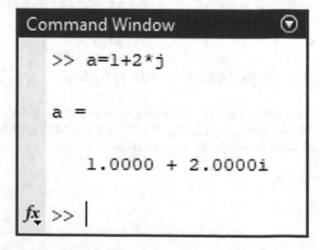

After running the commands of Figs. 1.4, 1.5 or 1.6, a new variable will be added to the workspace window (Fig. 1.7).

Fig. 1.7 Variable a is added to Workspace

Workspace	
Name ▲	Value
⊞ a	1.0000 + 2.0000i

The result of commands will not be shown if you put a semicolon at the end of commands (Fig. 1.8).

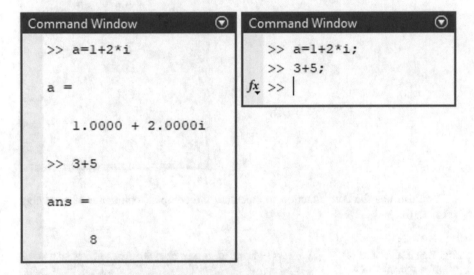

```
Command Window                    ⊙
   >> a=1+2*i

   a =

      1.0000 + 2.0000i

   >> 3+5

   ans =

      8
```

```
Command Window                    ⊙
      >> a=1+2*i;
      >> 3+5;
fx >> |
```

Fig. 1.8 Semicolon hides the results of commands

You can use the real and imag function to obtain the real and imaginary parts of a complex number (Fig. 1.9).

Fig. 1.9 Calculation of real
and imaginary parts with
real and imag commands

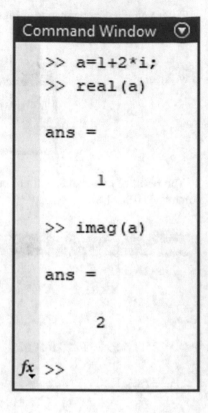

You can use the conj function to calculate the complex conjugate of a number
(Fig. 1.10).

Fig. 1.10 Calculation of
complex conjugate of a
number with conj command

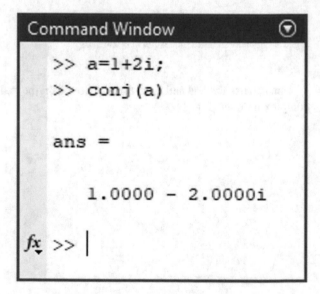

The point a = 1 + 2i is shown in Fig. 1.11. The polar form this point is $\sqrt{1^2 + 2^2}\, e^{\,j\tan^{-1}\frac{2}{1}}$. Value of $\sqrt{1^2 + 2^2}$ and $\tan^{-1}\frac{2}{1}$ is calculated with the aid commands shown in Fig. 1.12. You can calculate the magnitude and phase of a complex number with the aid of commands shown in Fig. 1.13 as well.

Fig. 1.11 Representation of 1 + 2i in the complex plane

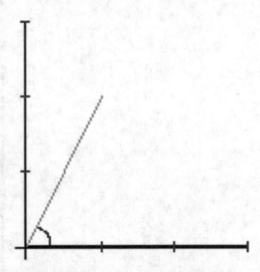

Fig. 1.12 Calculation of magnitude and angle of 1 + 2i

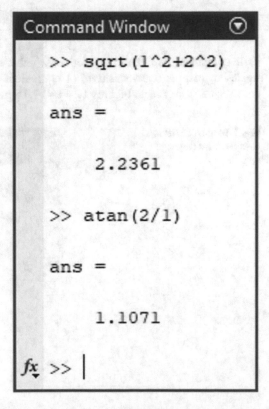

```
Command Window

>> sqrt(1^2+2^2)

ans =

    2.2361

>> atan(2/1)

ans =

    1.1071

fx >> |
```

Fig. 1.13 Calculation of magnitude and angle of 1 + 2i

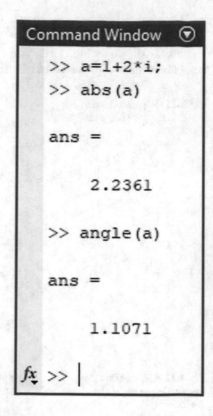

In computers, case sensitivity defines whether uppercase and lowercase letters are treated as distinct (case-sensitive) or equivalent (case-insensitive). MATLAB is a case sensitive language. Figures 1.14 and 1.15 proves this.

Fig. 1.14 MATLAB is a case sensitive language

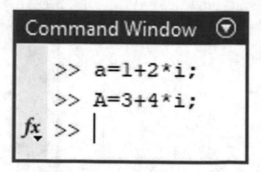

Fig. 1.15 a and A are two different variables

Workspace	⊙
Name ▲	Value
▦ a	1.0000 + 2.0000i
▦ A	3.0000 + 4.0000i

   MATLAB has a default variable named ans. This variable is used to save the results of commands when you don't define a variable. For instance, in Fig. 1.16 the result of multiplication will be put in the variable c however in Fig. 1.17, ans will be used to save the result of multiplication since the user didn't defined any variable.

Fig. 1.16 Variable c equals to product of a and b

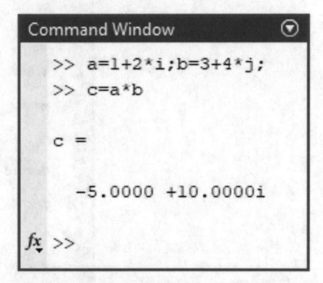

```
>> a=1+2*i;b=3+4*j;
>> c=a*b

c =

   -5.0000 +10.0000i
```

Fig. 1.17 Default variable ans saves the calculation result

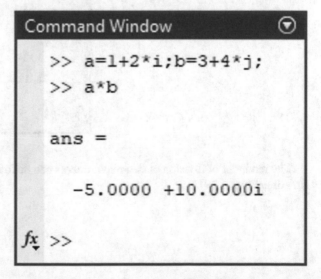

```
>> a=1+2*i;b=3+4*j;
>> a*b

ans =

   -5.0000 +10.0000i
```

Commands in Fig. 1.18 do some basic operations on two complex numbers. The percent symbol (%) is used for indicating a comment line.

Fig. 1.18 Basic operations in MATLAB

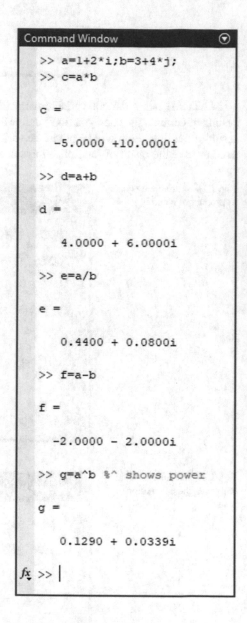

The reminder of division of two real numbers can be found using the mod or rem functions (Fig. 1.19).

Fig. 1.19 rem and mod
commands

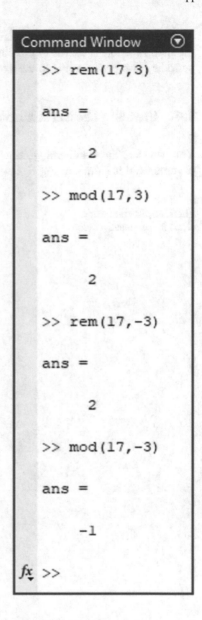

```
Command Window      ⊙

>> rem(17,3)

ans =

        2

>> mod(17,3)

ans =

        2

>> rem(17,-3)

ans =

        2

>> mod(17,-3)

ans =

       -1

fx >>
```

The concept of remainder after division is not uniquely defined, and the two functions mod and rem each compute a different variation. The mod function produces a result that is either zero or has the same sign as the divisor. The rem function produces a result that is either zero or has the same sign as the dividend.

Another difference is the convention when the divisor is zero. The mod function follows the convention that mod(a,0) returns a, whereas the rem function follows the convention that rem(a,0) returns NaN. NaN stands for "Not a Number" and is used to represents values that are not real or complex numbers. Expressions like 0/0 and inf/inf result in NaN.

Both of mod and rem functions have their uses. For example, in signal processing, the mod function is useful in the context of periodic signals because its output is periodic (with period equal to the divisor).

1.4   Clearing the Screen and Variables

You can clear the command window using the clc command (Fig. 1.20). Result of clc command is shown in Fig. 1.21.

Fig. 1.20 clc commands
clears the command window

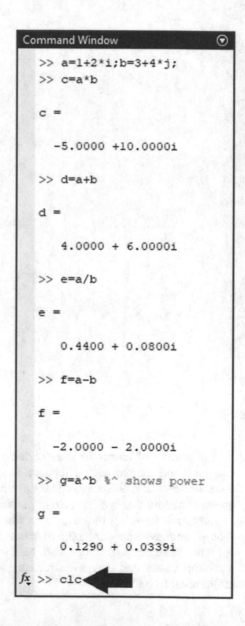

Fig. 1.21 Commands window is cleared with clc command

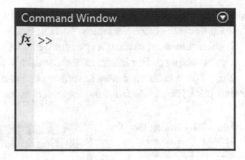

There is another way to clear the command window as well. You can right click on the command window and click the clear command window in the appeared window (Fig. 1.22).

Fig. 1.22 Clicking the clear command window is another way to clear the command window

Evaluate Selection	F9
Open Selection	Ctrl+D
Help on Selection	F1
Function Browser	Shift+F1
Show Function Browser Button	
Function Hints	Ctrl+F1
Cut	Ctrl+X
Copy	Ctrl+C
Paste	Ctrl+V
Select All	Ctrl+A
Find...	Ctrl+F
Print...	Ctrl+P
Print Selection...	
Page Setup...	
Clear Command Window	

Clearing the command window does not affect the variables (Fig. 1.23). If you want to clear all the variables, you need to use the command shown in Figs. 1.24 or 1.25. If you want to clear a specific variable, then you need to write its name after the clear command. For instance, the command shown in Fig. 1.26, clears the variable g only. There is another way to remove the variable g as well. You can right click on it and click the delete from the appeared menu (Fig. 1.27).

Fig. 1.23 Clearing the command window has no effect on the variables

Workspace	
Name ▲	Value
a	1.0000 + 2.0000i
b	3.0000 + 4.0000i
c	-5.0000 + 10.0000i
d	4.0000 + 6.0000i
e	0.4400 + 0.0800i
f	-2.0000 - 2.0000i
g	0.1290 + 0.0339i

Fig. 1.24 Deleting all the Workspace variables

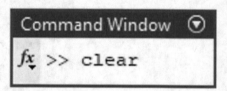

Fig. 1.25 Deleting all the Workspace variables

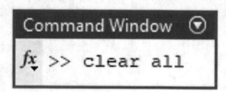

Fig. 1.26 Deleting the variable g of Workspace

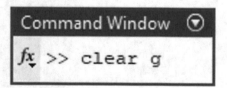

Fig. 1.27 Deleting the
variable g of Workspace

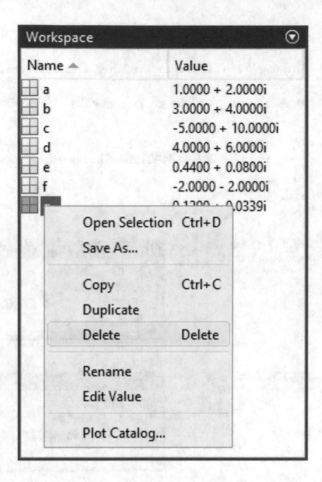

You can use the up and down arrows of the keyboard to access the previous
entered commands (Fig. 1.28).

Fig. 1.28 Accessing to the
previously entered
commands

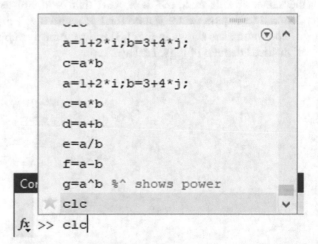

1.5 Basic Matrix Operations

Anything in MATLAB is a matrix. Even a number is considered as a 1×1 matrix.
Assume that you want to enter $A = [1\ 2\ 3], B = \begin{bmatrix} 4 \\ 5 \\ 6 \end{bmatrix}$ and $C = \begin{bmatrix} 7 & 8 & 9 \\ 10 & 11 & 12 + 2i \\ 13 & 14 & 16 \end{bmatrix}$ to MATLAB. The commands shown in Figs. 1.29 or 1.30 do this job.

Fig. 1.29 Defining the matrices A, B and C

```
Command Window                                              ⊙

    >> A=[1 2 3];
    >> B=[4;5;6];
    >> C=[7 8 9;10 11 12+2*i;13 14 16];
fx >> |
```

Fig. 1.30 Defining the matrices A, B and C

```
Command Window                                              ⊙

    >> A=[1,2,3];
    >> B=[4;5;6];
    >> C=[7,8,9;10,11,12+2*i;13,14,16];
fx >> |
```

You can read the element on the i'th row and j'th column of matrix X, by using the command X(i,j). If matrix X is $1 \times n$, then you can use X(i) instead of X(1,i). In the same way, If matrix X is $n \times 1$, then you can use X(i) instead of X(i,1). Generally, the term vector is used for $1 \times n$ and $n \times 1$ matrices.

The commands shown in Fig. 1.31, gives some example for reading the elements of matrices defined in Figs. 1.29 and 1.30.

Fig. 1.31 Accessing the
elements of the matrices

```
Command Window                                    ⊙

  >> A(1,2)

  ans =

        2

  >> A(2)

  ans =

        2

  >> C(1,3)

  ans =

        9

  >> C(2,3)

  ans =

        12.0000 + 2.0000i

fx >> |
```

The commands shown in Fig. 1.32, changes the element in the second row and third column of matrix C to 12 + 12i.

```
Command Window                                    ⊙

  >> C(2,3)=12+12*i;
  >> C

  C =

     7.0000 + 0.0000i    8.0000 + 0.0000i    9.0000 + 0.0000i
    10.0000 + 0.0000i   11.0000 + 0.0000i   12.0000 +12.0000i
    13.0000 + 0.0000i   14.0000 + 0.0000i   16.0000 + 0.0000i

fx >> |
```

Fig. 1.32 Changing the element on the second row and third column of C

You can use the size command to read the number of rows and columns (Fig. 1.33).

Fig. 1.33 Calculation of number of rows and columns of a matrix

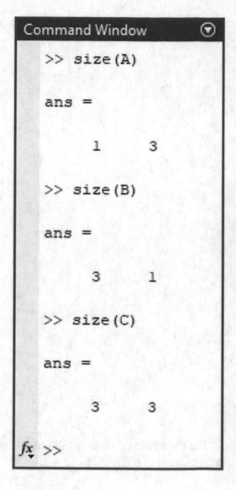

You can save the number of rows and number of columns in variables as well. The commands shown in Fig. 1.34 save the number of rows and columns of matrix A in RowA and ColumnB variables, respectively.

Fig. 1.34 Number of rows and columns of matrix A are saved in RowA and ColumnA, respectively

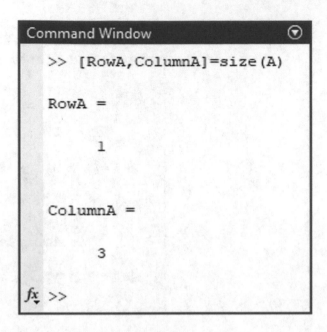

The multiplication, summation and subtraction of two matrices can be done with the aid of *, + and − operators, respectively. Note that the size of two matrices must be the same for summation and subtraction and the number of columns of the first matrix must be the same as the number of rows of the second matrix, otherwise the operation will not be done. MATLAB has elementwise operators as well. The element wise operator does the operation on the corresponding elements. For instance, in Fig. 1.35, [1 2 3].^[4 5 6]=[1^4 2^5 3^6]=[1 32 729] or [1 2 3].* [4 5 6]=[1*4 2*5 3*6]=[4 10 18].

Fig. 1.35 Elementwise
commands

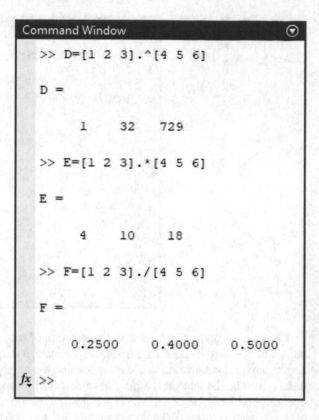

You can find the minimum and maximum of a vector by using the min and max
commands, respectively (Fig. 1.36).

Fig. 1.36 Calculation of
minimum and maximum of
vector A

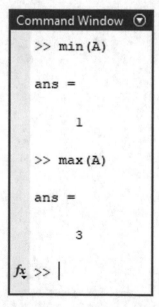

The min/max function can be applied to matrices as well. If the input of these functions is a m×n matrix, they return a n × 1 vector containing the minimum/maximum element from each column (Fig. 1.37). When the matrix is complex, the minimum/maximum is computed using the magnitude (Fig. 1.38). Remember that the magnitude of a complex number $a + bi = \sqrt{a^2 + b^2}$. In Fig. 1.38, the magnitude matrix is $\begin{bmatrix} 7 & 8 & 9 \\ 10 & 11 & 16.971 \\ 13 & 14 & 16 \end{bmatrix}$. So, the max function returns 12 + 12i as the maximum of the third column since 12 + 12i has the maximum magnitude in the third column.

Fig. 1.37 min and max commands can be applied to matrices as well

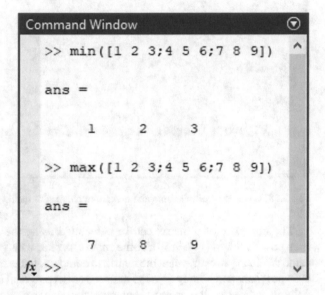

```
>> min([1 2 3;4 5 6;7 8 9])

ans =

       1       2       3

>> max([1 2 3;4 5 6;7 8 9])

ans =

       7       8       9
fx >>
```

```
Command Window                                                        ⊙

  >> C

 C =

    7.0000 + 0.0000i    8.0000 + 0.0000i    9.0000 + 0.0000i
   10.0000 + 0.0000i   11.0000 + 0.0000i   12.0000 +12.0000i
   13.0000 + 0.0000i   14.0000 + 0.0000i   16.0000 + 0.0000i

 >> min(C)

 ans =

        7      8      9

 >> max(C)

 ans =

   13.0000 + 0.0000i   14.0000 + 0.0000i   12.0000 +12.0000i

fx >>
```

Fig. 1.38 Calculation of minimum and maximum element of matrix C

The transpose of a matrix can be calculated using the single quote operator or transpose function (Fig. 1.39). If the matrix is real, then two methods produce the same results. However, when the matrix is complex, the results are not the same. The transpose function simply flips the matrix over its diagonal (i.e. switches the row and column indices of the matrix), however the single quote function calculates the complex conjugate of elements and then change the row and column indices.

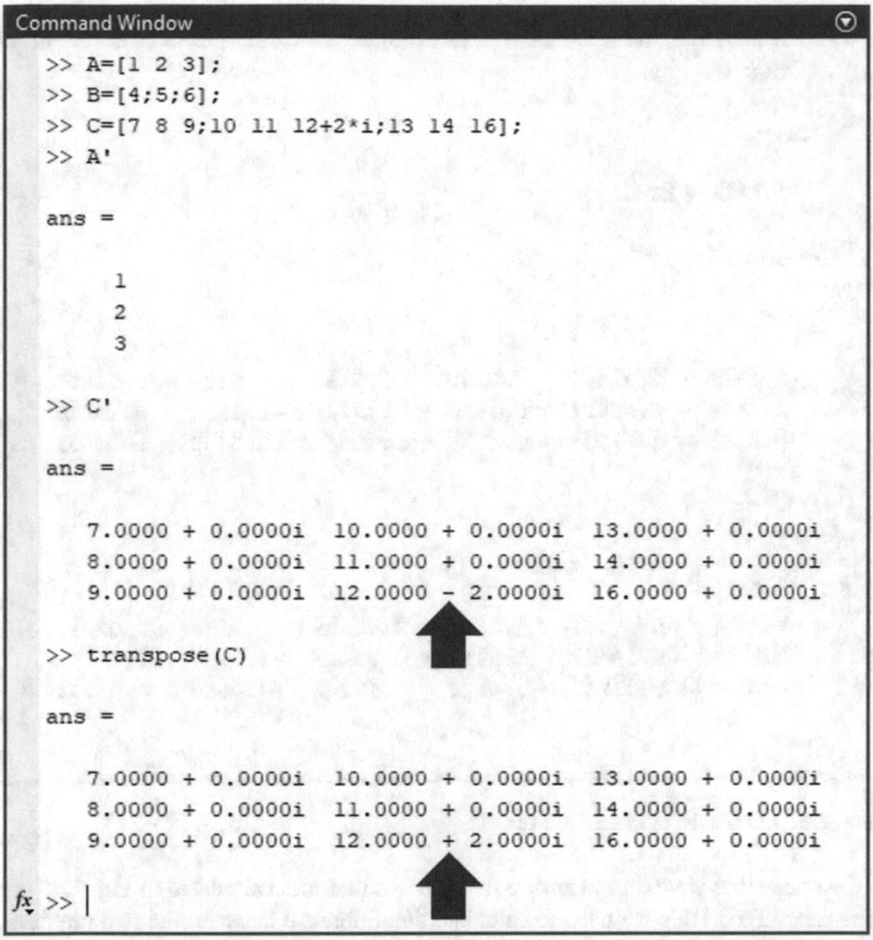

```
Command Window                                                    ⊙
  >> A=[1 2 3];
  >> B=[4;5;6];
  >> C=[7 8 9;10 11 12+2*i;13 14 16];
  >> A'

  ans =


      1
      2
      3

  >> C'

  ans =

      7.0000 + 0.0000i  10.0000 + 0.0000i  13.0000 + 0.0000i
      8.0000 + 0.0000i  11.0000 + 0.0000i  14.0000 + 0.0000i
      9.0000 + 0.0000i  12.0000 - 2.0000i  16.0000 + 0.0000i

  >> transpose(C)

  ans =

      7.0000 + 0.0000i  10.0000 + 0.0000i  13.0000 + 0.0000i
      8.0000 + 0.0000i  11.0000 + 0.0000i  14.0000 + 0.0000i
      9.0000 + 0.0000i  12.0000 + 2.0000i  16.0000 + 0.0000i

fx >> |
```

Fig. 1.39 Different methods of calculation of transpose of a matrix

You can use the det function to calculate the determinant of square matrix. The inverse of a matrix X can be calculated using the inv(X) or X^-1 commands.

```
Command Window                                              ⊙

 >> det(C)

ans =

  -3.0000 +12.0000i

 >> inv(C)

ans =

  -2.3529 - 0.0784i    0.0392 + 0.1569i    1.3137 - 0.0784i
   2.1176 - 0.1961i    0.0980 + 0.3922i   -1.2157 - 0.1961i
   0.0588 + 0.2353i   -0.1176 - 0.4706i    0.0588 + 0.2353i

 >> C^-1

ans =

  -2.3529 - 0.0784i    0.0392 + 0.1569i    1.3137 - 0.0784i
   2.1176 - 0.1961i    0.0980 + 0.3922i   -1.2157 - 0.1961i
   0.0588 + 0.2353i   -0.1176 - 0.4706i    0.0588 + 0.2353i

fx >>
```

Fig. 1.40 Calculation of inverse of matrix C

Assume that you want to ensure that the two matrices calculated in Fig. 1.40 are the same. To do this we need to calculate the difference between the two matrices. Figure 1.41 shows that the obtained results are the same.

Fig. 1.41 Calculation of difference between the C^-1 and inv(C) commands

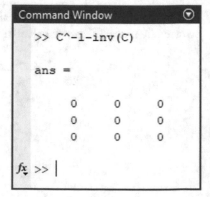

```
Command Window                          ⊙

 >> C^-1-inv(C)

ans =

        0        0        0
        0        0        0
        0        0        0

fx >> |
```

Assume that you want to solve the $\begin{bmatrix} 7 & 8 & 9 \\ 10 & 11 & 12+2i \\ 13 & 14 & 16 \end{bmatrix} x = \begin{bmatrix} 4 \\ 5 \\ 6 \end{bmatrix}$. The two commands shown in Fig. 1.42 can be used to calculate the vector x.

Fig. 1.42 Calculation of vector x

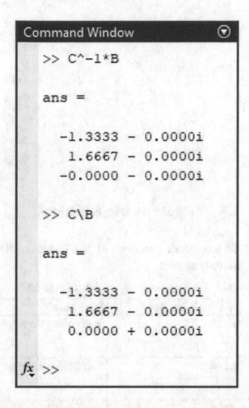

The Fig. 1.43 shows that the difference between the result of commands shown in Fig. 1.42. Note that 1e-15 mean 10^{-15}. So, the difference between the results is very small.

Fig. 1.43 Calculation of
difference between the C\B
and C^-1*B commands

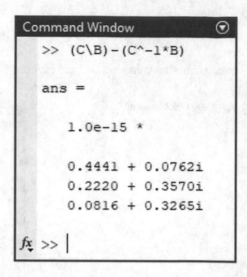

1.6   Trigonometric Functions

Trigonometric functions of MATLAB are listed in Table 1.2. x can be a complex
number as well.

Table 1.2 Trigonometric functions of MATLAB

MATLAB function	Description	Example
sin	$\sin(x)$. Takes the input in radians.	>>sin(pi/2) ans = 1
sind	$\sin(x)$. Takes the input in degrees.	>>sind(30) ans = 0.5000
asin	$\sin^{-1}(x)$. Returns the output in radians.	>>asin(0.5) ans = 0.5236
asind	$\sin^{-1}(x)$. Returns the output in degrees.	>>asind(.5) ans = 30.0000
cos	$\cos(x)$. Takes the input in radians.	>>cos(1) ans = 0.5403
cosd	$\cos(x)$. Takes the input in degrees.	>>cosd(60) ans = 0.5000
acos	$\cos^{-1}(x)$. Returns the output in radians.	>>acos(.5) ans = 1.0472
acosd	$\cos^{-1}(x)$. Returns the output in degrees.	>>acosd(.5) ans = 60.0000

(continued)

Table 1.2 (continued)

MATLAB function	Description	Example
tan	$\tan(x)$. Takes the input in radians.	>>tan(1) ans = 1.5574
tand	$\tan(x)$. Takes the input in degrees.	>>tand(60) ans = 1.7321
atan	$\tan^{-1}(x)$. Returns the output in radians.	>>atan(1) ans = 0.7854
atand	$\tan^{-1}(x)$. Returns the output in degrees.	>>atand(1) ans = 45
sec	$\sec(x)$. Takes the input in radians.	>>sec(1) ans = 1.8508
secd	$\sec(x)$. Takes the input in degrees.	>>secd(10) ans = 1.0154
asec	$\sec^{-1}(x)$. Returns the output in radians.	>>asec(30) ans = 1.5375
asecd	$\sec^{-1}(x)$. Returns the output in degrees.	>>asecd(30) ans = 88.0898
csc	$\csc(x)$. Takes the input in radians.	>>csc(1) ans = 1.1884
cscd	$\csc(x)$. Takes the input in degrees.	>>cscd(1) ans = 57.2987
acsc	$\csc^{-1}(x)$. Returns the output in radians.	>>acsc(1) ans = 1.5708
acscd	$\csc^{-1}(x)$. Returns the output in degrees.	>>acscd(1) ans = 90
cot	$\cot(x)$. Takes the input in radians.	>>cot(1) ans = 0.6421
cotd	$\cot(x)$. Takes the input in degrees.	>>cotd(30) ans = 1.7321
acot	$\cot^{-1}(x)$. Returns the output in radians.	>>acot(1) ans = 0.7854
acotd	$\cot^{-1}(x)$. Returns the output in degrees.	>>acotd(1) ans = 45

For instance, assume that we want to calculate for $\sin(x)^4 - 3\cos(x)$ for $x = 30°$. The commands shown in Figs. 1.44 or 1.45 do this.

Fig. 1.44 Calculation of sin $(x)^4 - 3\cos(x)$ for $x = 30°$

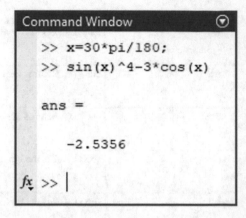

Fig. 1.45 Calculation of sin $(x)^4 - 3\cos(x)$ for $x = 30°$

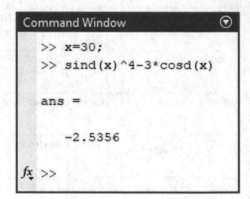

1.7   Hyperbolic Functions

Hyperbolic functions of MATLAB are listed in Table 1.3. x can be a complex number as well.

Table 1.3 Hyperbolic functions of MATLAB

Function	MATLAB representation
$\sinh(x)$	`sinh(x)`
$\cosh(x)$	`cosh(x)`
$\tanh(x)$	`tanh(x)`
$\coth(x)$	`coth(x)`
$\text{sech}(x)$	`sech(x)`
$\text{csch}^{-1}(x)$	`csch(x)`
$\sinh^{-1}(x)$	`asinh(x)`
$\cosh^{-1}(x)$	`acosh(x)`
$\tanh^{-1}(x)$	`atanh(x)`
$\coth^{-1}(x)$	`acoth(x)`
$\text{sech}^{-1}(x)$	`asech(x)`
$\text{csch}^{-1}(x)$	`acsch(x)`

1.8   Logarithmic and Exponential Function

Logarithmic and exponential functions of MATLAB are listed in Table 1.4. x can be a complex number as well.

Table 1.4 Logarithmic and exponential functions of MATLAB

MATLAB function	Description	Example

`exp`	Calculates e^x.	`>>exp(2)`
		 `ans =`
		 `7.3891`

`log`	Calculates the natural logarithm (with base of $e = 2.7182$).	`>>log`
		 `(2.7182)`
		 `ans =`
		 `1.0000`

`log10`	Calculates the common logarithm (with base of 10).	`>>log10`
		 `(100)`
		 `ans =`
		 `2`

`log2`	Calculates the binary logarithm with (with base of 2).	`>>log2(4)`
		 `ans =`
		 `2`

`sqrt`	Calculates the square root.	`>>sqrt`
		 `(16)`
		 `ans =`
		 `4`

`power`	Calculates the power function. Same as elementwise power operator (Fig. 1.46).	`>>power`
		 `(2,3)`
		 `ans =`
		 `8`

Fig. 1.46 Result of power
command

1.9 Rounding Functions

Rounding functions of MATLAB are shown in Table 1.5.

Table 1.5 Rounding function of MATLAB

MATLAB function	Expression	Example
fix	Rounds toward zero.	fix([-4.6 4.6]) ans = [-4 4]
floor	Round to negative infinity.	>>floor(8.4687) ans = 8
ceil	Round to positive infinity.	>>ceil(4.4) ans = 5
round	Round to nearest decimal or integer.	>>round(4.55) ans = 5

1.10 Colon Operator

The colon operator can be used for creating regularly spaced vectors, index into arrays, and define the bounds of a for loop. For instance, assume that you want to make a vector t from 0 to 1 with 0.1 s steps. The command shown in Fig. 1.47 makes the vecotor.

Fig. 1.47 Defining the
vector t with colon operator

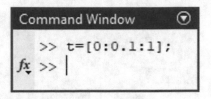

Values of vector t is shown in Fig. 1.48.

```
Command Window
>> t

t =

   Columns 1 through 10

        0    0.1000    0.2000    0.3000    0.4000    0.5000    0.6000    0.7000    0.8000    0.9000

   Column 11

   1.0000

fx >>
```

Fig. 1.48 Values of vector t defined in Fig. 1.47

The default step of colon operator equals to one. So, x = [3:10] and x = [3:1:10]
commands are the same, both of these commands define x = [3 4 5 6 7 8 9 10]. The
command shown in Fig. 1.49 defines a random matrix with 8 rows and 6 columns.
The A(5:8,:) command in Fig. 1.49, selects the elements in the 5th, 6th, 7th and 8th
columns.

```
Command Window                                                    ⊙

  >> A=rand(8,6)

 A =

      0.9619      0.8001      0.5797      0.0760      0.9448      0.3897
      0.0046      0.4314      0.5499      0.2399      0.4909      0.2417
      0.7749      0.9106      0.1450      0.1233      0.4893      0.4039
      0.8173      0.1818      0.8530      0.1839      0.3377      0.0965
      0.8687      0.2638      0.6221      0.2400      0.9001      0.1320
      0.0844      0.1455      0.3510      0.4173      0.3692      0.9421
      0.3998      0.1361      0.5132      0.0497      0.1112      0.9561
      0.2599      0.8693      0.4018      0.9027      0.7803      0.5752

  >> A(5:8,:)

 ans =

      0.8687      0.2638      0.6221      0.2400      0.9001      0.1320
      0.0844      0.1455      0.3510      0.4173      0.3692      0.9421
      0.3998      0.1361      0.5132      0.0497      0.1112      0.9561
      0.2599      0.8693      0.4018      0.9027      0.7803      0.5752

fx >> |
```

Fig. 1.49 Selection of 5th, 6th, 7th and 8th row of matrix A

The command A(:,4:6) in Fig. 1.50 selects all the elements in the 4th, 5th, 6th columns.

```
Command Window                                                              ⊙
  >> A=rand(8,6)


A =

      0.9619      0.8001      0.5797      0.0760      0.9448      0.3897
      0.0046      0.4314      0.5499      0.2399      0.4909      0.2417
      0.7749      0.9106      0.1450      0.1233      0.4893      0.4039
      0.8173      0.1818      0.8530      0.1839      0.3377      0.0965
      0.8687      0.2638      0.6221      0.2400      0.9001      0.1320
      0.0844      0.1455      0.3510      0.4173      0.3692      0.9421
      0.3998      0.1361      0.5132      0.0497      0.1112      0.9561
      0.2599      0.8693      0.4018      0.9027      0.7803      0.5752

  >> A(:,4:6)


ans =

      0.0760      0.9448      0.3897
      0.2399      0.4909      0.2417
      0.1233      0.4893      0.4039
      0.1839      0.3377      0.0965
      0.2400      0.9001      0.1320
      0.4173      0.3692      0.9421
      0.0497      0.1112      0.9561
      0.9027      0.7803      0.5752

fx >> |
```

Fig. 1.50 Selection of 4th, 5th and 6th column of matrix A

The command A(2:5,3:4) in Fig. 1.51 selects the elements A_{ij} where $2 \leq i \leq 5$ and $3 \leq j \leq 4$.

```
Command Window                                                    ⊙
  >> A=rand(8,6)                                                  ^

  A =

        0.9619     0.8001     0.5797     0.0760     0.9448     0.3897
        0.0046     0.4314     0.5499     0.2399     0.4909     0.2417
        0.7749     0.9106     0.1450     0.1233     0.4893     0.4039
        0.8173     0.1818     0.8530     0.1839     0.3377     0.0965
        0.8687     0.2638     0.6221     0.2400     0.9001     0.1320
        0.0844     0.1455     0.3510     0.4173     0.3692     0.9421
        0.3998     0.1361     0.5132     0.0497     0.1112     0.9561
        0.2599     0.8693     0.4018     0.9027     0.7803     0.5752

  >> A(2:5,3:4)

  ans =

        0.5499     0.2399
        0.1450     0.1233
        0.8530     0.1839
        0.6221     0.2400
fx >> |                                                          ∨
```

Fig. 1.51 Selection of A_{ij} where $2 \leq i \leq 5$ and $3 \leq j \leq 4$

The colon operator is used in for loops as well. For instance, the code in Fig. 1.52 counts the number of elements which are bigger or equal than 0.5.

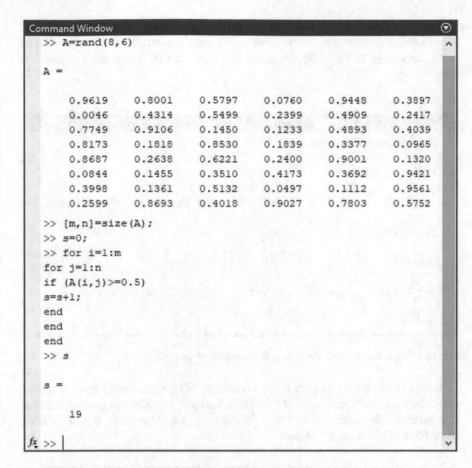

Fig. 1.52 Counting the number of elements which are equal or bigger than 0.5

1.11 Linspace and Logspace Commands

The linspace(X1, X2, N) generates N linearly spaced points between X1 and X2. MATLAB assumes N = 100 when it is not entered. For instance, the command in Fig. 1.53 generates a linear space from 1 to 10 which has 6 members. We expect the difference between two consecutive elements to be $\frac{10-1}{6-1} = 1.8$.

```
Command Window                                            ⊙
  >> X=linspace(1,10,6)

  X =

        1.0000    2.8000    4.6000    6.4000    8.2000   10.0000

fx >>
```

Fig. 1.53 Defining the variable X with linspace command

The difference between two consecutive elements can be calculated using the diff command (diff command can be used to calculate the derivative of a function as well). According to Fig. 1.54, the difference between two consecutive elements is 1.8.

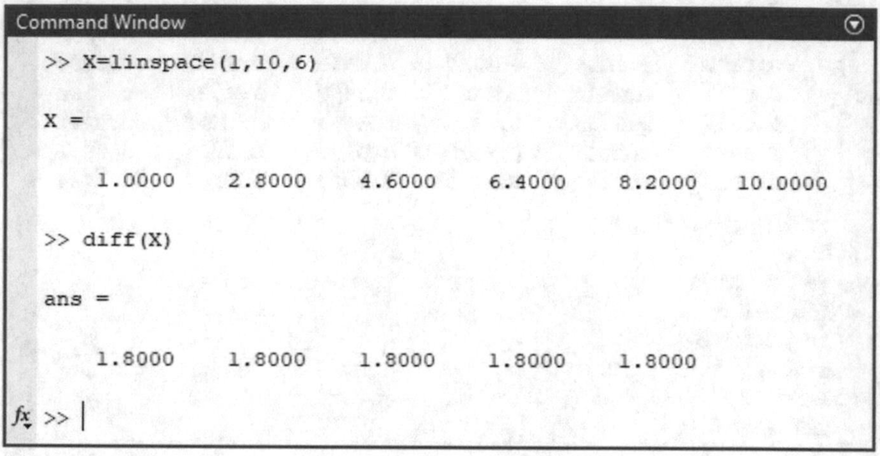

```
Command Window                                                    ⊙
  >> X=linspace(1,10,6)

  X =

       1.0000    2.8000    4.6000    6.4000    8.2000   10.0000

  >> diff(X)

  ans =

       1.8000    1.8000    1.8000    1.8000    1.8000

fx >> |
```

Fig. 1.54 Difference between the consecutive elements of vector X

logspace(X1, X2,N) generates a row vector of N logarithmically equally spaced points between 10^{X1} and 10^{X2}. MATLAB assumes $N = 100$ when it is not entered. For instance, the command in Fig. 1.55, makes a logarithmically equally spaced from 100 to 1000 with 5 members.

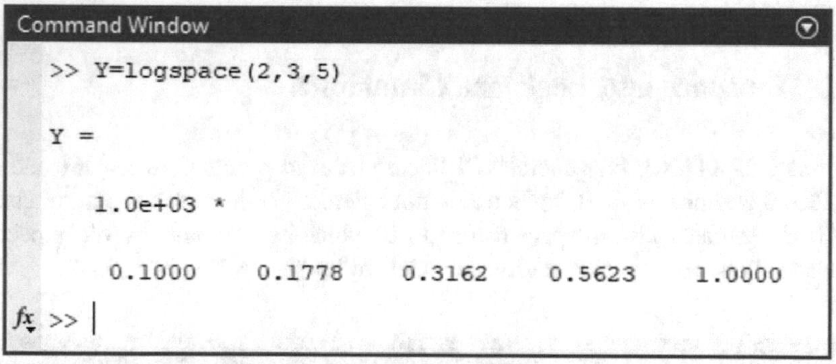

```
Command Window                                                    ⊙
  >> Y=logspace(2,3,5)

  Y =

     1.0e+03 *

       0.1000    0.1778    0.3162    0.5623    1.0000

fx >> |
```

Fig. 1.55 Defining the variable Y with logspace command

The ratio between any two consecutive elements are constant (Fig. 1.56). This ratio can be calculated from the $10^2 \times x^{5-1} = 10^3$ equation. Fig. 1.57 solves this equation. According to Fig. 1.57, the only positive answer of this equation is 1.7783.

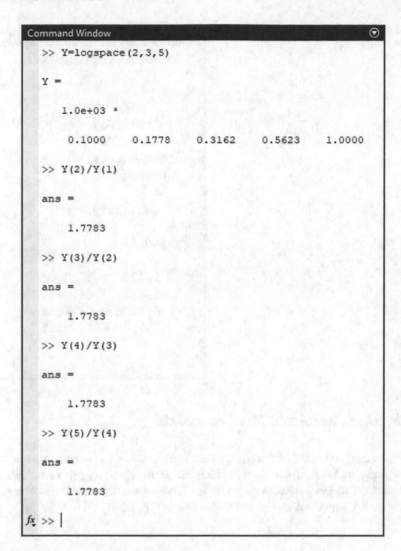

Fig. 1.56 Ratio between the consecutive elements of vector Y

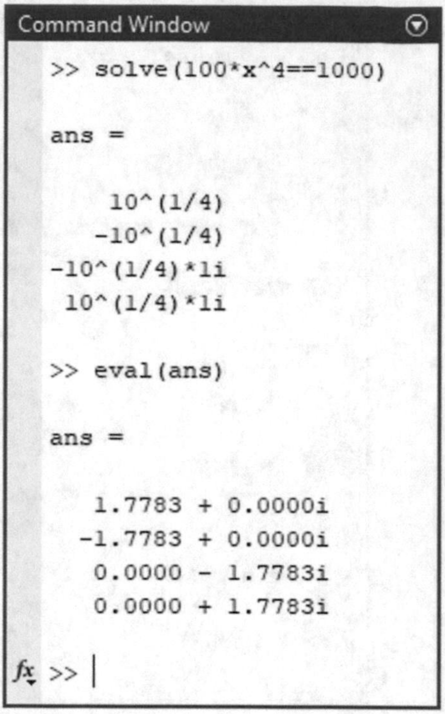

Fig. 1.57 Solution of
$10^2 \times x^{5-1} = 10^3$

```
Command Window

>> solve(100*x^4==1000)

ans =

    10^(1/4)
   -10^(1/4)
 -10^(1/4)*1i
  10^(1/4)*1i

>> eval(ans)

ans =

   1.7783 + 0.0000i
  -1.7783 + 0.0000i
   0.0000 - 1.7783i
   0.0000 + 1.7783i

fx >> |
```

1.12   Ones, Zeros and Eye Commands

The ones(m,n) command makes a m×n matrix which all of its elements are one. The zeros (m,n) makes a m×n matrix which all of its elements are zero. The eye (n) command makes a n×n identity matrix. The eye(m,n) command makes a m×n matrix with 1's on the diagonal and 0 elsewhere (Fig. 1.58).

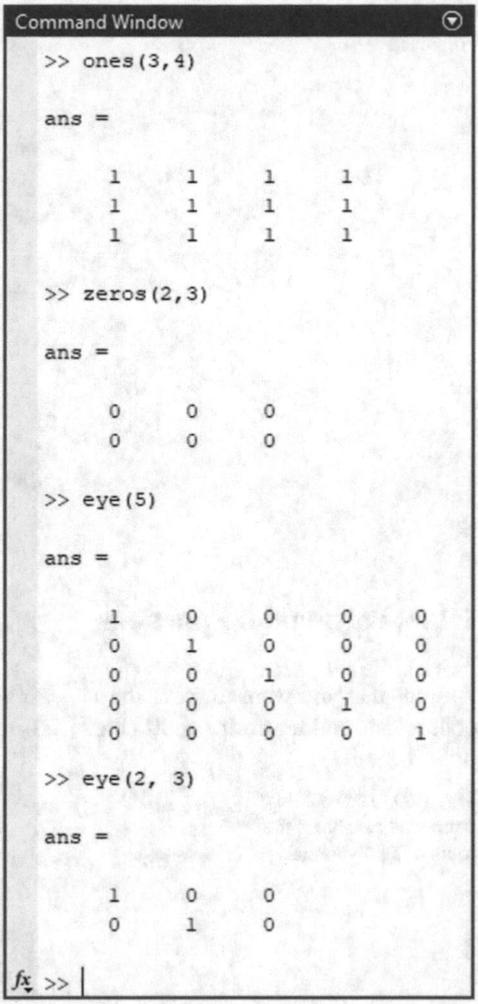

Fig. 1.58 Example for
ones, zeros and eye
commands

```
Command Window                                    ⊙

  >> ones(3,4)

  ans =

        1     1     1     1
        1     1     1     1
        1     1     1     1

  >> zeros(2,3)

  ans =

        0     0     0
        0     0     0

  >> eye(5)

  ans =

        1     0     0     0     0
        0     1     0     0     0
        0     0     1     0     0
        0     0     0     1     0
        0     0     0     0     1

  >> eye(2, 3)

  ans =

        1     0     0
        0     1     0

 fx >> |
```

1.13   Format Command

The format command permits you to set the output display format for command
window (Fig. 1.59).

Fig. 1.59 The format command

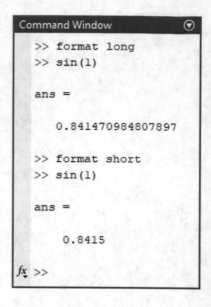

1.14   Polynomial Functions

Assume that you want to solve the $x^2 - 4x + 3 = 0$. The roots command takes the coefficients and return the roots (Fig. 1.60).

Fig. 1.60 The roots command calculates the roots of a polynomial

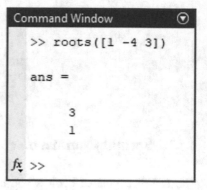

Assume that you want to find the polynomial which its roots are $-1+2i$, $-1-2i$ and 6. The poly function gives the coefficients of the polynomial which has that roots. According to Fig. 1.61, the polynomial is $p(x) = x^3 - 4x^2 - 7x - 30$.

Fig. 1.61 poly command
takes the roots of a
polynomial function and
gives its coefficients

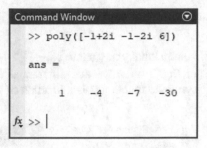

Assume that we want to calculate the value of $p(x) = x^3 - 4x^2 - 7x - 30$ at $x = 5$.
The polyval function can be used for this purpose (Fig. 1.62).

Fig. 1.62 polyval
command calculates the
value of a polynomial
function at a point

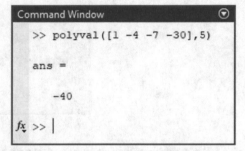

Assume that you want to calculate the product of $p_1(x) = x + 3$ and
$p_2(x) = x^2 + 5x + 6$. The commands shown in Fig. 1.63 calculates the product of
these two polynomials. According to Fig. 1.63, the product is
$p(x) = x^3 + 8x^2 + 21x + 18$.

Fig. 1.63 Calculation of
product of two polynomials
with the conv command

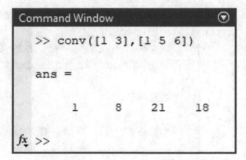

1.15   Solving the Ordinary Differential Equations

Assume that you want to solve $\frac{dy(t)}{dt} + 4y(t) = e^{-t}, y(0) = 1$. The commands shown in Fig. 1.64 solve this differential equation. Note that the first line defines the symbolic function y as a function of symbolic variable t.

Fig. 1.64 Solution of $\frac{dy(t)}{dt} + 4y(t) = e^{-t}, y(0) = 1$

```
Command Window                                              ⊙

    >> syms y(t)
    >> ode = diff(y)+4*y == exp(-t);
    >> cond = y(0) == 1;
    >> ySol(t) = dsolve(ode,cond)


    ySol(t) =


    exp(-t)/3 + (2*exp(-4*t))/3

fx >>
```

Let's study another example. Assume that we want to solve $2\frac{d^2y(t)}{dt^2} + \frac{dy(t)}{dt} + 11y(t) = e^{-t}, y(0) = 1, y'(0) = 0$. The commands shown in Fig. 1.65 solves this problem.

```
Command Window                                              ⊙

    >> syms y(t)
    >> eqn=2*diff(y,t,2)+diff(y,t)+11*y==exp(-t);
    >> D1y=diff(y,t);
    >> cond = [y(0)==1, D1y(0)==0];
    >> ySol=dsolve(eqn,cond)


    ySol =


    (exp(-t)*(29*sin((87^(1/2)*t)/4)^2 + 319*exp((3

fx >>
    <                                              >
```

Fig. 1.65 Solution of $2\frac{d^2y(t)}{dt^2} + \frac{dy(t)}{dt} + 11y(t) = e^{-t}, y(0) = 1, y'(0) = 0$

Assume that we want to see the graph of ySol for [0,6] time interval. You can use ezplot command (Fig. 1.66) to see the ySol graph (Fig. 1.67).

```
Command Window                                                    ⊙
    >> syms y(t)
    >> eqn=2*diff(y,t,2)+diff(y,t)+11*y==exp(-t);
    >> Dly=diff(y,t);
    >> cond = [y(0)==1, Dly(0)==0];
    >> ySol=dsolve(eqn,cond)

    ySol =

    (exp(-t)*(29*sin((87^(1/2)*t)/4)^2 + 319*exp((3

    >> ezplot(ySol,[0,6])
fx >>
    ⟨                                              ⟩
```

Fig. 1.66 Drawing the solution with ezplot

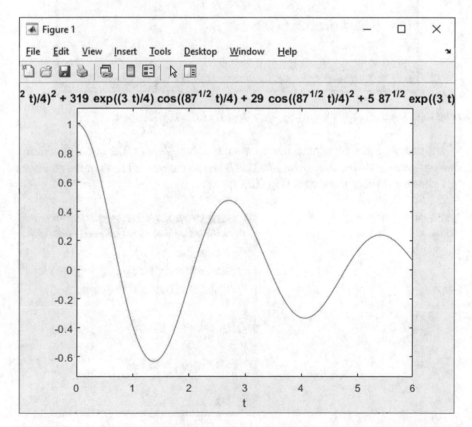

Fig. 1.67 Result of ezplot command

Assume that we want to solve $\frac{d^3y(x)}{dx^3} = \cos(2x) - y, y(0) = 1, y'(0) = 0, y''(0) = -1$. The commands shown in Fig. 1.68 solves this problem. You can use the simplify function to simplify the calculated result.

```
Command Window                                                    ⊙

  >> syms y(x)
  >> eqn=diff(y,x,2)==cos(2*x)-y;
  >> D1y=diff(y,x);
  >> D2y=diff(y,x,2);
  >> cond = [y(0)==1, D1y(0)==0, D2y(0)==0];
  >> ySol=dsolve(eqn,cond)


  ySol =


  (5*cos(x))/3 + sin(x)*(sin(3*x)/6 + sin(x)/2) -

  >> simplify(ySol)


  ans =


  1 - (8*sin(x/2)^4)/3

fx >> |

  <                                                            >
```

Fig. 1.68 Solution of $\frac{d^3y(x)}{dx^3} = \cos(2x) - y, y(0) = 1, y'(0) = 0, y''(0) = -1$

As another example, assume that we want to solve $\frac{dy(t)}{dt} = ty$. The initial conditions are not given in this problem. So, MATLAB shows the general form of the solution. C1 shows an arbitrary constant (Fig. 1.69).

Fig. 1.69 Solution of $\frac{dy(t)}{dt} = ty$

```
Command Window                                             ⊙

   >> syms y(t)
   >> eqn=diff(y,t)==t*y;
   >> ySol=dsolve(eqn)


   ySol =


   C1*exp(t^2/2)


  fx >>
```

1.16   Partial Fraction Expansion and Laplace Transform

You can use the residue command to calculate the partial fraction expansions. For instance, assume that we want to write the partial fraction expansion of $\frac{s+1}{s^2+5s+6}$. The commands shown in Fig. 1.70 calculates the residues at the poles. So, according to Fig. 1.70, the partial fraction expansion of $\frac{s+1}{s^2+5s+6}$ equals to $\frac{2}{s+3} + \frac{-1}{s+2}$.

Fig. 1.70 Result of residue command for $\frac{s+1}{s^2+5s+6}$

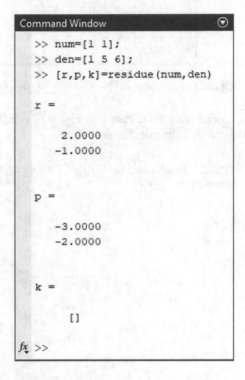

```
Command Window                              ⊙

  >> num=[1 1];
  >> den=[1 5 6];
  >> [r,p,k]=residue(num,den)

  r =

            2.0000
           -1.0000


  p =

           -3.0000
           -2.0000


  k =

              []

fx >>
```

Let's study another example. Assume that we want to write the partial fraction expansion of $\frac{1}{s^3(s-0.5)}$. Using the residue theory, $\frac{1}{s^3(s-0.5)} = \frac{-2}{s^3} + \frac{-4}{s^2} + \frac{-8}{s} + \frac{8}{s-0.5}$. Figure 1.71 shows that our result is correct.

```
Command Window                                    ⊙
   >> syms s
   >> simplify(-2/s^3-4/s^2-8/s+8/(s-0.5))

   ans =

   2/(s^3*(2*s - 1))

fx >>
```

Fig. 1.71 Verification of $\frac{-2}{s^3} + \frac{-4}{s^2} + \frac{-8}{s} + \frac{8}{s-0.5} = \frac{1}{s^3(s-0.5)}$

The commands shown in Fig. 1.72 calculates the residues. The residues are the same as hand calculations.

Fig. 1.72 Result of residue command for $\frac{1}{s^3(s-0.5)}$

```
Command Window                                    ⊙
   >> num=[1];
   >> den=[1 -.5 0 0 0];
   >> [r,p,k]=residue(num,den)

   r =

           8
          -8
          -4
          -2


   p =

       0.5000
            0
            0
            0


   k =

          []

fx >> |
```

The matrix k was a null matrix in both of the above examples. The matrix k is always null when the degree of the numerator is less than the denominator. When the degree of the numerator is equal or bigger than the denominator, it will not be null. For instance, assume $\frac{s^3+3s^2+7s+4}{s^2+2s+1}$. The commands shown in Fig. 1.73 calculates the partial fraction of this function. Note that matrix k is not null since the degree of numerator is bigger than denominator. According to Fig. 1.73, $\frac{s^3+3s^2+7s+4}{s^2+2s+1} = s+1+\frac{4}{(s+1)}+\frac{-1}{(s+1)^2}$.

Fig. 1.73 Result of residue command for $\frac{s^3+3s^2+7s+4}{s^2+2s+1}$

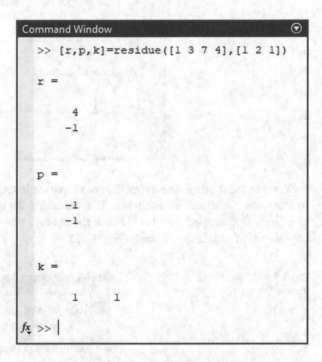

```
Command Window

>> [r,p,k]=residue([1 3 7 4],[1 2 1])

r =

        4
       -1


p =

       -1
       -1


k =

        1       1

fx >>
```

You can calculate the Laplace transform of function quite easily. The commands shown in Figs. 1.74 and 1.75 calculates the Laplace transform of e^{-6t} and $e^{-6t}\sin(\omega t)$, respectively.

Fig. 1.74 Calculation of the Laplace transform of e^{-6t}

```
Command Window

>> syms t
>> laplace(exp(-6*t))

ans =

1/(s + 6)

fx >>
```

Fig. 1.75 Calculation of the Laplace transform of $e^{-6t}\sin(\omega t)$

```
Command Window                                              ⊙

  >> syms w t
  >> laplace(exp(-6*t)*sin(w*t))


  ans =


  w/((s + 6)^2 + w^2)

  >> pretty(ans)
          w
  --------------
            2     2
  (s + 6)   + w

fx >> |
```

You can use the ilaplace command to calculate the inverse Laplace transform. For instance, the commands shown in Fig. 1.76 calculates the inverse Laplace transform of $\frac{s+1}{s^2+2s+1}$. Defining the variable F is not mandatory, you can calculate the inverse Laplace transform directly as well (Fig. 1.77).

Fig. 1.76 Calculation of inverse Laplace of $\frac{s+1}{s^2+2s+1}$

```
Command Window                                              ⊙

     >> syms s
     >> F=(s+1)/(s^2+2*s+2);
     >> ilaplace(F)


     ans =


     exp(-t)*cos(t)

fx >>
```

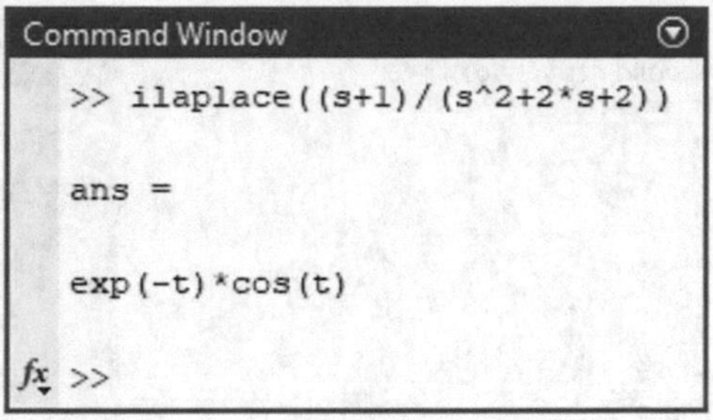

Fig. 1.77 Calculation of inverse Laplace of $\frac{s+1}{s^2+2s+1}$

```
Command Window                                    ⊙
  >> ilaplace((s+1)/(s^2+2*s+2))

  ans =

  exp(-t)*cos(t)

fx >>
```

You can calculate the Fourier/inverse Fourier transform with MATLAB as well (see the references for further study section).

1.17   Calculation of Limit, Derivative and Integral

Table 1.6 shows how to calculate the limit, derivative and integral using MATLAB. The syms command creates a symbolic variable for the calculations.

Table 1.6 Example for limit, diff and int functions of MATLAB

Mathematical expression	MATLAB representation
$\lim\limits_{x \to 0} \frac{\sin(x)}{x} = ?$	>>syms x >> limit(sin(x)/x,0)
$f(x) = \frac{x^3-x}{x^2+8}$ $\frac{df}{dx} = ?$	>>syms x >>diff((x^3-x)/(x^2+8))
$f(x,y) = x^3 + 3y^2 + 4xy$ $\frac{\partial f}{\partial y} = ?$	>>syms x y >>f=x^3+3*y^2+4*x*y >>diff(f,y)
$\int \frac{x^3-x}{x^2+8}\,dx = ?$	>>syms x >>int((x^3-x)/(x^2+8))
$\int_0^1 \frac{x^3-x}{x^2+8}\,dx = ?$	>>syms x >>int((x^3-x)/(x^2+8),x,0,1)

The command shown in Fig. 1.78 calculates the second derivative of function f with respect to variable x.

Fig. 1.78 Calculation of
second derivative of
$\frac{d^2(\sin(x)+\cos(y))}{dx^2}$

```
Command Window                              ⊙

    >> syms x y
    >> f=sin(x)+cos(y);
    >> diff(diff(f,x),x)


    ans =


    -sin(x)
fx >>
```

Sometimes it is difficult to read the MATLAB result since MATLAB shows the result in only one line. In these cases, you can use the pretty function which shows the result in a more readable manner (Fig. 1.79).

Fig. 1.79 Calculation of
integral with int command

```
Command Window                              ⊙

    >> syms x
    >> int((x^3-x)/(x^2+8))

    ans =

    x^2/2 - (9*log(x^2 + 8))/2

    >> pretty(ans)
     2                2
    x       9 log(x  + 8)
    --  -   --------------
     2                2

fx >> |
```

The simplify command, can be used to simplify the result of symbolic calculation (Fig. 1.80).

Fig. 1.80 simplify and
pretty commands

```
Command Window                          ⊙

  >> syms x
  >> simplify(1/(x+1)+1/(x-1))

  ans =

  (2*x)/(x^2 - 1)

  >> pretty(ans)
    2 x
  ------

   2
  x  - 1
fx >> |
```

1.18   MATLAB Editor

Command widow is not a suitable environment for writing long codes. It is better to
use MATLAB editor to write long codes. You can activate the editor by typing the
edit and press the enter key of the keyboard (Fig. 1.81). The MATLAB editor is
shown in Fig. 1.82.

Fig. 1.81 Running the
MATLAB editor

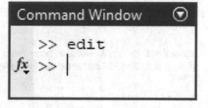

```
Command Window     ⊙

    >> edit
fx >> |
```

Fig. 1.82 MATLAB editor

The following commands, is a simple game. The user tries to guess a random number that is selected from 1 to 100. This code shows how to get input from the user, display suitable messages and how to use if command.

```
clc
clear all

maximumTry=5;
maximumNumber=100;

target=floor(maximumNumber*rand()); %the random target number
guess=input('Please enter your guess:');

noOfTrial=0;
while (guess~=target)
noOfTrial=noOfTrial+1;

if (guess>target)
 if (guess<0 ||noOfTrial>maximumTry)
 break
 end
 guess=input('Your guess is bigger than the target number. Enter new
guess:');
else
 if (guess<0||noOfTrial>maximumTry)
 break
 end
 guess=input('Your guess is smaller than the target number. Enter new
```

```
guess:');
end

end

if (guess==target)
 disp('Congratulation you win!')
end
disp ('Game finished.')
```

Enter the code to the MATLAB editor (Fig. 1.83) and save it by pressing the Ctrl+S (Fig. 1.84). The MATLAB codes are saved with the .m extension.

```
 1     clc
 2     clear all
 3
 4     maximumTry=5;
 5     maximumNumber=100;
 6
 7     target=floor(maximumNumber*rand()); %the random target number
 8     guess=input('Please enter your guess:');
 9
10     noOfTrial=0;
11    while (guess~=target)
12     noOfTrial=noOfTrial+1;
13
14     if (guess>target)
15         if (guess<0 ||noOfTrial>maximumTry)
16             break
17         end
18         guess=input('Your guess is bigger than the target number. Enter new guess:');
19     else
20         if (guess<0||noOfTrial>maximumTry)
21             break
22         end
```

Fig. 1.83   Code is entered to MATLAB editor

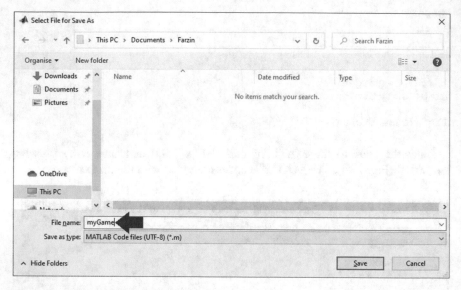

Fig. 1.84 Saving the entered code

Click the run button to run the program (Fig. 1.85). You can press the F5 key of your keyboard as well.

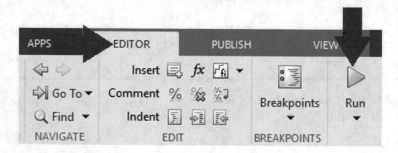

Fig. 1.85 Press the run button to run the entered code

The message box shown in Fig. 1.86 may appear after running the code. If it is appeared, click the add to path button.

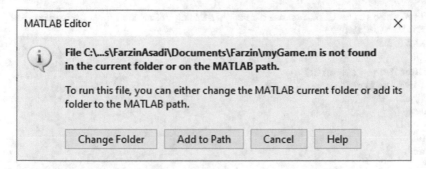

Fig. 1.86 MATLAB editor message

A sample run of the code is shown in Fig. 1.87.

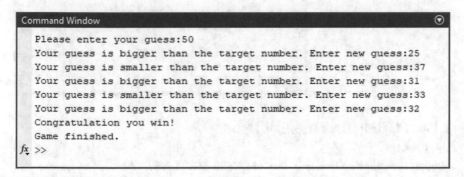

Fig. 1.87 Sample run of the entered code

The following program calculates the summation of even numbers from minNum to maxNum. This program uses a user defined function. Type this program in the MATLAB editor environment and save it with the name EvenNumberSummation. m. Run the program by pressing the F5 key. Figure 1.88 shows the output of this code.

```
%file: EvenNumberSummation.m
%this program claculates the summation of
%even numbers from minNum to maxNum.

sum=0;
minNum=0;
maxNum=10;

for n=minNum:maxNum
  sum=sum+n*isEven(n);
end
```

```
disp(strcat("sum of numbers from ",string(minNum)," to ",string
(maxNum)," is:"))
disp(sum)

function y=isEven(x)
if mod(x,2)==0
  y=1;
else
  y=0;
end
end
```

Fig. 1.88 Output of
the code

Command Window   ⌄

    >> EvenNumberSummation
    sum of numbers from 0 to 10 is:
        30

fx >> |

1.19   Plotting the Graph of Data

Plotting the graph of a data is very simple in MATLAB. You need to use the plot
command. The commands shown in Fig. 1.89 draws the graph of sin(x) for
$0 \leq x \leq 2\pi$. Result is shown in Fig. 1.90.

Fig. 1.89 Plotting the plot
of sin(x) for [0, 2π] interval

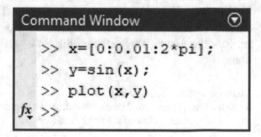

Command Window   ⌄

    >> x=[0:0.01:2*pi];
    >> y=sin(x);
    >> plot(x,y)
fx >>

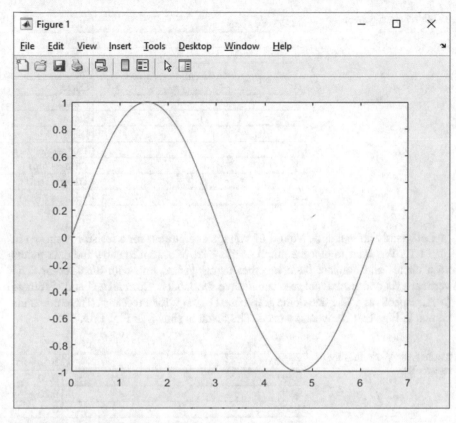

Fig. 1.90   Output of code in Fig. 1.89

You can add the operators shown in Tables 1.7, 1.8 and 1.9 to produce more custom plots.

Table 1.7   Types of lines

MATLAB command	Type of line
-	Solid
:	Dotted
-.	Dashdot
--	Dashed

Table 1.8   Colors

MATLAB command	Color
r	Red
g	Green
b	Blue
c	Cyan
m	Magenta
y	Yellow
k	Black
w	White

Table 1.9 Plot symbols

MATLAB command	Plot symbol
.	Point
+	Plus
*	Star
o	Circle
x	x-mark
s	Square
d	Diamond
v	Triangle (down)
^	Triangle (up)
<	Triangle (left)
>	Triangle (right)

Let's study an example. Values of voltage and current for a resistor is shown in Fig. 1.10. We want to plot the graph of this data. We want to show the data points with circles and connect them together using dashed line with black color. The vertical axis and horizontal axis must have the labels "Current(A)" and "Voltage (V)", respectively. The title of the graph must be "I–V for a resistor". The commands shown in Fig. 1.91 do what we need. The result is shown in Fig. 1.92.

Table 1.10 V–I values for resistor R1

V(volt)	I(Amper)
0.499	0.10
0.985	0.20
1.508	0.31
1.969	0.41
2.528	0.53
2.935	0.61
3.481	0.73
3.971	0.83
4.486	0.94
4.960	1.04
5.502	1.15
6.007	1.26
6.60	1.38

```
Command Window
>> V=[0.499 0.985 1.508 1.969 2.528 2.935 3.481 3.971 4.486 4.960 5.502 6.007 6.60];
>> I=[0.1 0.2 0.31 0.41 0.53 0.61 0.73 0.83 0.94 1.04 1.15 1.26 1.38];
>> plot(V,I,'--ko')
>> title('I-V for a resistor')
>> xlabel('Voltage(V)')
>> ylabel('Current(A)')
>> grid on
fx >>
```

Fig. 1.91 Drawing the plot of Table 1.10

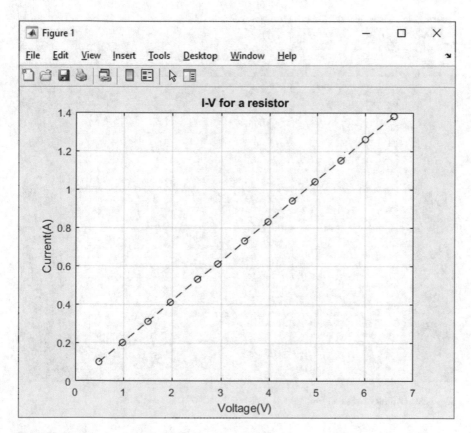

Fig. 1.92 Output of code in Fig. 1.91

The commands shown in Fig. 1.93 draws the I–V graph of Table 1.10. However, it uses red star for data points and solid black color for connecting the data points together.

```
Command Window                                                                    ⊙
  >> V=[0.499 0.985 1.508 1.969 2.528 2.935 3.481 3.971 4.486 4.960 5.502 6.007 6.60];
  >> I=[0.1 0.2 0.31 0.41 0.53 0.61 0.73 0.83 0.94 1.04 1.15 1.26 1.38];
  >> plot(V,I,'k',V,I,'r*'),grid minor
fx >>
```

Fig. 1.93 Drawing the plot of Table 1.10

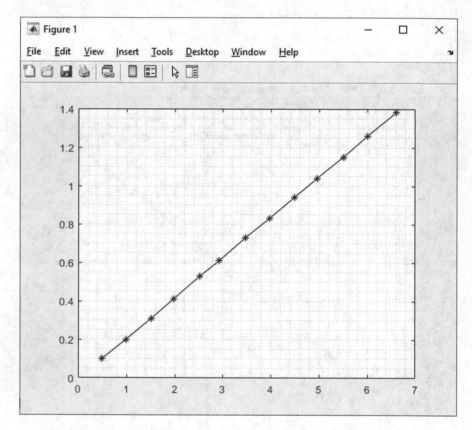

Fig. 1.94 Output of code in Fig. 1.93

Figure 1.94 doesn't have any labels and title. You can add the desired labels and titles to it using the insert menu (Fig. 1.95).

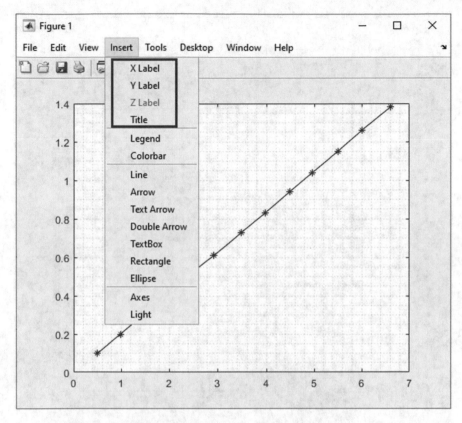

Fig. 1.95 Addition of title and labels to the axis

You can copy the drawn graph to the clipboard easily with the aid of edit> copy figure (Fig. 1.96). After copying the graph to the clipboard you can easily paste it in programs like MS Word® by pressing Ctrl+C.

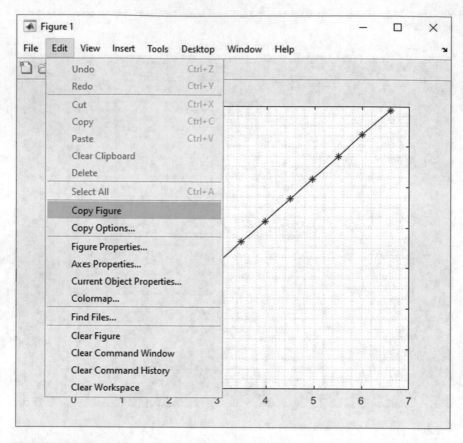

Fig. 1.96 Copying the drawn figure to the clipboard

You can save the drawn graph as a graphical file as well. To do this, click use the file>save as (Fig. 1.97). After clicking the file>save as, the save as window is appeared. Select the desired output file format from the save as type drop down list (Fig. 1.98).

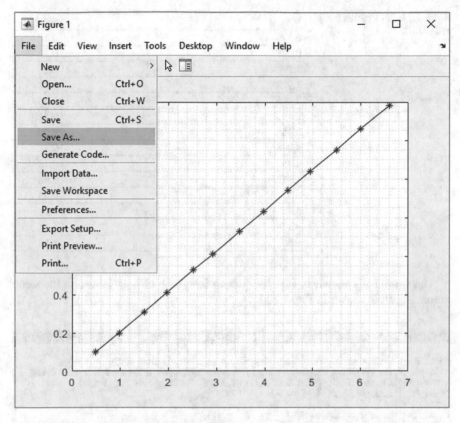

Fig. 1.97   Saving the graph as a graphical file

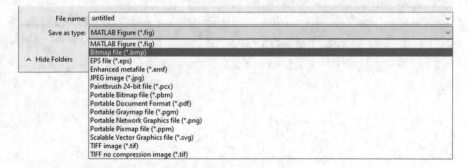

Fig. 1.98   Selection of desired type of file

Sometimes you need to show two or more datasets on the same graph. You need to use the hold on command to show two or more graphs simultaneously. Assume that we have another dataset (Table 1.11) and we want to show both datasets (Tables 1.10 and 1.11) on the same graph.

Table 1.11 V–I values for
resistor R2.

V(volt)	I(Amper)
0.579	0.10
0.978	0.17
1.598	0.28
1.976	0.34
2.496	0.43
2.953	0.51
3.458	0.60
4.068	0.71
4.450	0.78
4.917	0.86
5.35	0.93
5.75	1.01
6.37	1.11
6.60	1.15

The commands shown in Fig. 1.99 draws the graph of both tables on the same
graph. Result is shown in Fig. 1.100.

```
Command Window                                                                    ⊙
    >> V1=[0.499 0.985 1.508 1.969 2.528 2.935 3.481 3.971 4.486 4.960 5.502 6.007 6.60];
    >> I1=[0.1 0.2 0.31 0.41 0.53 0.61 0.73 0.83 0.94 1.04 1.15 1.26 1.38];
    >> V2=[0.579 0.978 1.598 1.976 2.496 2.953 3.458 4.068 4.450 4.917 5.35 5.75 6.37 6.60];
    >> I2=[0.1 0.17 0.28 0.34 0.43 0.51 0.60 0.71 0.78 0.86 0.93 1.01 1.11 1.15];
    >> plot(V1,I1,'b',V1,I1,'r*')
    >> hold on
    >> plot(V2,I2,'k',V2,I2,'r+')
    >> grid minor
    >> xlabel('Voltage (V)')
    >> ylabel('Current (A)')
    >> title('Comparison of I-V plot of R1 and R2')
fx >>
```

Fig. 1.99 Drawing the Tables 1.10 and 1.11 on the same graph

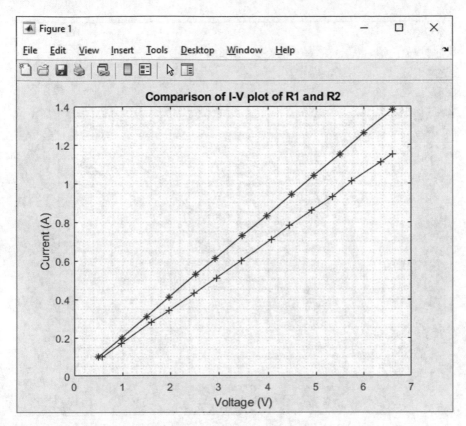

Fig. 1.100 Output of code in Fig. 1.99

You can use the insert> legend (Fig. 1.101) to show which graph belongs to which resistor.

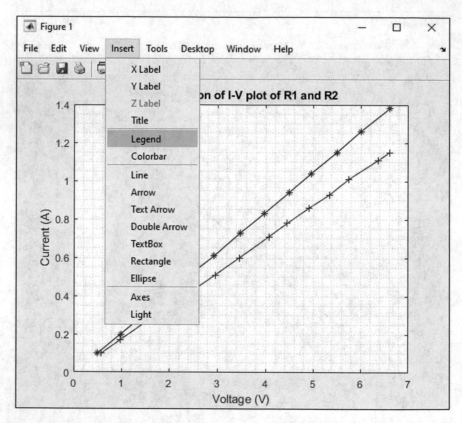

Fig. 1.101   Addition of legend to the graph

After clicking the insert> legend, the legend shown in Fig. 1.102 will be added to the graph.

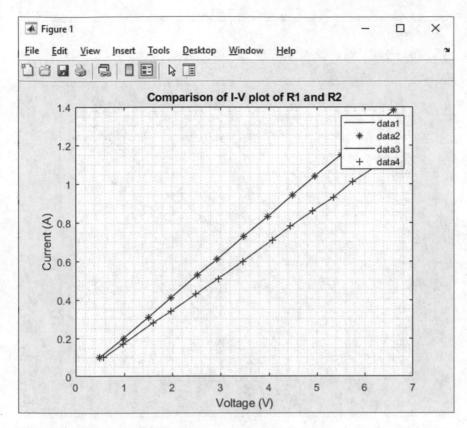

Fig. 1.102 Legend is added to the graph

Double the data1 in the legend box (Fig. 1.102) and change it to the desired text. Repeat this for data 2, data3 and data4. You can move the legend box by clicking on it, holding down the mouse button and dragging it to the desired location (Fig. 1.103). You can even right click on the legend box and use the predefined locations (Fig. 1.104).

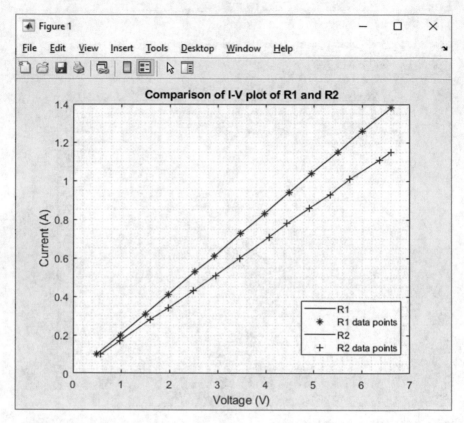

Fig. 1.103 Customized legend

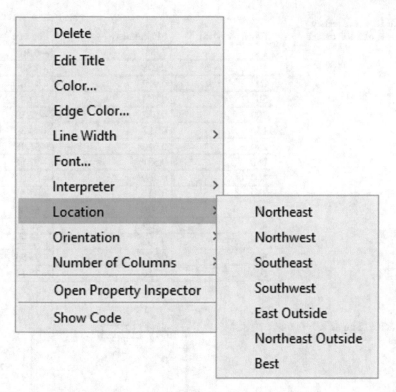

Fig. 1.104 Default locations for legend

The graph of I–V for the studied resistors uses linear axis. If you want to draw the frequency response graphs, then you need to use logarithmic axis. Let's study an example. Assume the frequency response given in Table 1.12. This table shows the frequency response of the circuit shown in Fig. 1.105.

Table 1.12 Frequency response of a RC circuit

Frequency(Hz)	Magnitude($\left|\frac{V_o(j\omega)}{V_{in}(j\omega)}\right|$)	Phase($\angle\frac{V_o(j\omega)}{V_{in}(j\omega)}$)
1	1.000	−0.36°
10	0.998	−3.60°
20	0.992	−7.16°
50	0.954	−17.44°
100	0.847	−32.13°
150	0.728	−43.30°
200	0.623	−51.48°
250	0.537	−57.51°
300	0.469	−62.05°
350	0.414	−65.54°
400	0.370	−68.30°
450	0.333	−70.51°
500	0.303	−72.34°
550	0.278	−73.85°
600	0.256	−75.14°

Fig. 1.105 Simple RC circuit

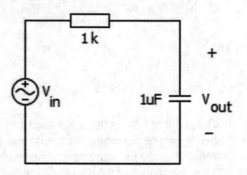

The commands shown in Fig. 1.106, draws the frequency response of the data in Table 1.12. Result is shown in Fig. 1.107.

```
Command Window
>> f=[1 10 20 50 100 150 200 250 300 350 400 450 500 550 600];
>> Amp=[1 0.998 0.992 0.954 0.847 0.728 0.623 0.537 0.469 0.414 0.370 0.333 0.303 0.278 0.256];
>> Phase=-[0.36 3.6 7.16 17.44 32.13 43.30 51.48 57.51 62.05 65.54 68.30 70.51 72.34 73.85 75.14];
>> subplot(211),semilogx(f,20*log10(Amp)),grid minor
>> title('Frequency responce of the RC circuit')
>> xlabel('Freq(Hz.)')
>> ylabel('Amplitude (dB)')
>> subplot(212),semilogx(f,Phase),grid minor
>> xlabel('Freq(Hz.)')
>> ylabel('Phase(Degrees)')
fx >> |
```

Fig. 1.106 Drawing the plot of Table 1.12

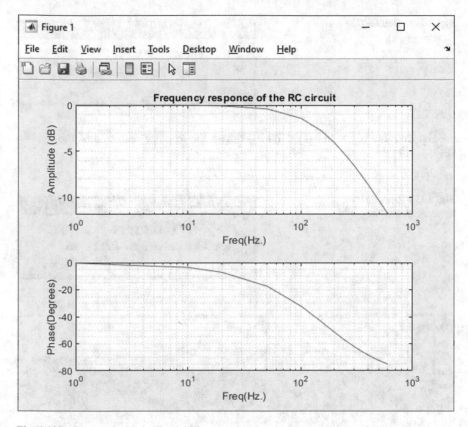

Fig. 1.107 Output of code in Fig. 1.106

1.20   Impulse Response, Step Response and Frequency Response of Dynamical Systems

Impulse response, step response and frequency response of dynamical systems can be drawn easily with MATLAB. Assume we want to draw the aforementioned responses for $H(s) = \frac{100}{s+6s+100}$. The commands shown in Figs. 1.108 or 1.109 defines the given system.

Fig. 1.108 Defining the $H(s) = \frac{100}{s+6s+100}$

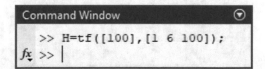

Fig. 1.109 Defining the
$H(s) = \frac{100}{s+6s+100}$

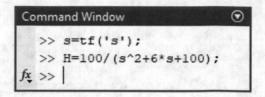

The impulse command (Fig. 1.110) draws the impulse response of the system (Fig. 1.111).

Fig. 1.110 Impulse
response of $H(s) = \frac{100}{s+6s+100}$

```
Command Window
   >> H=tf([100],[1 6 100]);
   >> impulse(H), grid on
fx >>
```

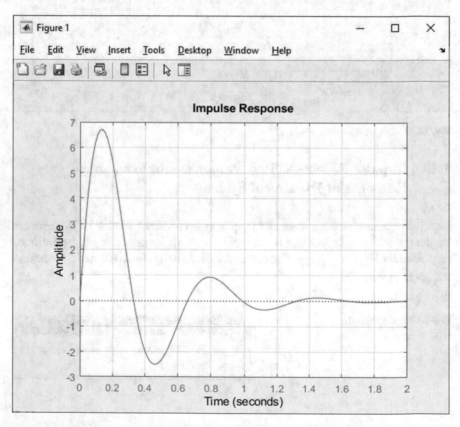

Fig. 1.111 Output of code in Fig. 1.110

You can add a cursor or cursors to the graph by clicking on the graph (Fig. 1.112). You can remove the added cursor by right clicking on it and click the delete.

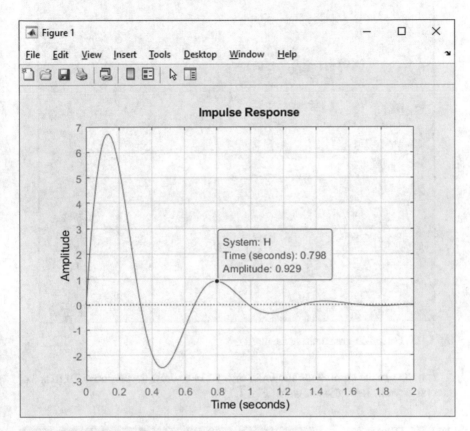

Fig. 1.112 Addition of cursor to the graph

The graph shown in Fig. 1.112 shows the impulse response for [0 s, 2 s] interval. Assume that you need the response for [0 s, 3 s] interval. Simply double click on the graph and enter the desired range (Fig. 1.113).

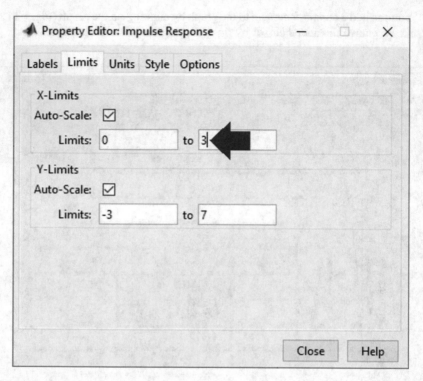

Fig. 1.113 Determining the limits of the plot

You can use the commands shown in Fig. 1.114 to draw the response for [0 s, 3 s] interval as well. Result is shown in Fig. 1.115.

Fig. 1.114 Drawing the impulse response for the desired time interval

```
Command Window                        ⊙
    >> time=[0:0.01:3];
    >> impulse(H,time), grid on
fx >>
```

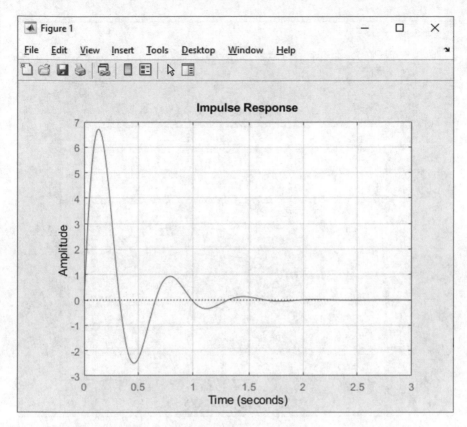

Fig. 1.115 Output of code in Fig. 1.114

You can measure the impulse response characteristics by right clicking on the graph and selecting the characteristics (Fig. 1.116).

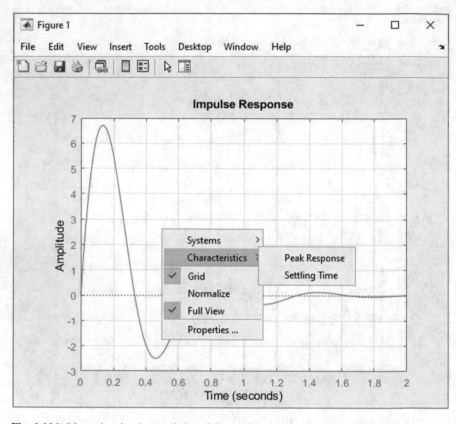

Fig. 1.116 Measuring the characteristics of the graph

The step command (Fig. 1.117) draws the step response of the system (Fig. 1.118).

Fig. 1.117 Drawing the step response of H(s) = $\frac{100}{s+6s+100}$

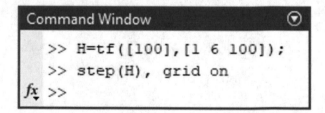

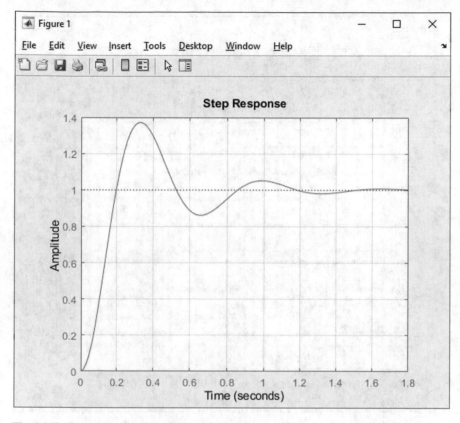

Fig. 1.118 Output of code shown in Fig. 1.117

You can add cursors to the graph, change the shown interval of the graph and measure the response characteristics similar to what described for the impulse response case. The bode command (Fig. 1.119) draws the frequency response of the system (Fig. 1.120).

Fig. 1.119 Drawing the bode diagram of $H(s) = \frac{100}{s+6s+100}$

```
Command Window
    >> H=tf([100],[1 6 100]);
    >> bode(H), grid on
fx >> |
```

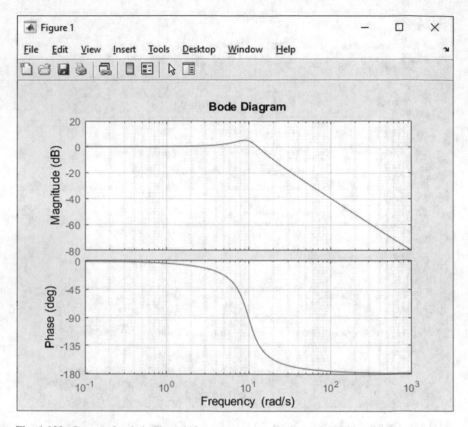

Fig. 1.120 Output of code in Fig. 1.119

Note that the bode command uses rad/s for the horizontal axis. You can change it to Hz if you prefer. Double click on the graph in order to change the unit of the horizontal axis. Go to the units tab in the opened window and select the desired unit (Fig. 1.121).

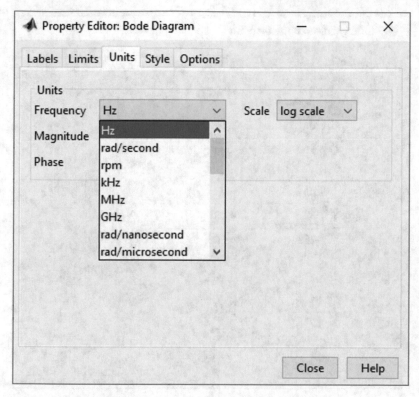

Fig. 1.121   Selection of Hz for horizontal axis

Use the limits tab to change the shown interval of the graph (Fig. 1.122).

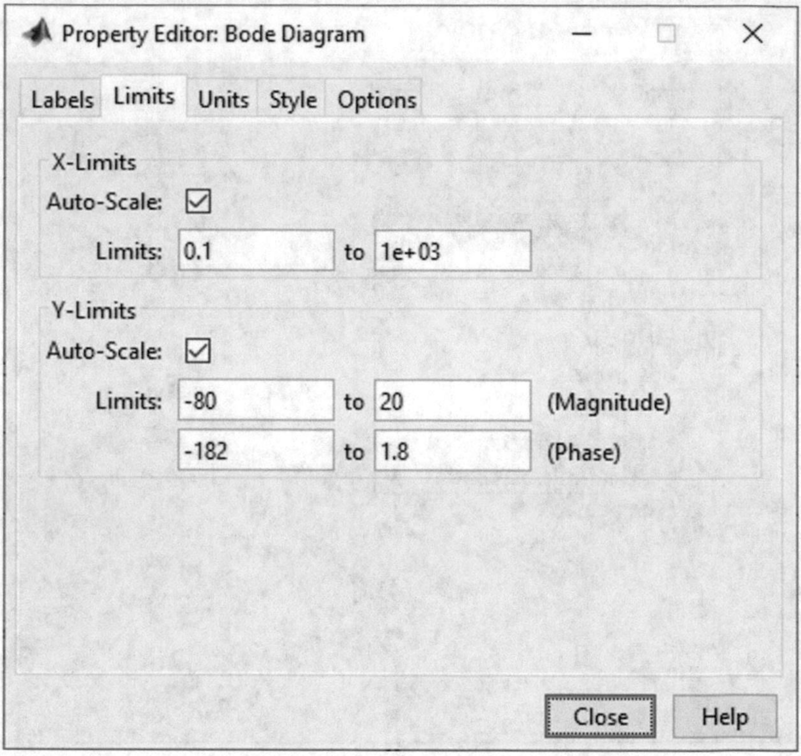

Fig. 1.122 Selection of desired interval for the drawn graph

You can use the code shown in Fig. 1.123 to draw the bode diagram in the desired range as well. This code draws the bode diagram for $\left[2\pi \times 0.1\ \frac{\text{Rad}}{\text{s}}, 2\pi \times 100\frac{\text{Rad}}{\text{s}}\right]$ range.

Command Window

```
>> H=tf([100],[1 6 100]);
>> wmin=2*pi*0.1;
>> wmax=2*pi*100;
>> w=logspace(log10(wmin),log10(wmax));
>> bode(H,w), grid on
```

Fig. 1.123 Drawing the bode diagram of $\frac{100}{s+6s+100}$ for $\left[2\pi \times 0.1\ \frac{\text{Rad}}{\text{s}}, 2\pi \times 100\frac{\text{Rad}}{\text{s}}\right]$ interval

You can add cursors to the graph and measure the response characteristics similar to what described for the impulse response case.

Assume that you want to calculate the bode diagram at $\omega = 15.3 \frac{Rad}{s}$. The commands shown in Figs. 1.124 or 1.125 calculate the value of H(s) at $\omega = 15.3 \frac{Rad}{s}$. Figure 1.126 shows that the obtained results are correct.

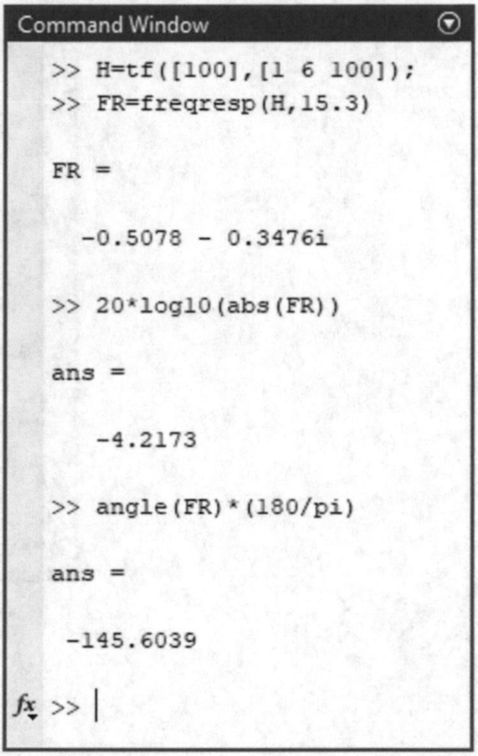

Fig. 1.124 Calculation of frequency response of $\frac{100}{s+6s+100}$ at $\omega = 15.3 \frac{Rad}{s}$

```
Command Window                              ⊙

  >> H=tf([100],[1 6 100]);
  >> FR=freqresp(H,15.3)

  FR =

    -0.5078 - 0.3476i

  >> 20*log10(abs(FR))

  ans =

    -4.2173

  >> angle(FR)*(180/pi)

  ans =

   -145.6039

fx >> |
```

Fig. 1.125 Calculation of
frequency response of
$\frac{100}{s+6s+100}$ at $\omega = 15.3 \frac{Rad}{s}$

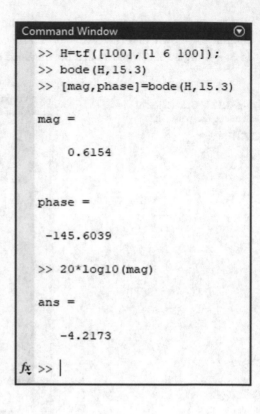

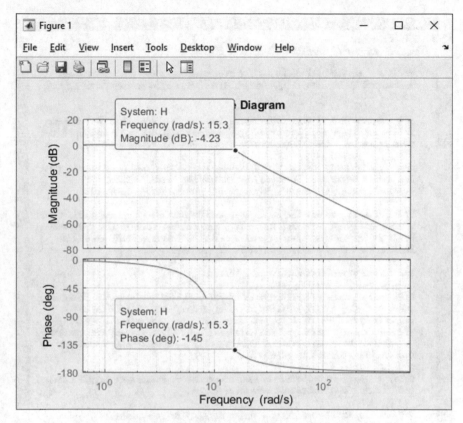

Fig. 1.126 Calculation of frequency response of $\frac{100}{s+6s+100}$ at $\omega = 15.3 \ \frac{Rad}{s}$

1.21   Getting Help in MATLAB

You can use the help command to get more information about the commands studied in this chapter. For instance, assume that you need more information about the freqresp command. The help freqresp command brings the document of the command for you (Fig. 1.127).

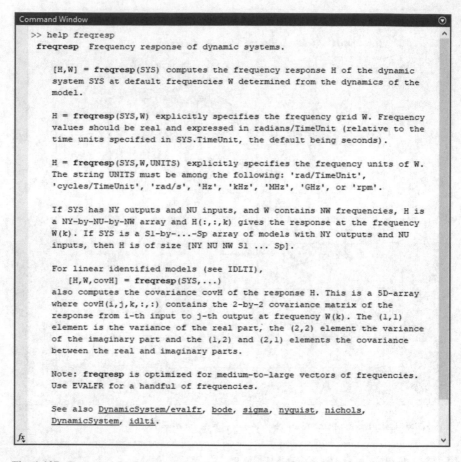

Fig. 1.127 Document for freqresp command

There are other ways to get help in MATLAB as well. For instance, if you press the F1 key of the keyboard, the window shown in Fig. 1.128 appears. Click the open help browser.

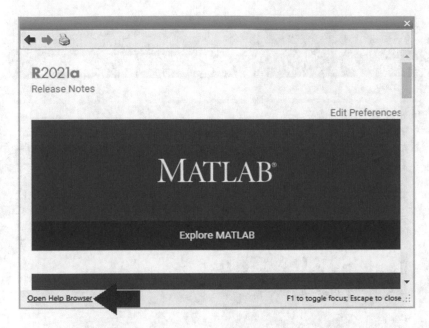

Fig. 1.128 MATLAB help window

After clicking the open help browser (Fig. 1.128), the window shown in Fig. 1.129 appears. Enter the search term into the search documentation box. As you type the search term, a list of related topics appears below the search box. If you click on them, then that topic will be shown.

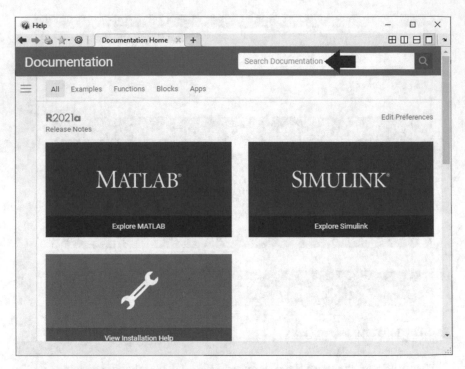

Fig. 1.129 Enter the search term to the search documentation box

Let's study another way to get help. On the left side, inside the command window, there is something visible in small fonts, and that is fx (Fig. 1.130). If you click on it (or press Shift+F1), a drop-down search bar gets opened (Fig. 1.131). The appeared list is the name of the products we have installed. Click on any of the products to get a list of all the related functions.

Fig. 1.130 The fx button

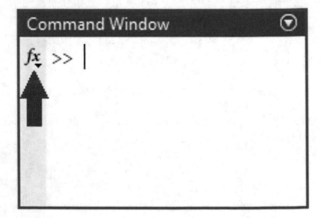

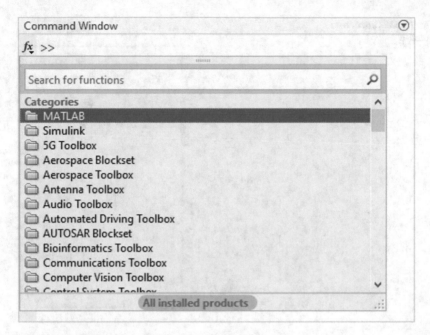

Fig. 1.131 The window which appears after clicking the fx button

1.22 Exercises

1. Do the following calculations with MATLAB:

   (a) $\sin(1+2i)$ **(b)** e^{1+2i} **(c)** $Log_{10}(1 + 2i)$ **(d)** $Ln_{10}(1 + 2i)$ **(e)** $\frac{1+2i}{3+4i}$ **(f)** $\sin(60°)$ **(g)**
   $\cos(\frac{\pi}{3})$ **(h)** $\sqrt{8^2 + 2^2}$ **(i)** $\sqrt[4]{8^2 + 2^2}$

2. Solve the $x^4 + 2x^3 + 4x + 1 = 0$ with MATLAB.
3. Write a MATLAB program to find the steady state value of current drawn from
   the source. $V_1 = 110\sqrt{2}\sin(2\pi \times 60t)$ (Fig. 1.132).

Fig. 1.132 Circuit of
exercise 3

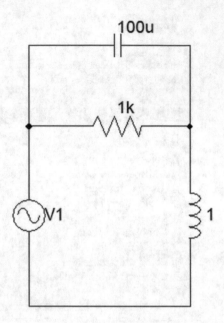

4. Use the dsolve command to find the current of circuit shown in Fig. 1.133. $V_1 = 10 + 25 \sin (2\pi \times 60t)$. Initial conditions are $V_C = 10 \ V$ and $i_L = 0 \ A$.

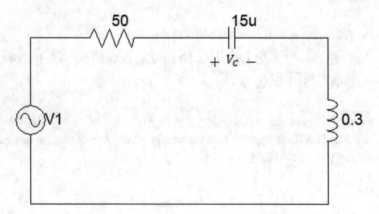

Fig. 1.133 Circuit of exercise 4

5. Calculate the transfer function between input voltage source V1 and capacitor voltage Vc, i.e. $\frac{V_1(s)}{V_c(s)}$. Use MATLAB to draw the step response of the system
6. Write a MATLAB program which calculates the product of odd numbers from 1 to 20.
7. Draw the graph of following functions $e^{-x} \sin (3x)$ for $[0, 5]$. Use the graph to find its maximum value.

Further Readings

1. S. Chapman, *MATLAB Programming for Engineers*, 6th edn. (Cengage, 2019)
2. J.O. Attia, *Electronics and Circuit Analysis Using MATLAB*, 2nd edn. (CRC Press, 2007)
3. B. Hahn, D. Valentine, *Essential MATLAB for Engineers and Scientists*, 7th edn. (Academic Press, 2019)
4. H. Moore, *MATLAB for Engineers*, 5th edn. (Pearson, 2017)
5. K. Ogata, *MATLAB for Control Engineers* (Pearson, 2007)
6. Fourier transform with MATLAB: https://www.mathworks.com/help/symbolic/sym.fourier.html (Visiting date: 15.07.2021)
7. Mathematical Functions of MATLAB: https://www.mathworks.com/help/symbolic/mathematical-functions.html (Visiting date: 15.07.2021)

Chapter 2
Simulation of Electric Circuits with Multisim™

2.1 Introduction

Multisim™ is industry-standard SPICE simulation and circuit design software for analog, digital, and power electronics in education and research. Multisim has an intuitive, user friendly interface. Simulation results are quite close to labaratory results. Printed Circuit Board (PCB) of circuits analyzed in Multisim can be designed with the aid of Ultiboard™.Ultiboard is printed circuit board design and layout software that integrates seamlessly with Multisim to accelerate PCB prototype development.

This chapter introduces the Multisim and shows how it can be used to analyze electric circuits.

2.2 Multisim Environment

Multisim environment is shown in Fig. 2.1. It consists of 7 sections. Section 2.1 contains the software menus. Section 2.2 contains the shortcut for elements that you need in your schematic. Section 2.3 is used to manage various elements in the schematic. For instance, you can see the tree that shows the files in the design. Section 2.4 is used to add some comments to the schematic. Section 2.5 (schematic editor) is used to draw the circuit schematic. Section 2.6 contains the measurement devices. Section 2.7 is used to show the error message and warnings to the user.

© The Author(s), under exclusive license to Springer Nature Switzerland AG 2022   91
F. Asadi, *Essential Circuit Analysis using NI Multisim™ and MATLAB®*,
https://doi.org/10.1007/978-3-030-89850-2_2

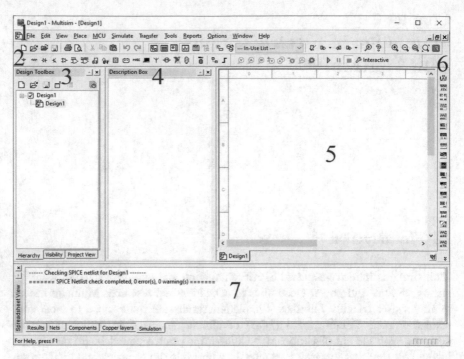

Fig. 2.1 Multisim environment

By default, the schematic editor has some dots on it (Fig. 2.2). In order to remove the dots, click the options> sheet properties (Fig. 2.3) and uncheck the show grid box (Fig. 2.4).

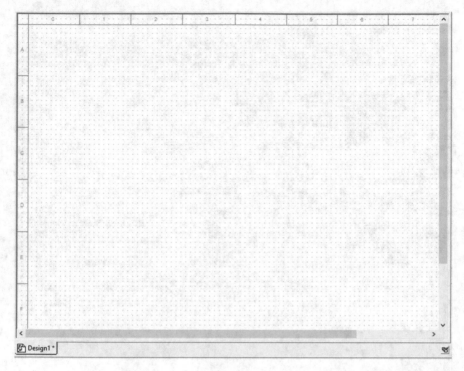

Fig. 2.2   Schematic editor

Fig. 2.3   Sheet properties

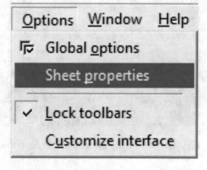

Fig. 2.4 Removing the grid points from the schematic editor

2.3   Opening a New File

Like any other Windows program, you can open a new file by clicking the paper icon (Fig. 2.5) or clicking the File> New (Fig. 2.6).

Fig. 2.5 New file icon

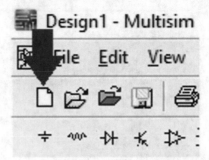

Fig. 2.6 Opening a new file

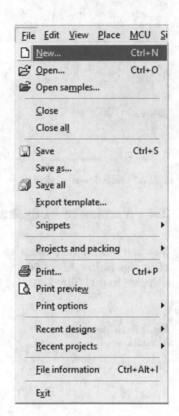

After clicking the paper icon or clicking the File> New, the window shown in Fig. 2.7 will appear. Simply click the Create button and your new file is ready.

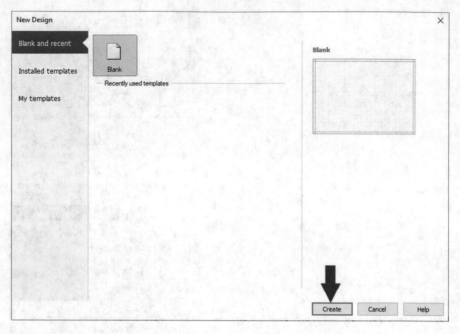

Fig. 2.7 New design window

2.4 Version of Multisim

You can use the Help menu in order to learn the version of your software. Simply click the Help> About Multisim (Fig. 2.8) and Multisim shows the version of software to you (Fig. 2.9).

Fig. 2.8 Help> About Multisim

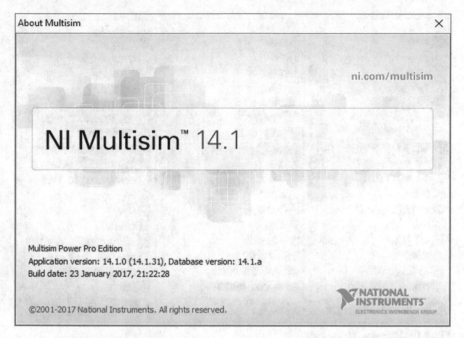

Fig. 2.9   About Multisim window

2.5   Multisim Components

Multisim has many components which are divided into different categories. These categories are shown in Figs. 2.10, 2.11, 2.12 and 2.13.

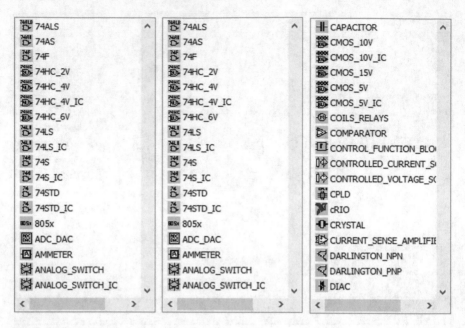

Fig. 2.10 Multisim components

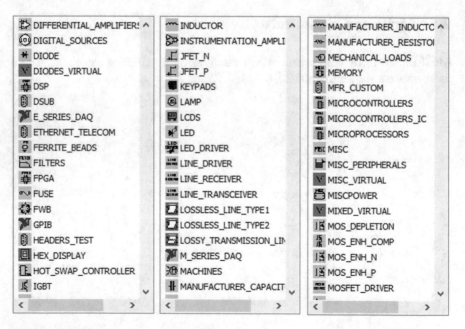

Fig. 2.11 Multisim components

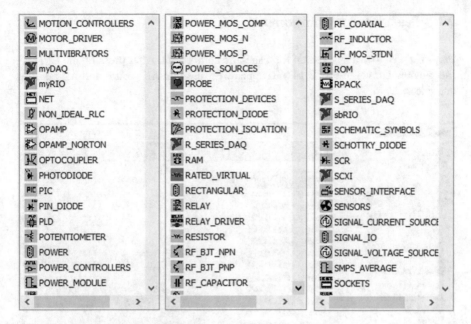

Fig. 2.12   Multisim components

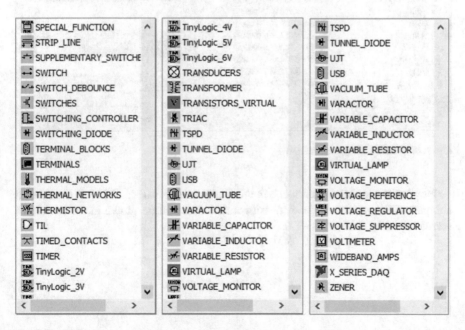

Fig. 2.13   Multisim components

2.6   Search for a Component

Press Ctrl+W in order to search for a specific component. After pressing the Ctrl+W, the window shown in Fig. 2.14 will appear. Ensure that All groups and All families are selected.

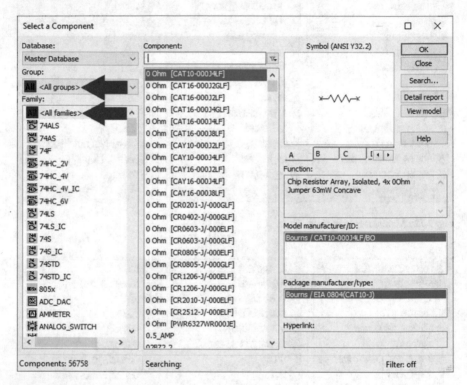

Fig. 2.14 Select a component window

Now write the component name in the Component box (Fig. 2.15). If the entered component is available in the Multisim database, it will be shown in the bottom of search box (Fig. 2.16).

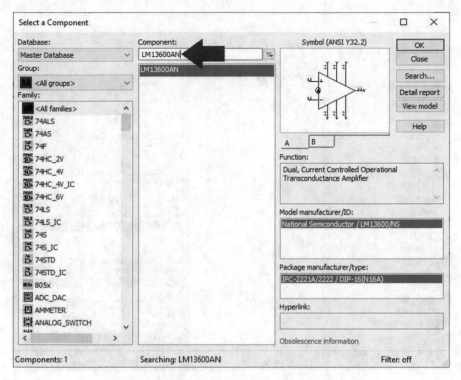

Fig. 2.15   Searching for LM13600AN

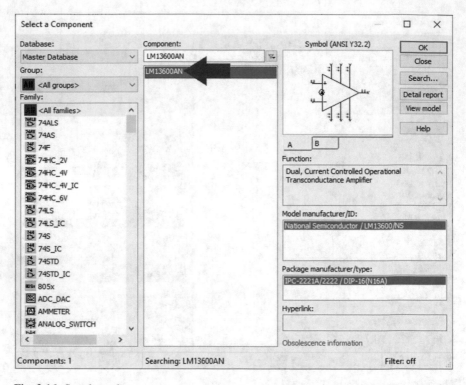

Fig. 2.16 Search result

2.7 Sample Simulations

Multisim has many ready to use sample simulations. Click the open samples icon (Fig. 2.17) in order to open the examples folder.

Fig. 2.17 Open
samples icon

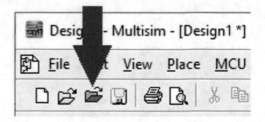

After clicking the sample simulations, the window shown in Fig. 2.18 will appear. Now you can select the category which you want. For instance, the MCU folder contains the sample simulations related to microcontrollers.

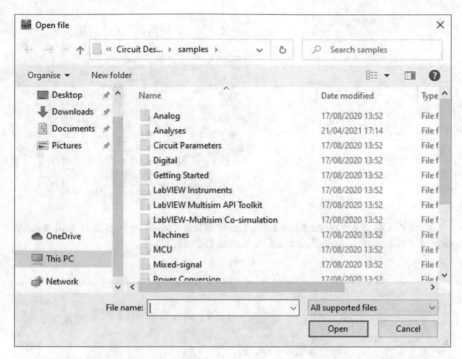

Fig. 2.18 Open file window

2.8　Example 1: A Simple Resistive Voltage Divider

The simple resistive circuit shown in Fig. 2.19 is analyzed in this section.

Fig. 2.19 Schematic of
example 1

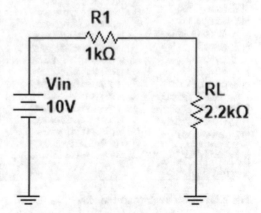

In order to do this, open a new file, then click the place basic icon shown in Fig. 2.20. You can press the Ctrl+W instead of clicking the place basic icon as well.

Fig. 2.20 Place basic icon

After clicking the place basic icon, the window shown in Fig. 2.21 will appear. Select 1 kΩ resistor and click the OK button (Fig. 2.22).

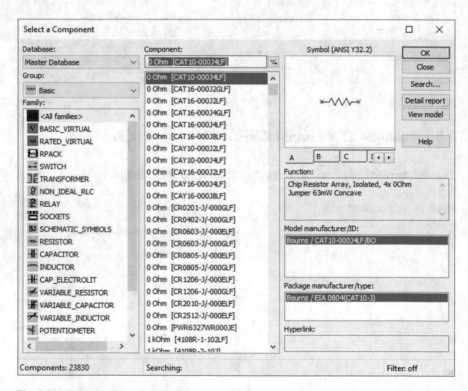

Fig. 2.21 Select a component window

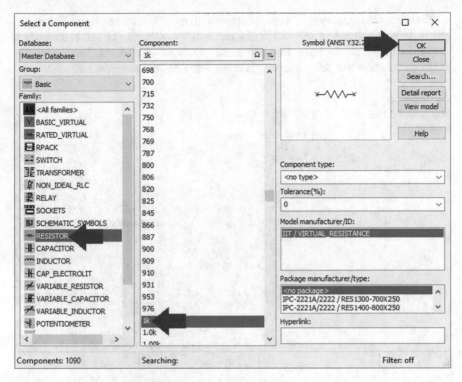

Fig. 2.22 Selection of 1 kΩ resistor

Now click on the schematic in order to add the 1 kΩ resistor to it (Fig. 2.23). The component will be rotated if you press the Ctrl+R before clicking on the schematic. After clicking and adding the component to the schematic, window shown in Fig. 2.22 will appear again automatically and permits you to select and add a new component quickly. If you don't want to add any component, click the close button.

Fig. 2.23 Addition of resistor to the schematic editor

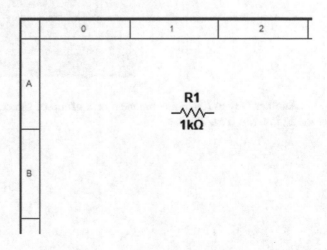

There are other methods to rotate a component, as well. You can right click on the component or a group of selected components and use the flip or rotate commands shown in Fig. 2.24. In order to select more than one component, hold down the mouse left button and draw a rectangle around the desired components.

Fig. 2.24 Flip/rotate commands

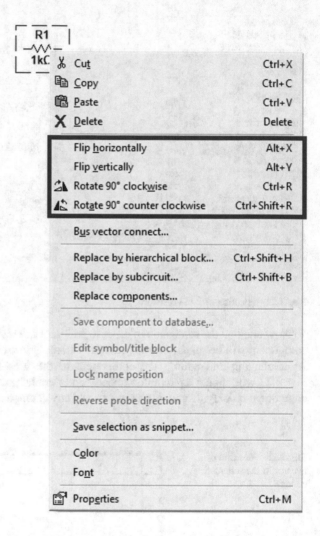

Another way to rotate a component or a group of selected components is to use the Edit menu (Fig. 2.25).

Fig. 2.25 Flip/rotate commands

Now add another resistor to the schematic (Fig. 2.26).

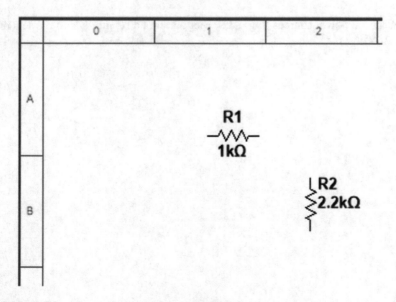

Fig. 2.26 Addition of second resistor to the schematic editor

After addition of resistor R2, select a component window will be opened again. Go to the sources section (Fig. 2.27).

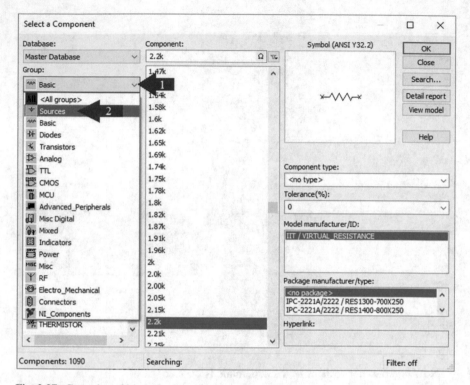

Fig. 2.27 Ground symbol can be found in the source group

Select the power sources (Fig. 2.28) section and add 2 grounds to the schematic (Fig. 2.29).

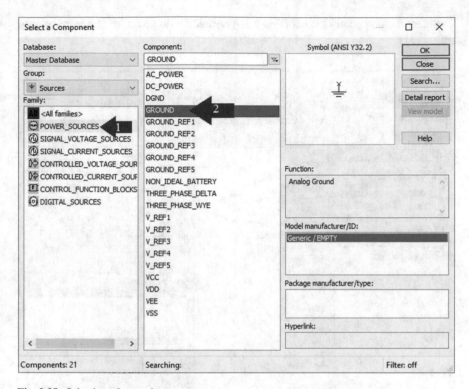

Fig. 2.28   Selection of ground

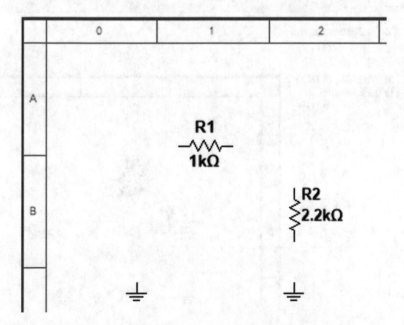

Fig. 2.29   Addition of two ground symbols to the schematic editor

Select the dc power element (Fig. 2.30) and add a dc power source to the circuit (Fig. 2.31).

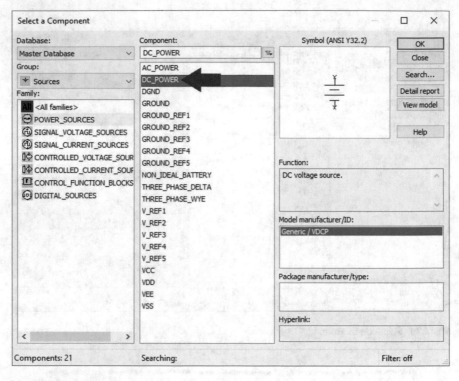

Fig. 2.30 DC power element

Fig. 2.31 Addition of DC power to the schematic editor

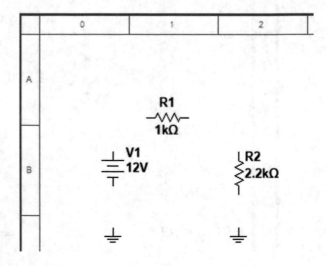

Double click on the V1 and use the RefDes box in the Label tab to rename the source to Vin (Fig. 2.32).

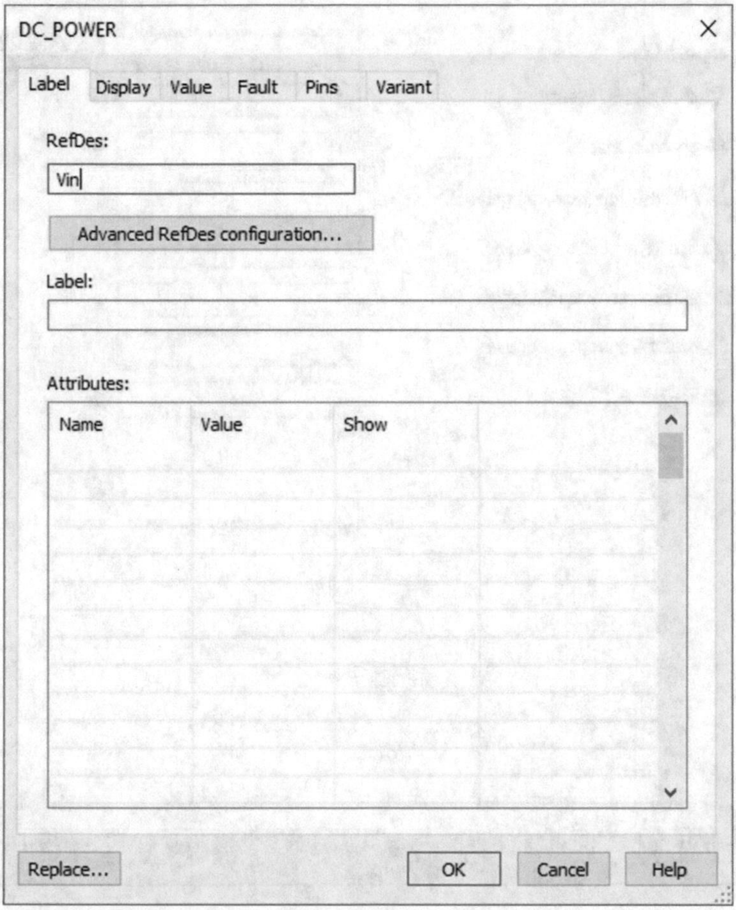

Fig. 2.32 Entering the name of DC power component

Now go to the value tab and use the voltage (V) box to enter the desired voltage (Fig. 2.33).

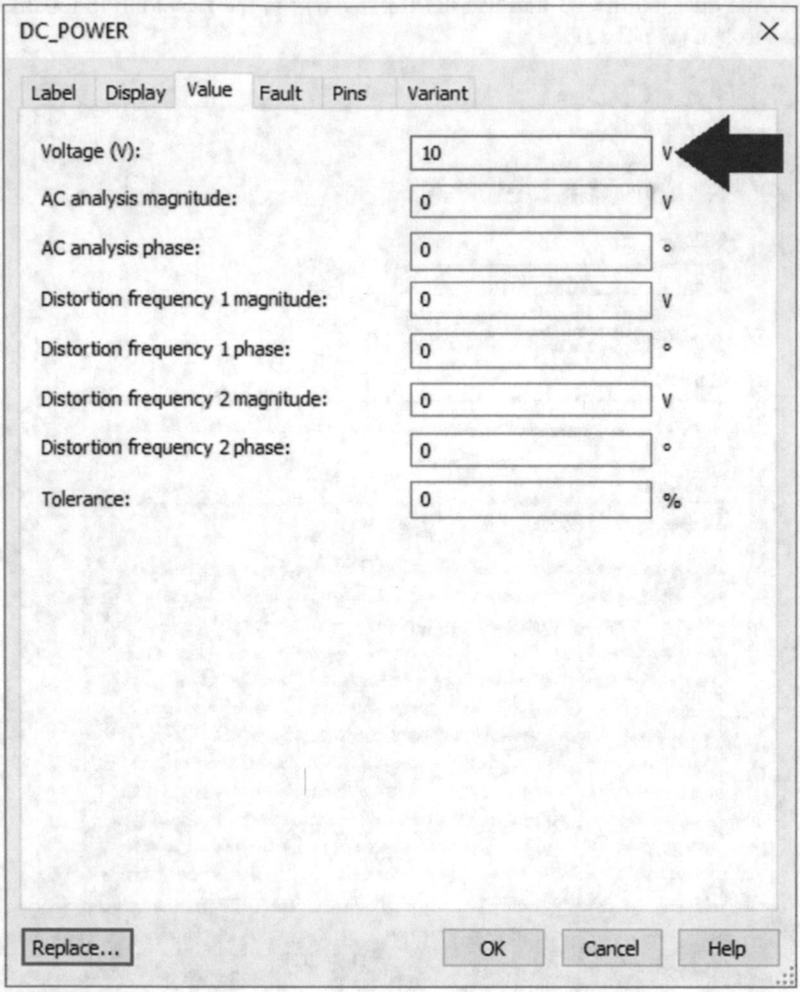

Fig. 2.33 Entering the voltage value

If you click the help button, you can see the help page related to that tab. For instance, if you click the help button in Fig. 2.32, the page shown in Fig. 2.34 will appear. If you click the help button in Fig. 2.33, the page shown in Fig. 2.35 will appear.

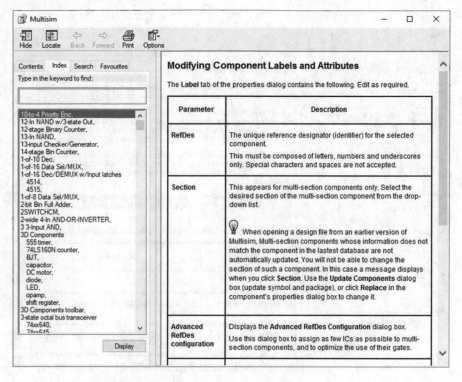

Fig. 2.34 Help page of label tab

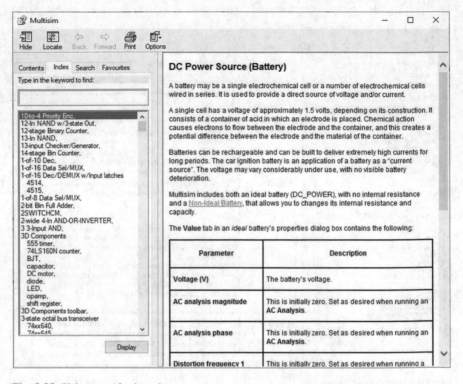

Fig. 2.35 Help page of value tab

Double click on the resistor R2 and rename it to RL (Fig. 2.36).

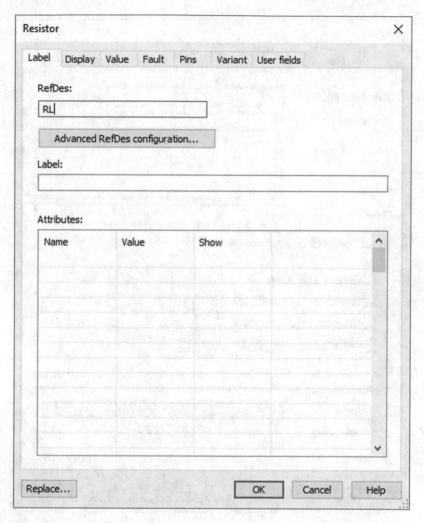

Fig. 2.36 Entering the name of resistor

You can change the resistor value by going to the value tab and entering the new value to the resistance box (Fig. 2.37). When a value is entered, you can use the prefixes shown in Table 2.1.

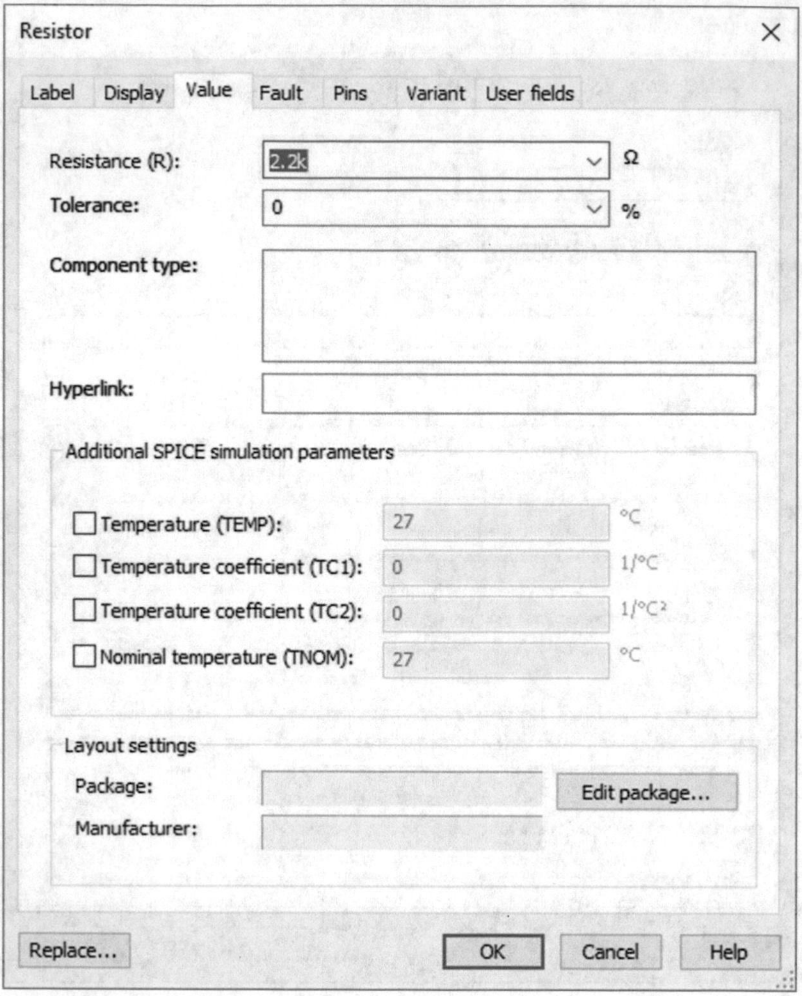

Fig. 2.37 Entering the value of resistor

Table 2.1 Available prefixes in Multisim

Prefix	Meaning
M	Mega
k	Kilo
m	Mili
u	Micro
n	Nano
p	Pico

Now the schematic must looks like Fig. 2.38. If you bring the mouse pointer close to the components terminals, the mouse pointer changes to crosshair. After seeing the crosshair, click the mouse left button and move the mouse pointer toward the destination terminal. When you reach the destination terminal, click it and a wire is drawn between the terminals. Use this technique to connect the components together (Fig. 2.39).

Fig. 2.38 Required components are added to the schematic editor

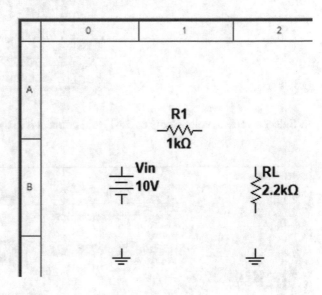

Fig. 2.39 Connecting the components together

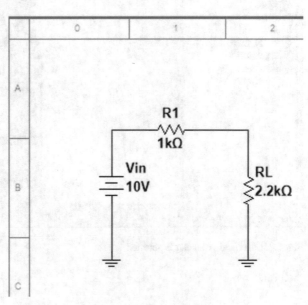

Now the circuit is complete. We need a voltmeter to see the resistor RL voltage. Click the place indicator icon (Fig. 2.40) in order to add a voltmeter to the schematic.

Fig. 2.40 Place
indicator icon

Select a vertical voltmeter (Fig. 2.41) and connect it to the resistor RL (Fig. 2.42).

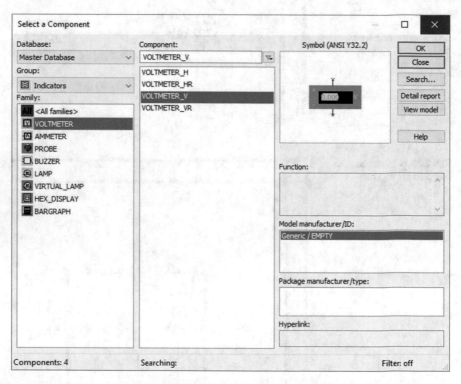

Fig. 2.41 Vertical voltmeter is selected

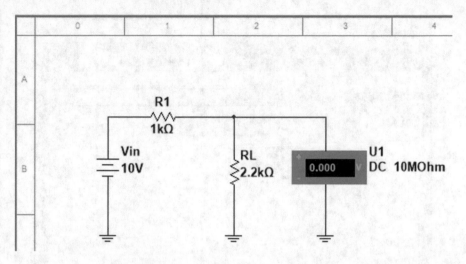

Fig. 2.42 Addition of vertical voltmeter to the schematic

You can set the desired internal resistance for the voltmeter by double clicking the voltmeter and entering the desired value to the resistance box. Ensure that DC is selected for the mode box. The DC mode measures the average value of signal (Fig. 2.43).

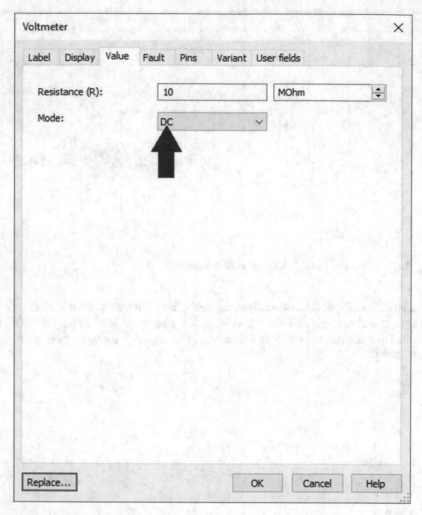

Fig. 2.43 Selection of DC mode

Now run the simulation by clicking the play icon (Fig. 2.44) or press the F5 button. Ensure that interactive is selected. If you see something else is selected instead of interactive, click on it and select Interactive simulation in the opened window. Then click the save button (Fig. 2.45).

Fig. 2.44 Running the simulation

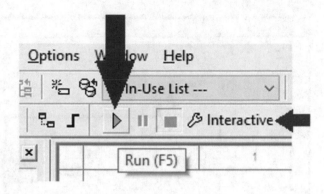

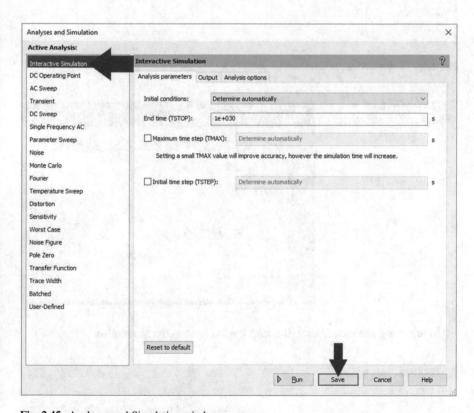

Fig. 2.45 Analyses and Simulation window

Simulation result is shown in Fig. 2.46. The obtained result can be verified easily with the simple calculation shown in Fig. 2.47.

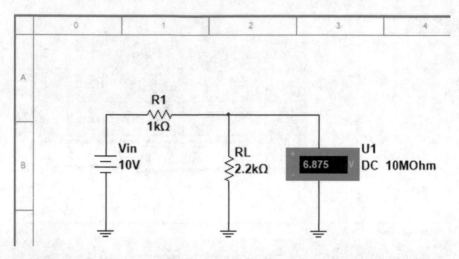

Fig. 2.46 Simulation result. Load voltage is 6.875 V

Fig. 2.47 MATLAB
calculation of output voltage

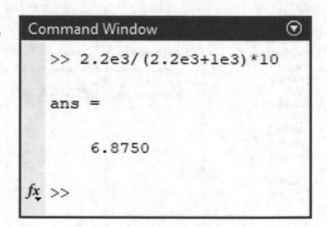

After seeing the result, click the stop button to stop the simulation (Fig. 2.48).

Fig. 2.48 The stop button

Assume that you want to measure the current of circuit. We need to put an ammeter in series to the source to measure the circuit current. Select the wire that connects the source to resistor R1 by left clicking on it (Fig. 2.49). Now press the delete key of keyboard to remove this wire (Fig. 2.50).

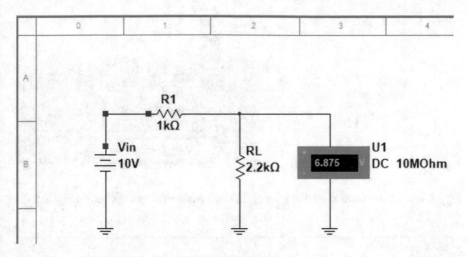

Fig. 2.49   Selection of the wire

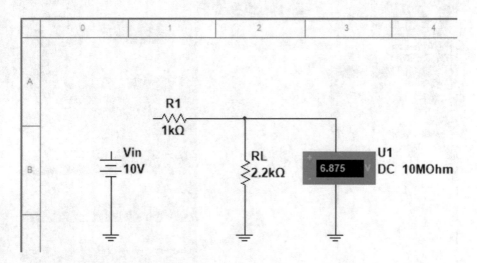

Fig. 2.50   Removing the selected wire

Click the place indicator button (Fig. 2.51).

Fig. 2.51 Place indicator
button

Select a horizontal ammeter (Fig. 2.52) and connect it to the circuit (Fig. 2.53).

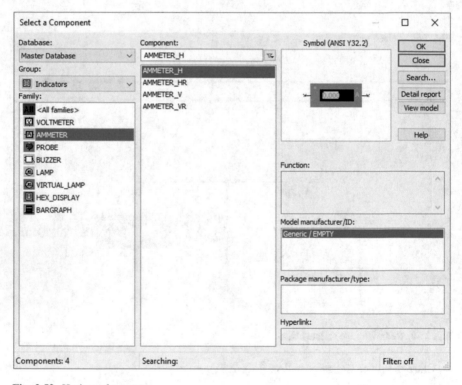

Fig. 2.52 Horizontal ammeter

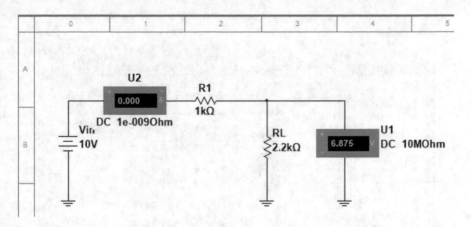

Fig. 2.53 Connecting the horizontal ammeter to the circuit

You can set the internal resistance of ammeter by double clicking the ammeter and use the resistance box. Ensure that DC is selected for the mode box (Fig. 2.54).

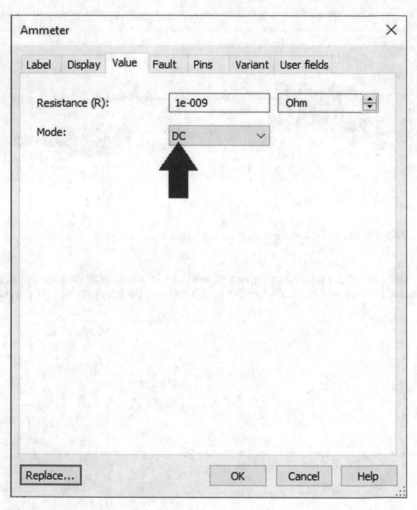

Fig. 2.54 Selection of DC mode

After running the simulation, the result shown in Fig. 2.55 is obtained. The obtained result can be verified with the simple calculation shown in Fig. 2.56.

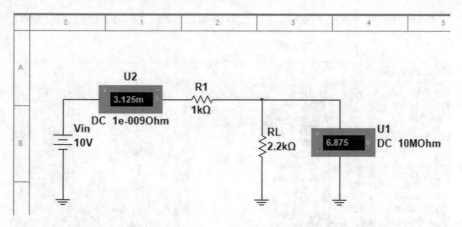

Fig. 2.55 Simulation result

Fig. 2.56 MATLAB code
to calculate the circuit
current

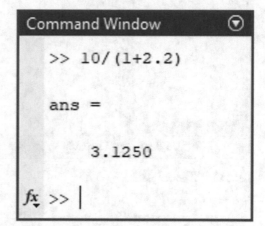

2.9   Example 2: DC/AC Mode of Voltmeter/Ammeter

Let's add an AC source in series to the DC source of Example 1. Existence of AC
source adds an AC component to the voltages/currents of circuit and helps to see the
difference between DC and AC voltmeter/ammeter.

Click the place source (Fig. 2.57) in order to add an AC source to the schematic of
Example 1.

Fig. 2.57 Place sources
button

Select the AC power element (Fig. 2.58) and add it to the schematic (Fig. 2.59).

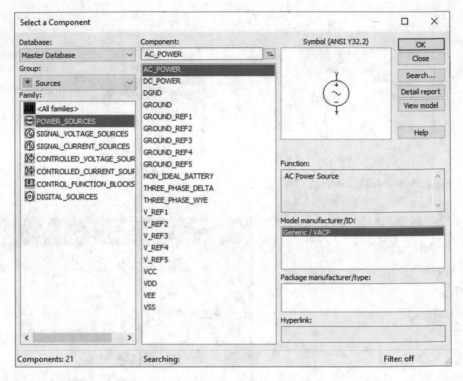

Fig. 2.58 AC power element

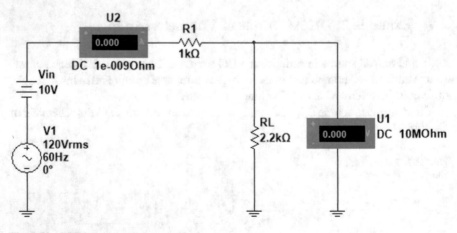

Fig. 2.59 Addition of AC power to the circuit

Double click the AC source and enter the values shown in Fig. 2.60. After entering the values, click the OK button. The schematic must be updated to what shown in Fig. 2.61.

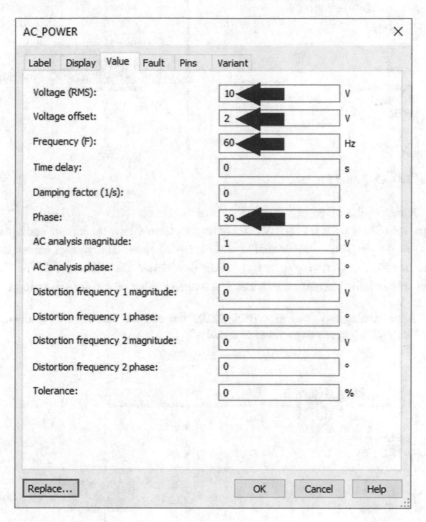

Fig. 2.60 AC power settings

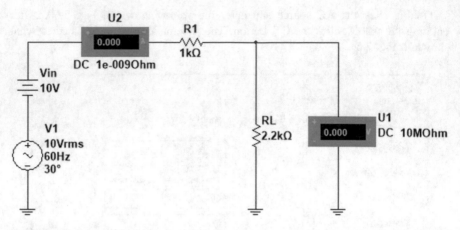

Fig. 2.61 Completed schematic

Now, the input voltage of circuit is $v(t) = 10 + 2 + 10\sqrt{2} \sin\left(2\pi \times 60 \times t + 30°\right)$ V. According to Ohm's law, the circuit current is $i(t) = \frac{v(t)}{R_1+R_L} = 3.75 + 4.4194 \sin\left(2\pi \times 60 \times t + 30°\right)$mA. The voltage across the resistor RL is $v_{RL}(t) = \frac{R_L}{R_1+R_L} \times v(t) = 8.25 + 9.723 \sin\left(2\pi \times 60 \times t + 30°\right)$. The average value of current is 3.75 mA and average value of resistor RL voltage is 8.25 V.

After running the simulation (Fig. 2.62), you can see that the DC voltmeter/ammeter shows the average value of signals.

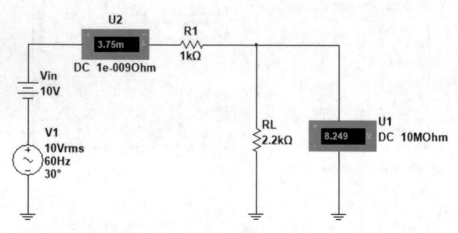

Fig. 2.62 Result of simulation

Now double click the voltmeter and ammeter blocks. Then use the value tab to select the AC mode (Fig. 2.63). Result of simulation with AC voltmeter and ammeter is shown in Fig. 2.64. The AC mode measures the Root Mean Square (RMS) of AC component of the signal (Appendix A reviews the RMS value of a signal). When the average value of a signal is subtracted from the signal itself, the resultant is the AC component. For instance the AC component of i-$(t) = 3.75 + 4.4194 \sin (2\pi \times 60 \times t + 30°)$mA is $4.4194 \sin (2\pi \times 60 \times t + 30°)$ mA and the AC component of $v_{RL}(t) = 8.25 + 9.723 \sin (2\pi \times 60 \times t + 30°)$ is $9.723 \sin (2\pi \times 60 \times t + 30°)$.

The RMS of AC component of current is $\frac{4.4194\ mA}{\sqrt{2}} = 3.125$ mA and the RMS of AC component of resistor RL voltage is $\frac{9.723\ V}{\sqrt{2}} = 6.8752$ V. These numbers are the same as the simulation result (Fig. 2.64).

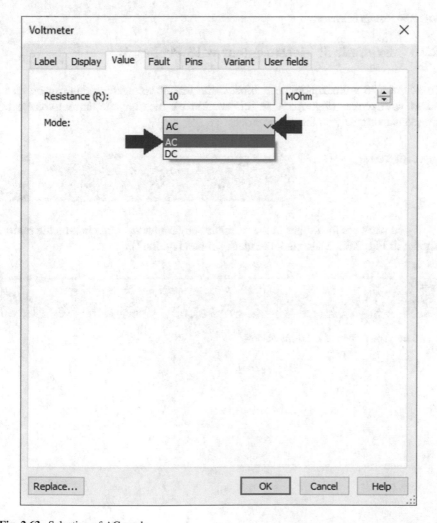

Fig. 2.63 Selection of AC mode

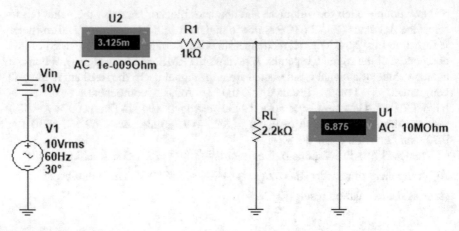

Fig. 2.64 Result of simulation

2.10 Example 3: Measurement with Probes

In addition to voltmeter/ammeter blocks, the circuit voltage/currents can be measured with probes (Fig. 2.65). In this section we measure the current/voltage of previous example with the aid of probes.

Fig. 2.65 Probes

If you can't see the probes in the Multisim environment, right click on the region shown in Fig. 2.66. Then click the place probe (Fig. 2.67).

Fig. 2.66 Top section of Multisim window

Fig. 2.67 Ensure that Place
probes is check

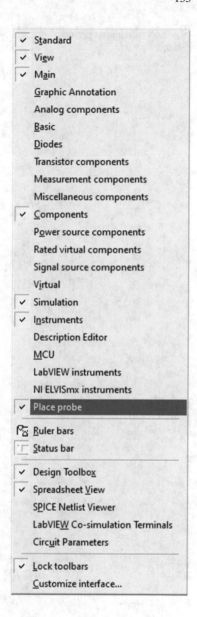

Before starting the measurement, click the probe settings (Fig. 2.68) and ensure
that instantaneous and periodic is selected (Fig. 2.69).

Fig. 2.68 Simulate> probe
settings

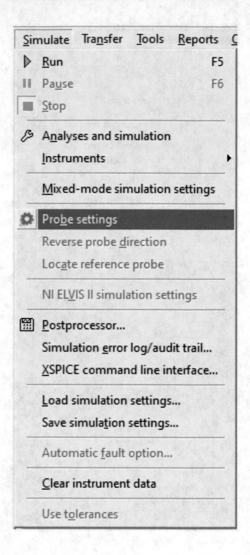

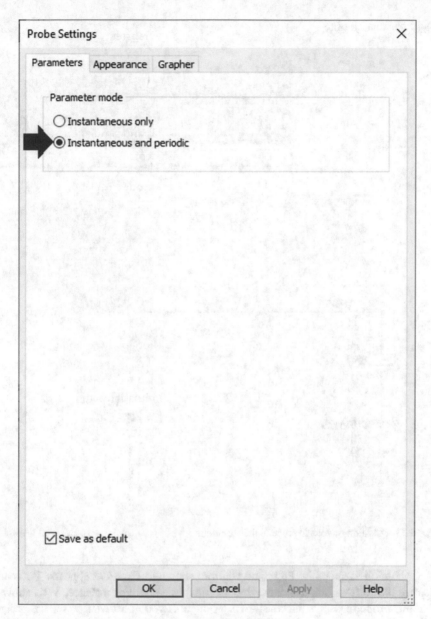

Fig. 2.69 Ensure that Instantaneous and periodic is checked

Add two voltage probes (Fig. 2.70) to the schematic of previous example (Fig. 2.71). Note that voltage probe measures the voltage of a node with respect to ground. You can move the box which is connected to the probe (Fig. 2.71) by clicking on it, holding down the mouse left button and dragging the box to desired location. You can remove the probes by clicking on them and pressing the delete key of your keyboard.

Fig. 2.70 Voltage probe

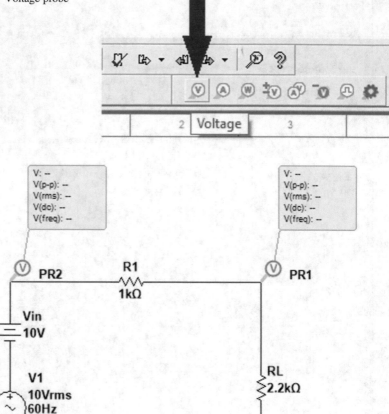

Fig. 2.71 Addition of voltage probe to the schematic

Double click the probe PR1. The window shown in Fig. 2.72 appears. You can change the RefDes box to a more understandable name. For instance, VRL shows that this probe measures the resistor RL voltage. If you don't want to see the probe name on the schematic, select the hide RefDef.

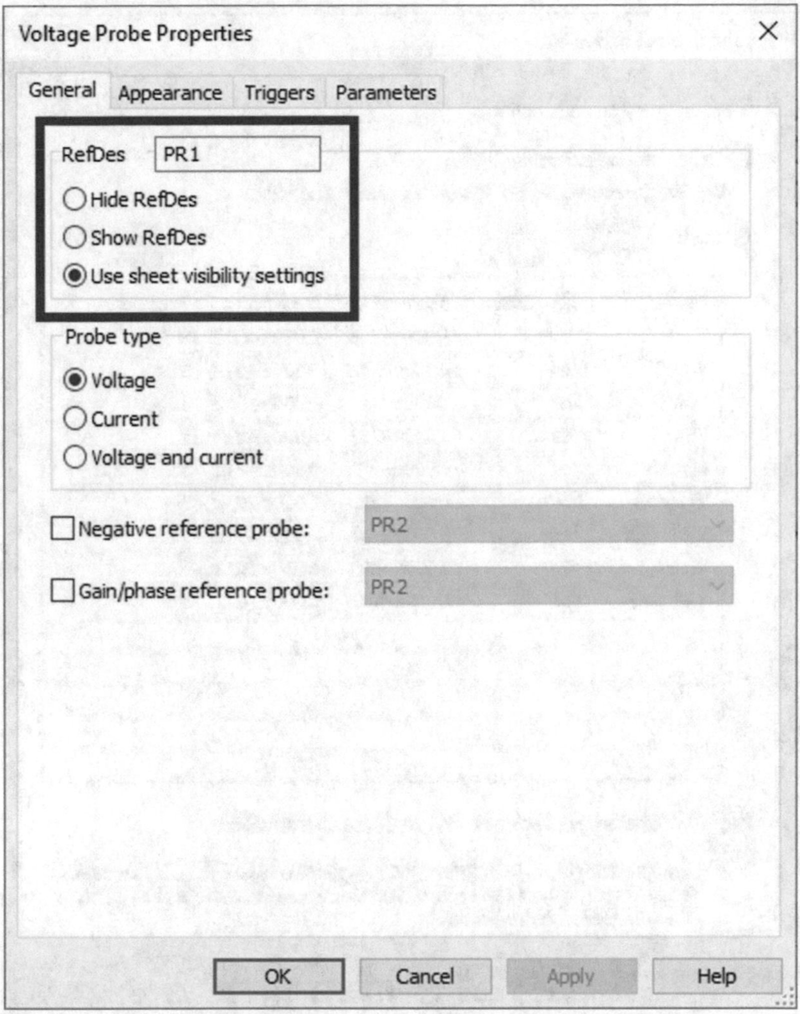

Fig. 2.72   General tab of voltage probe properties window

With the aid of parameters tab of voltage probe properties window (Fig. 2.73), you can select the measurements that you want to be shown. In order to do this, click the custom and use the show column. If you don't want a measurement to be shown, change the show column value for that row to No. You can set the desired precision for each measurement with the aid of precision column.

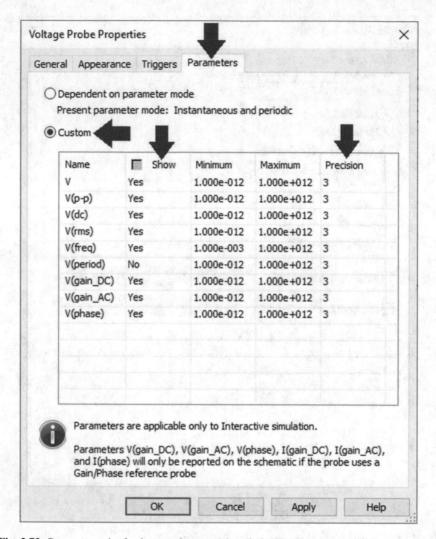

Fig. 2.73 Parameters tab of voltage probe properties window

Now run the simulation. The result shown in Fig. 2.74 is obtained. According to Fig. 2.74, the instantaneous value of resistor RL voltage (at the instant that figure is captured) is 11.9, the peak-peak value of resistor RL voltage is 19.4 V, the RMS value of resistor RL voltage 10.7 V, the DC component value of resistor RL voltage 8.25 V and the frequency of the resistor RL voltage is 60 Hz.

Lets check these results. The resistor RL voltage found in the previous example is $v_{RL}(t) = 8.25 + 9.723 \sin (2\pi \times 60 \times t + 30°)$. The peak-peak of this voltage is $2 \times 9.723 = 19.446$. The RMS of this function is $\sqrt{(8.25)^2 + \left(\frac{9.723}{\sqrt{2}}\right)^2} = 10.7392\ \text{V}$. The DC component (average value) is 8.25 V and the frequency of this function is 60 Hz. The reading of probe PR2 can be verified in the same way.

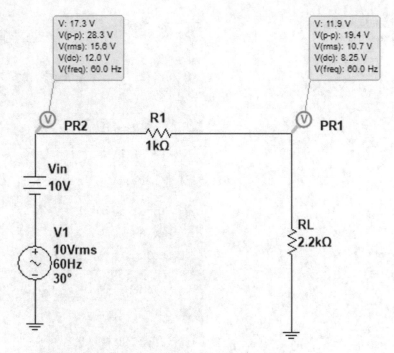

V: 17.3 V
V(p-p): 28.3 V
V(rms): 15.6 V
V(dc): 12.0 V
V(freq): 60.0 Hz

V: 11.9 V
V(p-p): 19.4 V
V(rms): 10.7 V
V(dc): 8.25 V
V(freq): 60.0 Hz

PR2

R1
1kΩ

PR1

Vin
10V

V1
10Vrms
60Hz
30°

RL
2.2kΩ

Fig. 2.74 Simulation result

After stopping the simulation, the measurement values will be cleared from the boxes that connected to the probes. If you want to clear the measured values, click the clear instrument data from simulate menu (Fig. 2.75).

Fig. 2.75 Simulate> clear
instrument data

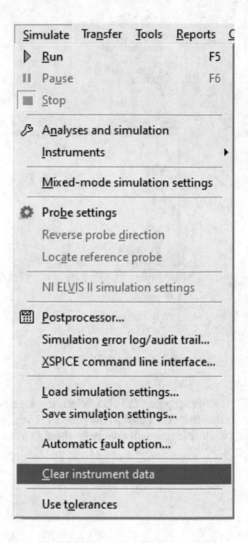

Now add a current probe (Fig. 2.76) to the circuit (Fig. 2.77). The current probe
has an arrow on it. The current that flows in the direction shown on current probe is
assumed to be positive.

Fig. 2.76 Current probe

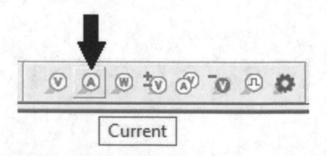

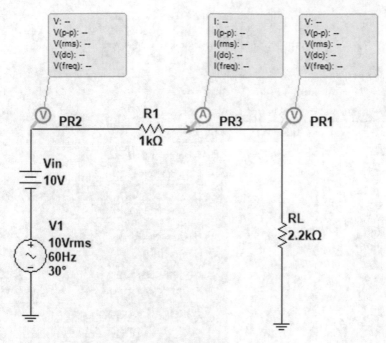

Fig. 2.77   Addition of current probe to the schematic

If you want to change the direction of current probe arrow, right click it and then click the reverse probe direction (Fig. 2.78).

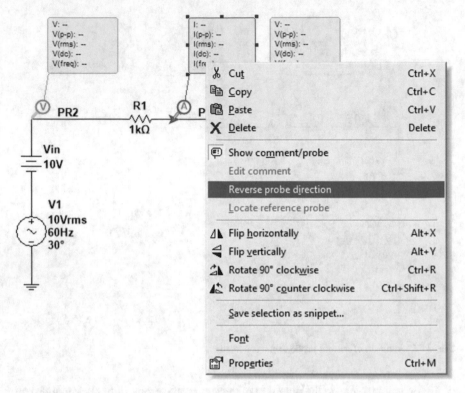

Fig. 2.78 Reversing the current probe direction

After running the simulation, the result shown in Fig. 2.79 is obtained. Let's verify the current probe reading. In the previous example, the circuit current found to be $i(t) = 3.75 + 4.4194 \sin(2\pi \times 60 \times t + 30°)$mA. The peak-peak of this current it $2 \times 4.4194 = 8.8388$ mA. The RMS is $\sqrt{(3.75)^2 + \frac{1}{2}(4.4194)^2} = 4.8814$ mA. The DC (average) value is 3.75 mA and the frequency of this function is 60 Hz.

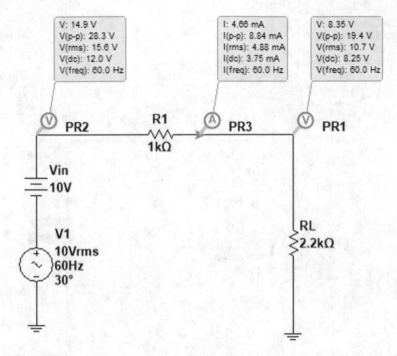

V: 14.9 V
V(p-p): 28.3 V
V(rms): 15.6 V
V(dc): 12.0 V
V(freq): 60.0 Hz

I: 4.66 mA
I(p-p): 8.84 mA
I(rms): 4.88 mA
I(dc): 3.75 mA
I(freq): 60.0 Hz

V: 8.35 V
V(p-p): 19.4 V
V(rms): 10.7 V
V(dc): 8.25 V
V(freq): 60.0 Hz

PR2   R1   PR3   PR1
         1kΩ

Vin
10V

V1
10Vrms
60Hz
30°

RL
2.2kΩ

Fig. 2.79 Simulation result

Sometime you need to measure the voltage and current of a node simultaneously. In such cases there is no need to use two separate probes, i.e. a voltage and a current probe. You can do the measurement with a voltage and current probe (Fig. 2.80). For instance, assume that you want to measure the voltage and current of resistor RL. In order to do this, add a voltage and current probe to the resistor RL (Fig. 2.81). After running the simulation, the result shown in Fig. 2.82 is obtained. The V and I shows the instantaneous values of voltage and current, respectively.

Fig. 2.80 Voltage and current probe

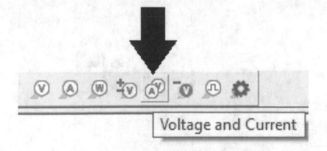

Voltage and Current

Fig. 2.81 Addition of voltage and current probe to the circuit

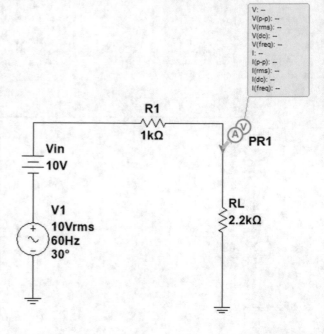

Fig. 2.82 Simulation result

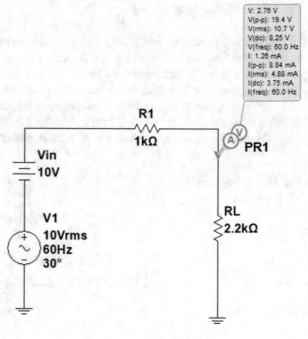

2.11   Example 4: Average Power Measurement with Power Probe

You can measure the average power of circuit with power probe (Fig. 2.83) easily.

Fig. 2.83 Power probe

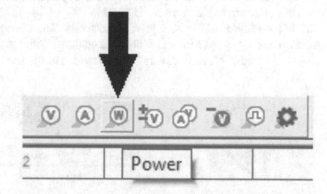

Assume that we want to measure the average power of components in the previous example circuit. In order to do this, connect a power probe to each element (Fig. 2.84).

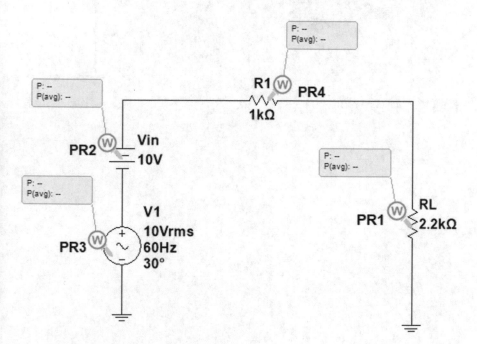

Fig. 2.84 Addition of power probe to the schematic

After running the simulation, the result shown in Fig. 2.85 is obtained. P shows the instantaneous value of power and P(avg) shows the average value of power. Note that the average power value for sources are negative. This shows that sources supply power to the circuit. Positive average power values show the consumption of power.

Let's check the results. The RMS value of circuit current is 4.88 mA. So, the resistor R1 and R2 power is 1 kΩ × (4.88 mA)^2 = 23.8 mW and 2.2 kΩ × (4.88 mA)^2 = 52.4 mW, respectively. The average power for DC source and the average power for AC source is calculated with the aid commands shown in Figs. 2.86 and 2.87, respectively. These results are the same as Fig. 2.85.

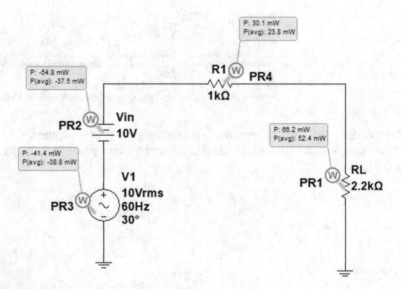

Fig. 2.85 Simulation result

Fig. 2.86 MATLAB
calculations

```
Command Window                                    ⊙
   >> f=60;T=1/f;w=2*pi*f;
   >> syms t
   >> v=2+10*sqrt(2)*sin(w*t+pi/6);
   >> i=(v+10)/(1e3+2.2e3);
   >> pDC=10*i;
   >> P_DC_avg=1/T*int(pDC,0,T)

   P_DC_avg =

   3/80

   >> eval(P_DC_avg)

   ans =

       0.0375

fx >> |
```

Fig. 2.87 MATLAB
calculations

```
Command Window                                    ⊙
   >> f=60;T=1/f;w=2*pi*f;
   >> syms t
   >> v=2+10*sqrt(2)*sin(w*t+pi/6);
   >> i=(v+10)/(1e3+2.2e3);
   >> p=v*i;
   >> Pavg=1/T*int(v*i,0,T)

   Pavg =

   31/800

   >> eval(Pavg)

   ans =

       0.0387

fx >>
```

According to the conservation of energy, the summation of instantaneous and average powers must be zero. Figures 2.88 and 2.89 show the summation of instantaneous and average powers for Fig. 2.85, respectively. The result is small however it is not zero. The reason is the limited precision of probes. If you like you can increase the precision of probes by double clicking on them and increase the number in the precision column (Fig. 2.90). This will increase the precision of shown numbers and the summation of powers will be closer to zero.

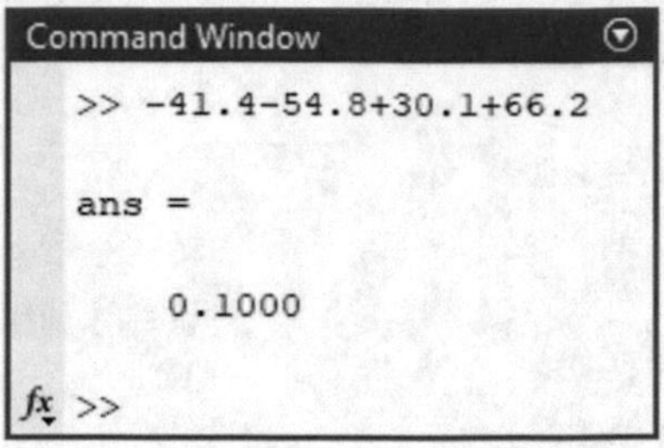

Fig. 2.88 Summation of
instantaneous power is
0.1 W

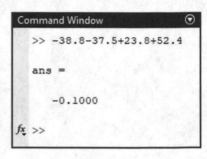

Fig. 2.89 Summation of
average powers is 0.1 W

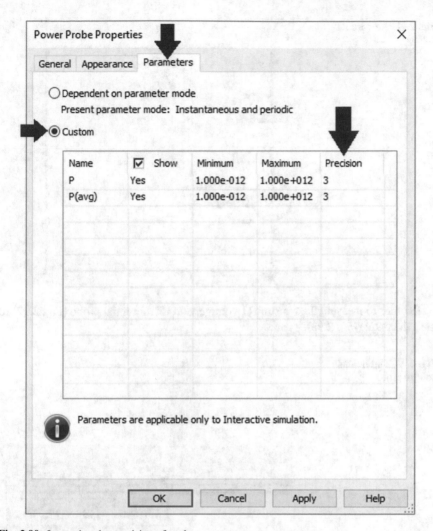

Fig. 2.90 Increasing the precision of probe

2.12   Example 5: Differential Probe

Assume that we want to measure the resistor R1 voltage (Fig. 2.91). The voltage probe can't be used since we want to measure with respect to a point which is not ground.

Fig. 2.91 Resistor R1
voltage

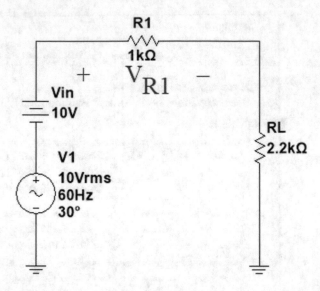

In order to measure the resistor R1 voltage, add a differential probe (Fig. 2.92) to the schematic (Fig. 2.93).

Fig. 2.92 Differential
voltage probe

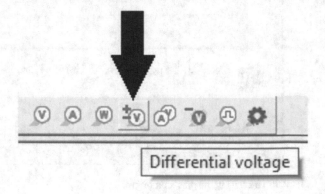

Fig. 2.93 Addition of differential voltage probe to the schematic

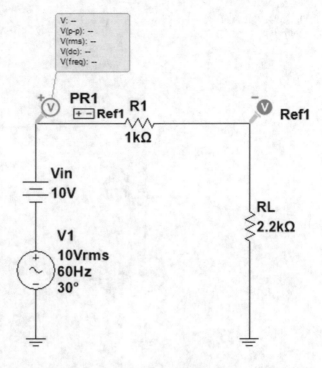

Run the simulation. The result shown in Fig. 2.94 is obtained. Let's check this result. The circuit current is found to be $i(t) = 3.75 + 4.4194 \sin (2\pi \times 60 \times t + 30^{\circ})$ mA. According to Ohm's law, the resistor voltage is be $v_{R1}(t) = R_1 \times i(-t) = 3.75 + 4.4194 \sin (2\pi \times 60 \times t + 30^{\circ})$ V. The peak-peak, RMS and DC values are calculated in Fig. 2.95.

Fig. 2.94 Result of
simulation

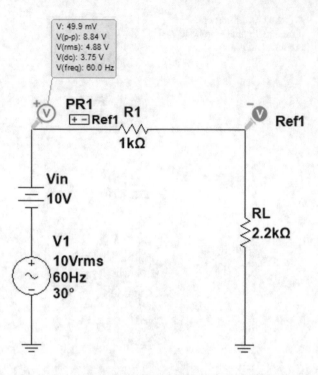

Fig. 2.95 MATLAB
calculations

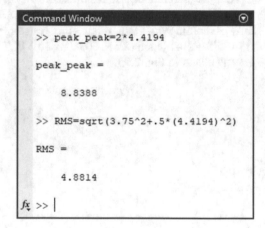

2.13 Example 6: RMS Measurement in Presence
of Harmonics

In this example, measurement of AC/DC voltmeters/ammeters is compared with the
probe measurements. Assume the schematic shown in Fig. 2.96.

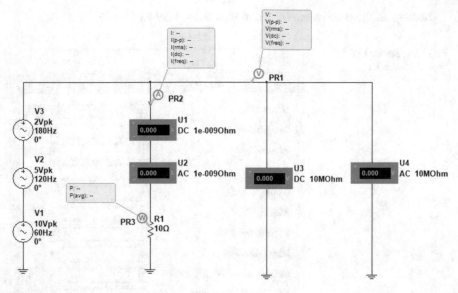

Fig. 2.96 Schematic of example 6

This schematic uses the ac voltage source element (Fig. 2.97).

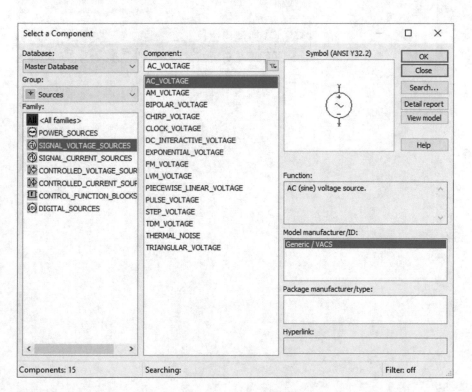

Fig. 2.97 AC voltage source

Settings of the sources are shown in Figs. 2.98, 2.99 and 2.100.

Fig. 2.98 Settings of AC voltage source V1

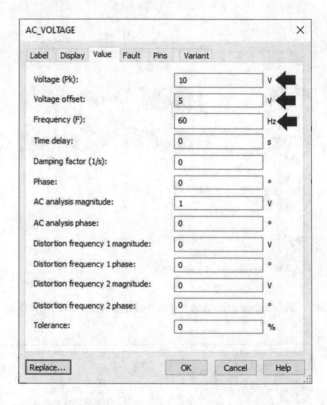

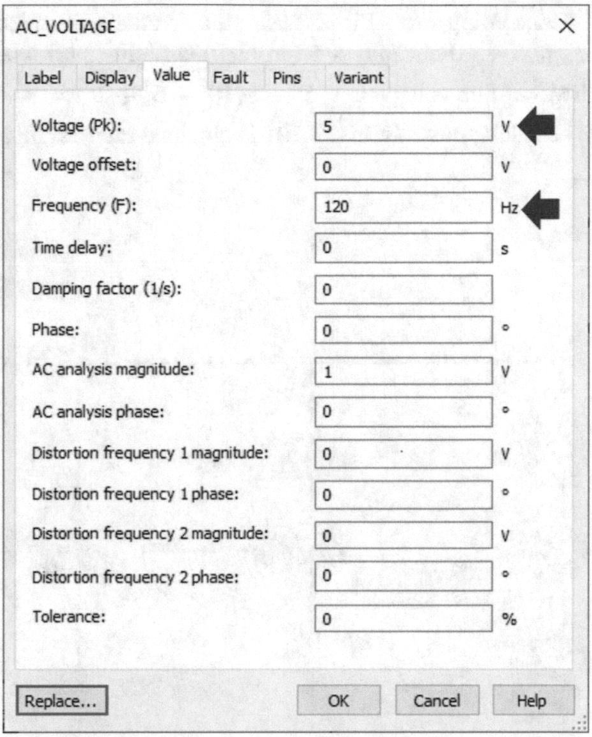

Fig. 2.99 Settings of AC voltage source V2

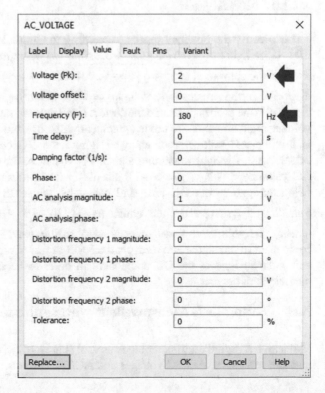

Fig. 2.100 Settings of AC voltage source V3

According to Fig. 2.96, the voltage entered to the circuit is v-$(t) = 5 + 10 \sin (\omega t) + 5 \sin (2\omega t) + 2 \sin (3\omega t)$ where $\omega = 2\pi \times 60 = 120\pi$. RMS of this voltage is $\sqrt{5^2 + \frac{1}{2}\left(10^2 + 5^2 + 2^2\right)} = 9.4604$ V. Now run the circuit. The result shown in Fig. 2.101 is obtained after running the circuit.

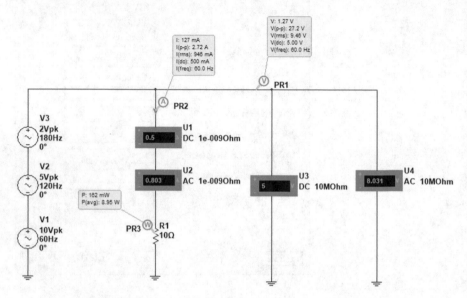

Fig. 2.101 Simulation result

Let's verify the obtained results. According to Ohm's law, the circuit current is $i(t) = 0.5 + 1 \sin (\omega t) + 0.5 \sin (2\omega t) + 0.2 \sin (3\omega t)$ where $\omega = 2\pi \times 60 = 120\pi$. The RMS of this current is $\sqrt{5^2 + \frac{1}{2}\left(1^2 + 0.5^2 + 0.2^2\right)} = 0.946$ A. The average power dissipated in the resistor is $R \times I_{RMS}^2 = 10 \times 0.946^2 = 8.9492$ W. According to Fig. 2.101, the probes measured the correct RMS values, however the RMS values of voltmeter and ammeter are not in agreement with our hand calculations. The reason is that the AC voltmeter or ammeter ignore the DC component of signal and calculate the RMS of remaining signal. For instance, the signal which enters the AC voltmeter U4 is $v(t) = 5 + 10 \sin (\omega t) + 5 \sin (2\omega t) + 2 \sin (3\omega t)$. The AC voltmeter ignores the DC term and calculates the RMS of $\hat{v}(t) = 10 \sin (\omega t) + 5 \sin (2\omega t) + 2 \sin (3\omega t)$ which equals to $\sqrt{\frac{1}{2}\left(10^2 + 5^2 + 2^2\right)} = 8.0312$ V. A real (cheap) AC voltmeters and ammeters work in this way, i.e., they ignore the DC component of the signal and measure the RMS of rest of signal. Multisim voltmeter and ammeter ignores the DC component in order to be similar to the real world measurement devices.

2.14 Example 7: Measurement with Multimeter

Multisim has digital multimeter block as well (Fig. 2.102). You can use this block to measure resistance, AC/DC voltages and AC/DC currents.

Fig. 2.102 Digital
multimeter

If you can't see digital multimeter block in the right side of screen, right click on the area shown in Fig. 2.103. Then check the instruments (Fig. 2.104). You can even add a digital multimeter to the schematic by using the simulate> instruments as well (Fig. 2.105).

Fig. 2.103 Top section of Multisim window

Fig. 2.104 Ensure that
Instruments is checked

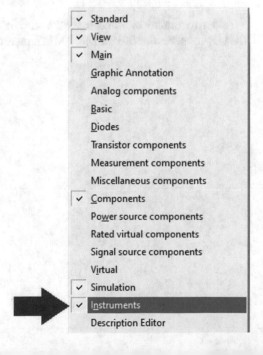

Fig. 2.105 Simulate>
Instruments

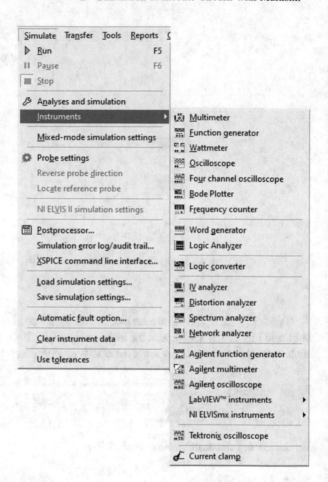

Add two digital multimeter to the circuit of previous example (Fig. 2.106).
XMM1 measures the current and XMM2 measures the voltage.

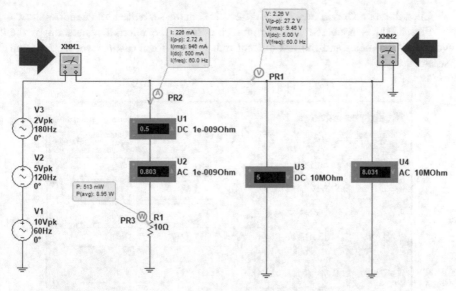

Fig. 2.106 Simulation result

Double click on the digital multimeter block. The window shown in Fig. 2.107 will appear. The button with sinusoidal wave shows the AC mode measurement and the button with straight line shows the DC mode measurements. In DC mode, the digital multimeter measures the average value of the signal and in the AC mode it measures the RMS of signal without DC component.

Fig. 2.107 Multimeter window

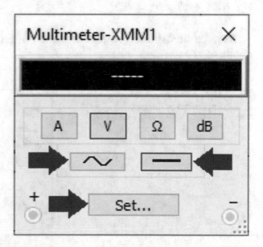

Click the set button in Fig. 2.107. After clicking the set button, the window shown in Fig. 2.108 is appeared and permits you to set the internal resistance of the voltmeter, ammeter and the test current that is used for measurement in the Ohm section.

Multimeter Settings			×
Electronic setting			
Ammeter resistance (R):	10	μΩ	
Voltmeter resistance (R):	1	GΩ	
Ohmmeter current (I):	10	nA	
dB relative value (V):	774.597	mV	
Display setting			
Ammeter overrange (I):	1	GA	
Voltmeter overrange (V):	1	GV	
Ohmmeter overrange (R):	10	GΩ	
OK		Cancel	

Fig. 2.108 Multimeter settings

Run the simulation. The results shown in Figs. 2.109, 2.110, 2.111 and 2.112 is obtained.

Fig. 2.109 Average
(DC component) of current

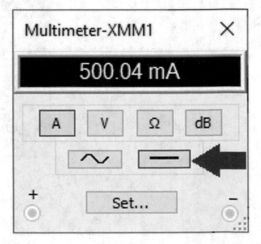

Fig. 2.110 RMS of AC
component of the current

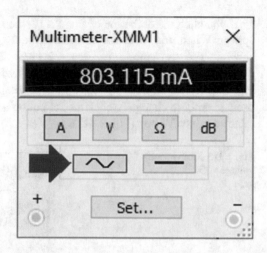

Fig. 2.111 Average value
of voltage

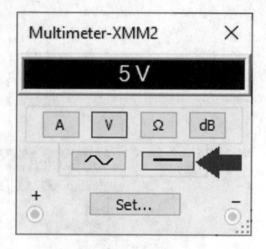

Fig. 2.112 RMS of AC
component of voltage

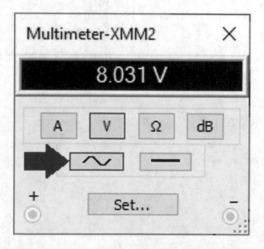

If you click the db button of Fig. 2.112, then the result shown in Fig. 2.113 will appear. When you press the db button, the $20 \log_{10}\left(\frac{V}{0.774597}\right)$ is calculated and is shown. For instance, in Fig. 2.112, the voltmeter reads 8.031 V. $20 \log_{10}\left(\frac{8.031}{0.774597}\right) = 20.314$ dB (Fig. 2.114) which is exactly the value shown in Fig. 2.113. The number 0.774597 in the $20 \log_{10}\left(\frac{V}{0.774597}\right)$ can be changed by the user. In order to change it, click the set button and enter the desired value in the dB relative value (V) box (Fig. 2.115).

Fig. 2.113 Conversion of voltage to db

Fig. 2.114 Calculation of $20 \log_{10}\left(\frac{8.031}{0.774597}\right)$

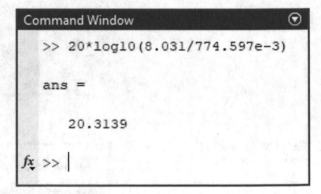

Fig. 2.115 Multimeter
settings window

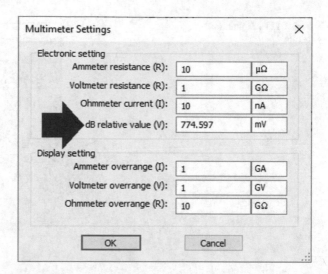

You can use the ohmmeter to measure the resistance of a series of connected resistors. For instance, consider schematic shown in Fig. 2.116. After running the simulation, the result shown in Fig. 2.117 is obtained. Figure 2.118 verifies the obtained result.

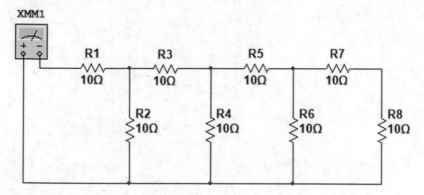

Fig. 2.116 Measurement of resistance with multimeter

Fig. 2.117 Simulation
result

Fig. 2.118 Verification of obtained result

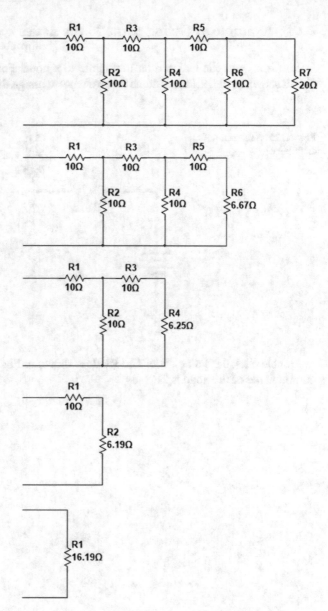

2.15   Example 8: Giving name to the nodes

In Multisim, you can assign a desired name to a node. For instance, consider the circuit shown in Fig. 2.119. Assume that we want to assign the name V2 to the center node.

Fig. 2.119 Schematic of example 8

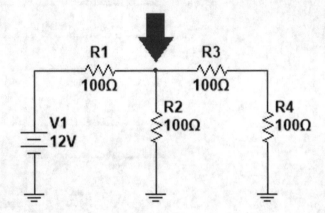

Double click the center node. The window shown in Fig. 2.120 is appeared. The current name of the node is 2.

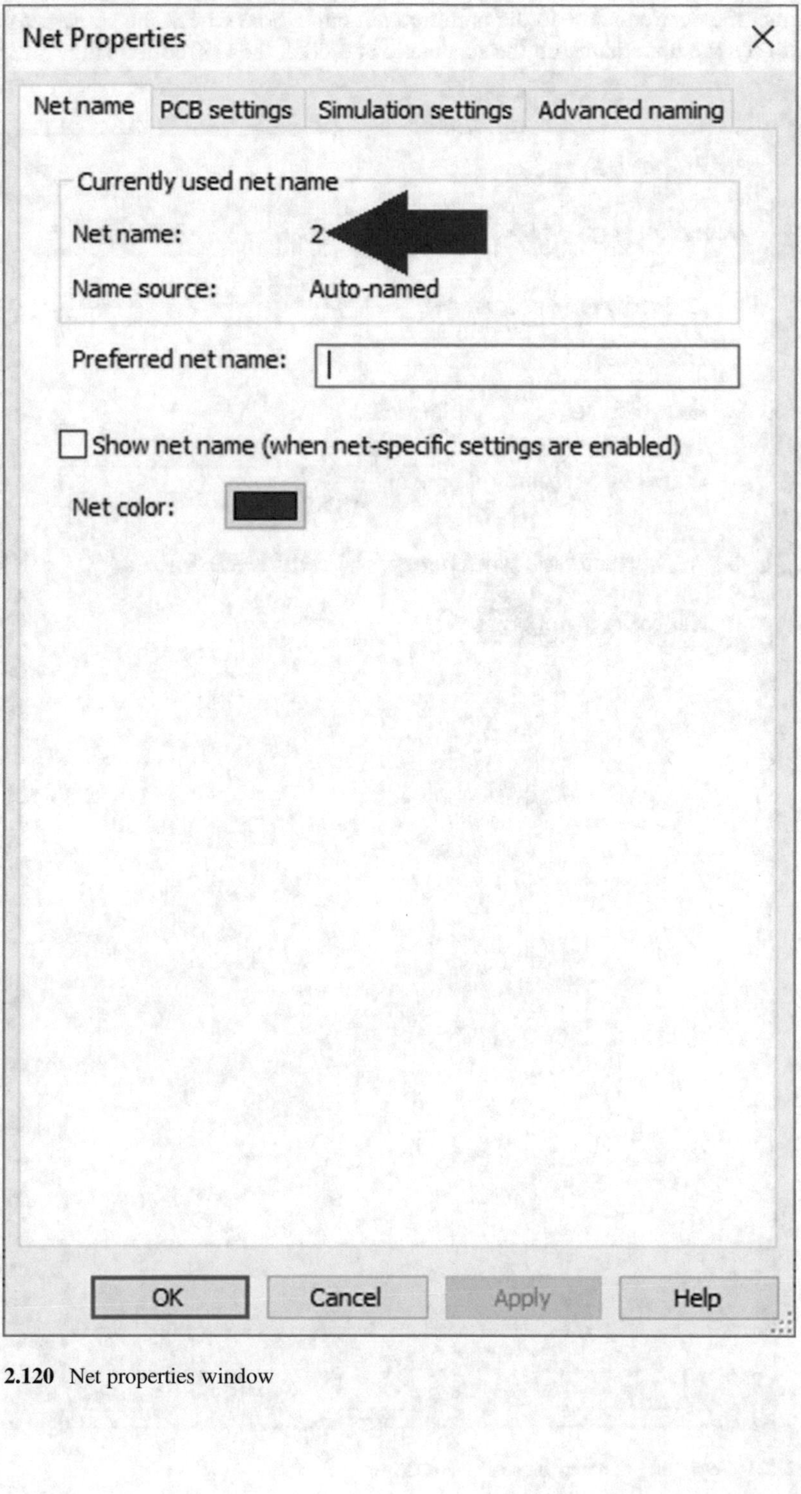

Fig. 2.120 Net properties window

Enter the desired name to the preferred net name box. Check the show net name box to see the node name on the schematic and click the OK button (Fig. 2.121).

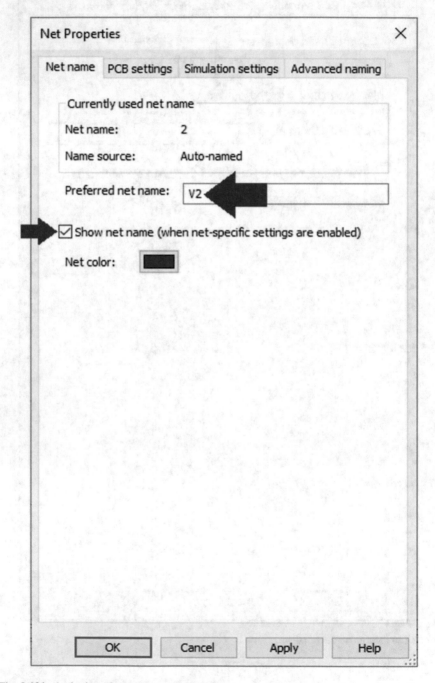

Fig. 2.121 Assigning of a new name to the node

Now the schematic changes to what shown in Fig. 2.122.

Fig. 2.122 Name of the
node is shown on the
schematic

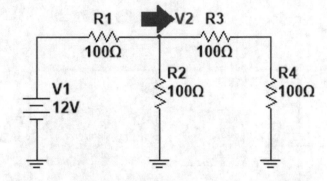

2.16 Example 9: Ground Element

PSIM has different symbol for ground connection (Fig. 2.123) however all of them
are the same from the simulation point of view. In order to understand this subject,
assume the schematic shown in Fig. 2.124.

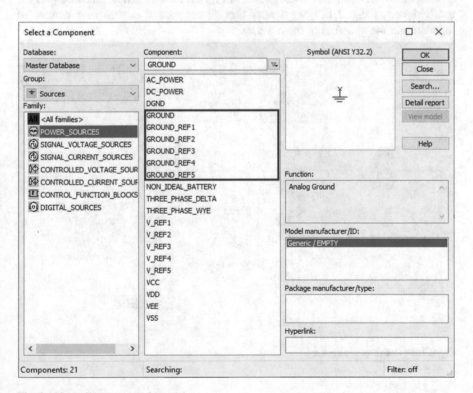

Fig. 2.123 Different types of grounds

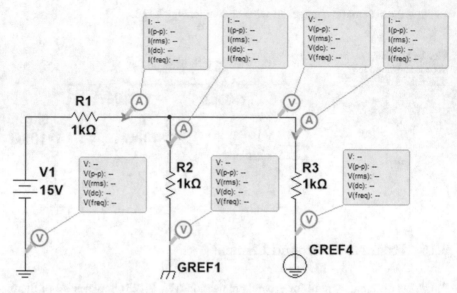

Fig. 2.124 Test simulation

Run the simulation. The result shown in Fig. 2.125 is obtained. According to the result shown in Fig. 2.125, we deduce that all the ground elements are connected together and the potential of them are zero.

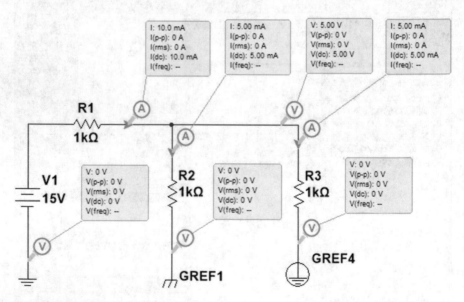

Fig. 2.125 Result of simulation

2.17 Example 10: Junction Tool

Assume that you want to connect the two nodes that is shown in Fig. 2.126.

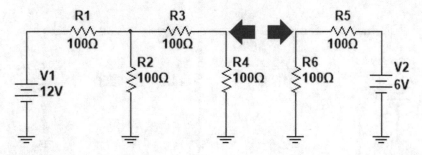

Fig. 2.126 Schematic of example 10

Press the Ctrl+J or click the Junction tool (Fig. 2.127) in order to activate the junction tool.

Fig. 2.127 Place> junction

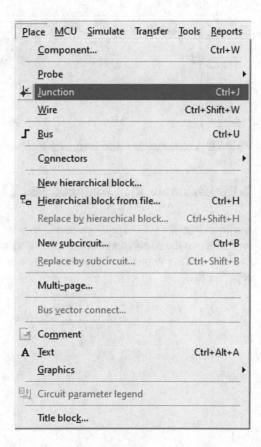

Click on one of the nodes. This add a small dot to the wire (Fig. 2.128). Now you can bring your mouse close to the added dot and when it changed to crosshair, you can click it and connect it to the destination node (Fig. 2.129).

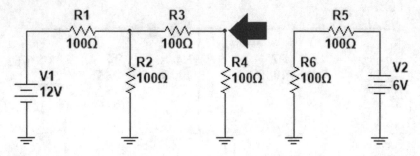

Fig. 2.128 Addition of a junction to the schematic

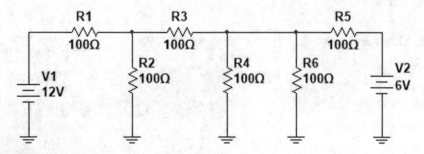

Fig. 2.129 Connecting the added junction to the other part of the circuit

2.18   Example 11: Comment Block

Multisim permits you to add comments to your circuits. The comments don't have any effect on the simulation however it is a valuable tool to add some description to the circuit. One way to add comments to the circuit is to use the comment tool (Fig. 2.130).

Fig. 2.130 Place>
comment

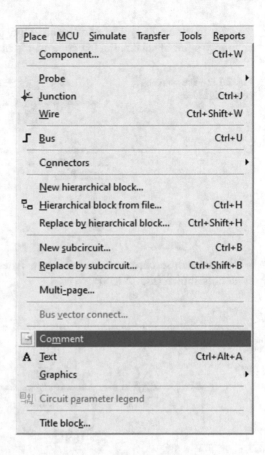

Activate the comment tool by clicking the place> comment. Then click on the schematic and type what you want (Fig. 2.131).

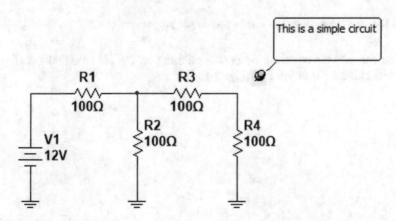

Fig. 2.131 Entering the comment content

After finishing the typing, click on an empty point of schematic. After clicking on an empty point of schematic, the comment is hidden (Fig. 2.132).

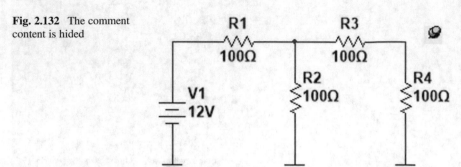

Fig. 2.132 The comment content is hided

Put the mouse cursor on the pin icon. After few seconds, the comment block content is shown (Fig. 2.133).

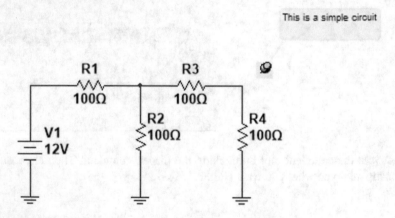

Fig. 2.133 The comment content is shown when you put the cursor on it

You can edit the comment by double clicking on the pin icon and entering the new text into the Comment box (Fig. 2.134).

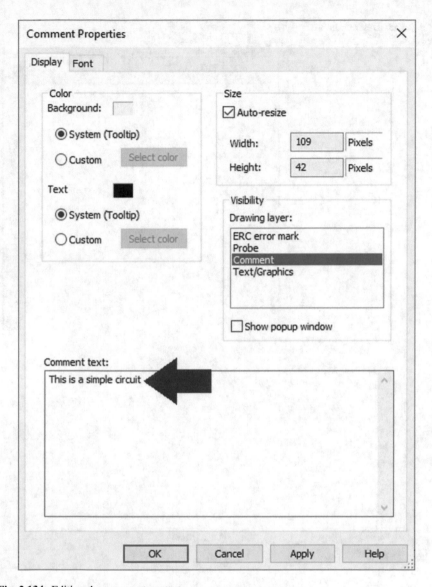

Fig. 2.134 Editing the comment content

2.19   Example 12: Circuit Description Box

The circuit description box (Fig. 2.135) is another way to add some comment to the circuit. You can also place bitmaps, sound and video in the description box. If you can't see the description box, check the description box in the view menu (Fig. 2.136).

Fig. 2.135 Description box
window

Fig. 2.136 Ensure that
description box is checked

Click the title block editor (Fig. 2.137) in order to start editing the title block.

Fig. 2.137 Tools> title
block editor

After clicking the title block editor, the environment shown in Fig. 2.138 will be opened. Use this environment to type the comment. After typing the comment, click the close button (Fig. 2.139) to finish editing and returning to the Multisim environment.

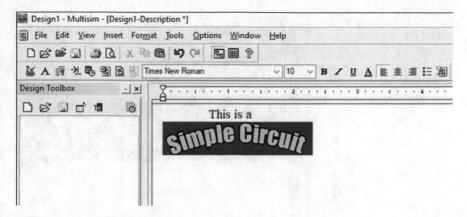

Fig. 2.138   Entering the desired content

Fig. 2.139   Closing the title
block editor

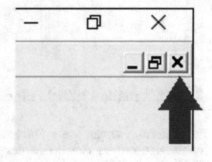

Now, the typed comment is shown in the description box (Fig. 2.140).

Fig. 2.140 The entered content is added to the description box

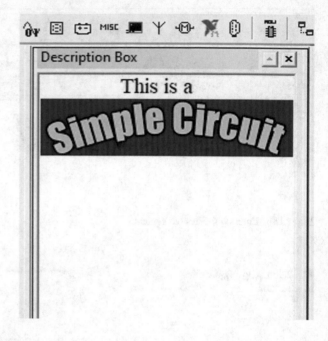

2.20   Example 13: Title Block

In Multisim, you can add a title block to the schematic. The title block shows information such as circuit title, designer name, review date, etc. Click the title block (Fig. 2.141) to add a title block to the schematic.

Fig. 2.141 Place> title
block

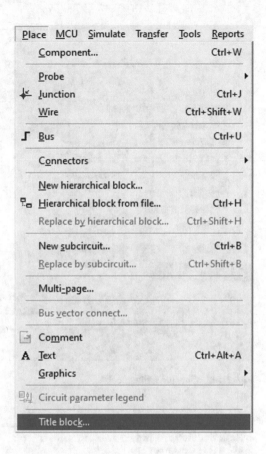

After clicking the title block, the open window shown in Fig. 2.142 will appear and permits you to select a block from the available blocks. For instance, select default.tb7 and click the open button (Fig. 2.143). Then click on the desired empty location of the schematic to add the title block to there.

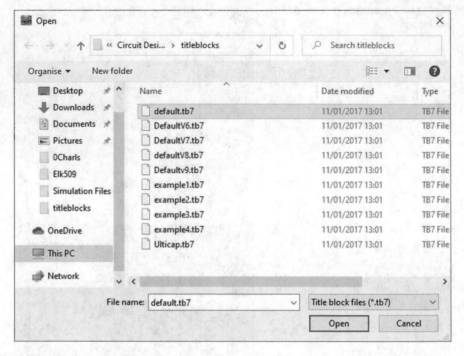

Fig. 2.142 Opening a title block

National Instruments 801-111 Peter Street Toronto, ON M5V 2H1 (416) 977-5550		NATIONAL INSTRUMENTS™
Title: Multiviberator	Desc.: A	
Designed by: B	Document No: C	Revision: D
Checked by: E	Date: F	Size: A
Approved by: G	Sheet 1 of 1	

Fig. 2.143 default.tb7 title block

If you double click on the title block, the window shown in Fig. 2.144 appears. Now, you can enter desired text for each field of title block.

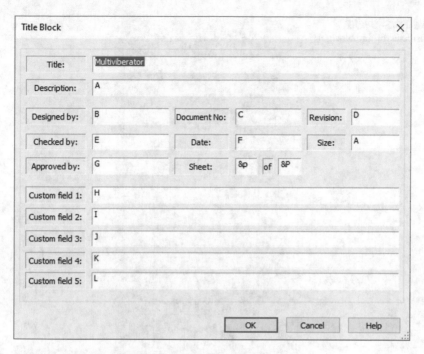

Fig. 2.144 Settings of default.tb7 title block of Fig. 2.143

You can move the title block to desired location. In order to do this, left click on the title block and without releasing the button, drag the title block to the desired location. You can set the title block position by left clicking on it and select one of the options in the edit>title block position as well (Fig. 2.145).

Fig. 2.145 Default
positions for title block

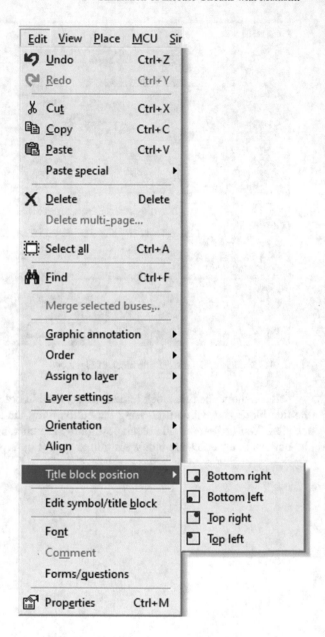

2.21 Example 14: Exporting the Schematic as a Graphical File

You can export the drawn schematics as a graphical file. This is very useful when you want to write a report and you need to add the circuit schematic to the report. Click the capture screen area to start capturing the schematic (Fig. 2.146).

Fig. 2.146 Tools> capture screen area

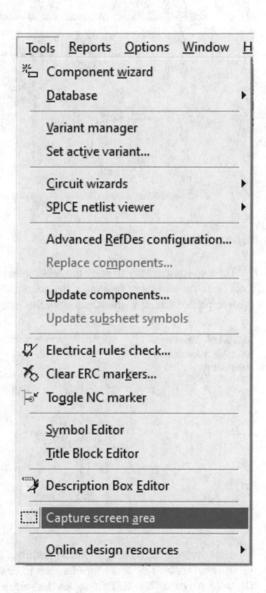

After clicking the tools> > capture screen area, a rectangle appears on the screen and permits you to select the capture area. Use the small dots to move the borders of capturing area. When the rectangle surrounded all of the desired area, click the copy button (shown with big arrow in Fig. 2.147) to copy the schematic into the clipboard. Now you can paste it in the desired text editor.

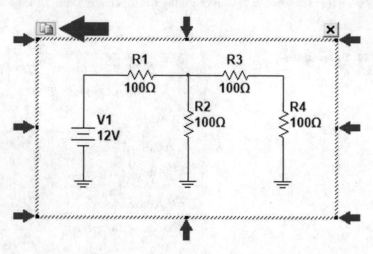

Fig. 2.147 Determining the capture area

There is another way to copy the schematic into the clipboard as a graphical file. Left click on an empty region of the schematic and draw a rectangle around the desired part of circuit without releasing the mouse left button (Fig. 2.148).

Fig. 2.148 Drawing a rectangle around the components

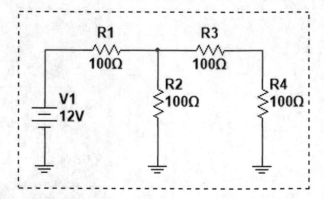

After drawing the rectangle which surrounds the desired part of circuit, release the mouse left button. This cause the elements to be selected (Fig. 2.149). Now, press the Ctrl+C keys of keyboard. This copies the selected circuit into the clipboard and you can paste it in any text editor using the Ctrl+V.

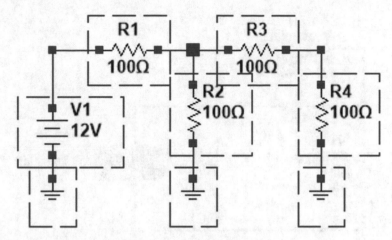

Fig. 2.149 Components inside the rectangle of Fig. 2.148 are selected

2.22 Example 15: Voltage Division in AC Circuits

In this example, we study the voltage division in the circuit of Fig. 2.150. The voltage source V1 has the equation $V_1(t) = 120\sqrt{2}\sin\left(2\pi \times 60 \times t + 45°\right)$.

Fig. 2.150 Schematic of example 15

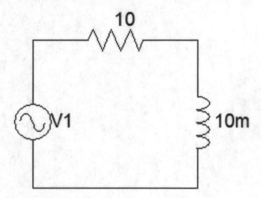

Figure 2.151 shows the required schematic.

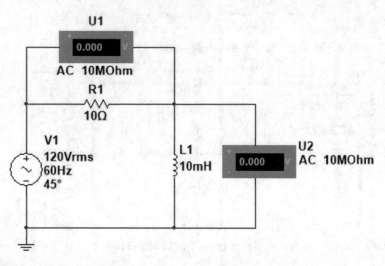

Fig. 2.151 Multisim equivalent of Fig. 2.150

Settings of voltage source V1 is shown in Fig. 2.152.

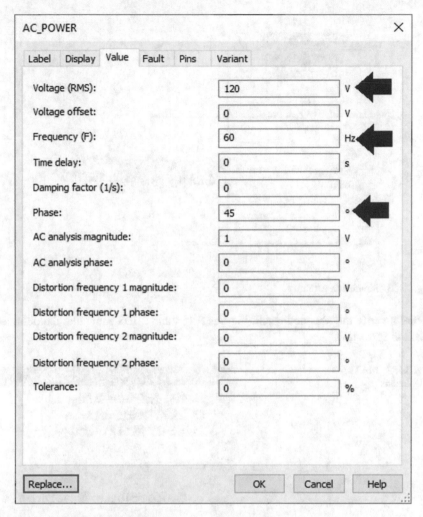

Fig. 2.152 Settings of voltage source V1

Run the simulation. The result shown in Fig. 2.153 is obtained.

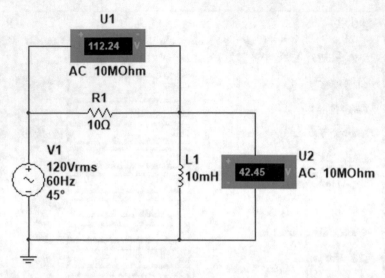

Fig. 2.153 Simulation result

Let's verify the obtained result. Figures 2.154 and 2.155 show that the obtained result is correct.

Fig. 2.154 MATLAB
calculations

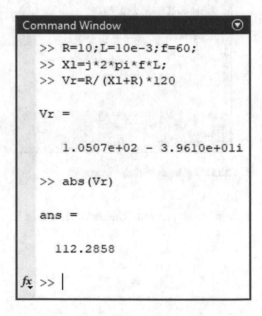

Fig. 2.155 MATLAB
calculations

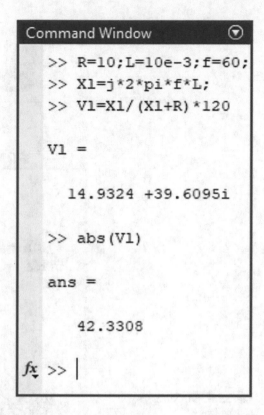

```
>> R=10;L=10e-3;f=60;
>> X1=j*2*pi*f*L;
>> V1=X1/(X1+R)*120

V1 =

   14.9324 +39.6095i

>> abs(V1)

ans =

   42.3308
```

You can use a current probe to measure the circuit current (Fig. 2.156). The RMS of circuit current is 11.2 A. The commands shown in Fig. 2.157 shows that Multisim result is correct.

Fig. 2.156 Simulation
result

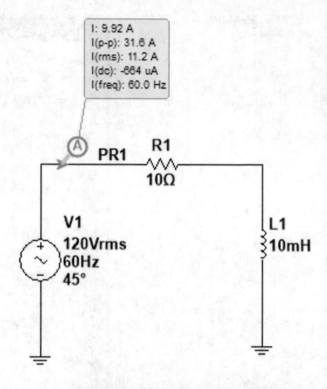

Fig. 2.157 MATLAB
calculations

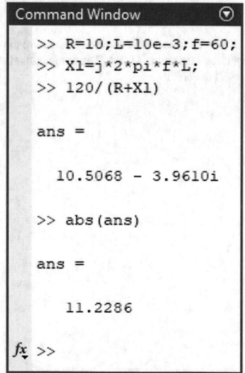

Now, assume that we want to analyze the circuit shown in Fig. 2.158. There is no need to draw the schematic of this circuit from scratch. You can double click on the inductor in Fig. 2.156 and use the replace button (Fig. 2.159).

Fig. 2.158 Simple RC circuit

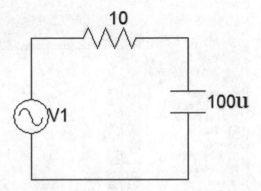

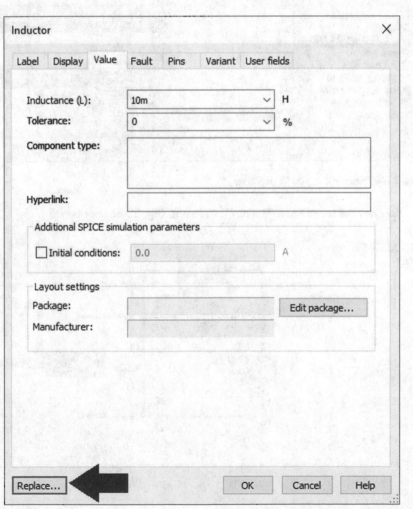

Fig. 2.159 Replace button

After clicking the replace button, the select a component window (Fig. 2.160) appears and you can select the required capacitor.

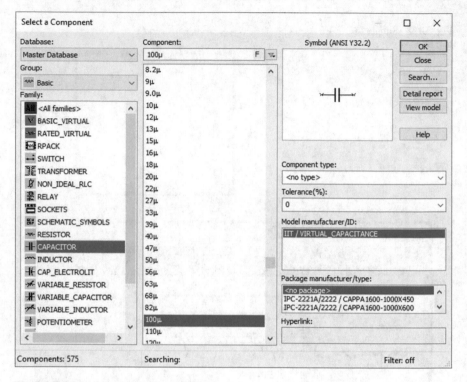

Fig. 2.160 Selection of 100 μF capacitor

Now run the simulation. The result shown in Fig. 2.161 is obtained.

Fig. 2.161 Simulation result

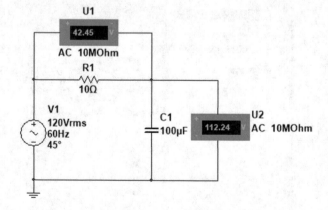

Let's verify the Multisim result. The calculation shown in Figs. 2.162 and 2.163 shows that the Multisim result is correct.

Fig. 2.162 MATLAB calculations

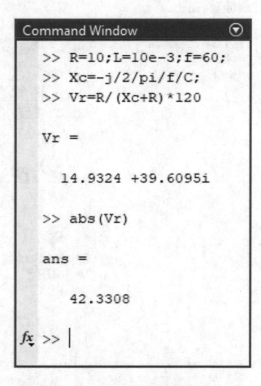

Fig. 2.163 MATLAB calculations

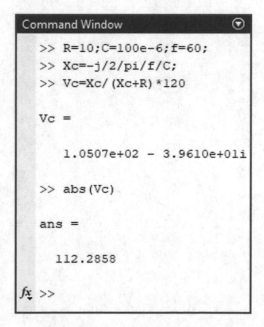

You can use a current probe to measure the circuit current (Fig. 2.164). The RMS of circuit current is 12 A. The commands shown in Fig. 2.165 shows that Multisim result is correct.

Fig. 2.164 Simulation result

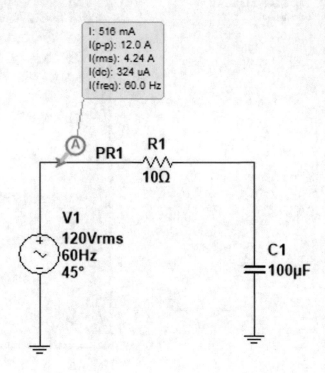

Fig. 2.165 MATLAB calculations

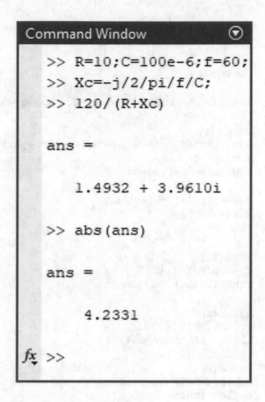

2.23 Example 16: Power Waveforms

The power of components in Fig. 2.166 is studied in this example.

Fig. 2.166 Schematic of example 16

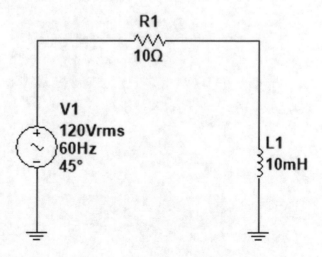

The following code calculates and draws the power waveforms for each element.

```
clc
clear all

R=10; L=10e-3; f=60; w=2*pi*f; Vm=120*sqrt(2); phi=pi/4;

syms i(t)
ode = L*diff(i,t)+R*i == Vm*sin(w*t+phi);
cond = i(0) == 0;
iSol(t) = dsolve(ode,cond);

P=Vm*sin(w*t+phi)*iSol;
PR=R*iSol^2;
PL=P-PR;

figure(1)
ezplot(P,[0 0.05])
title('Total load power')

figure(2)
ezplot(PR,[0 0.05])
title('Resistor power')

figure(3)
ezplot(PL,[0 0.05])
title('Inductor power')
```

After running the code, the result shown in Figs. 2.167, 2.168 and 2.169 is obtained.

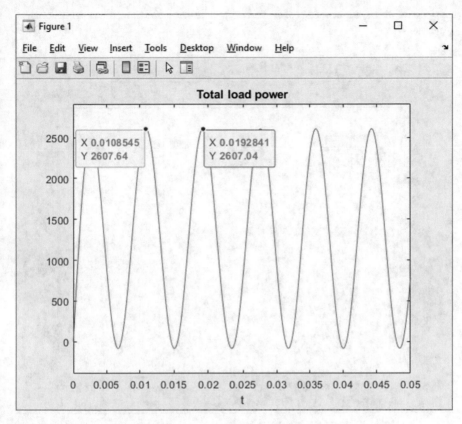

Fig. 2.167 Summation of instantaneous power resistor and inductor

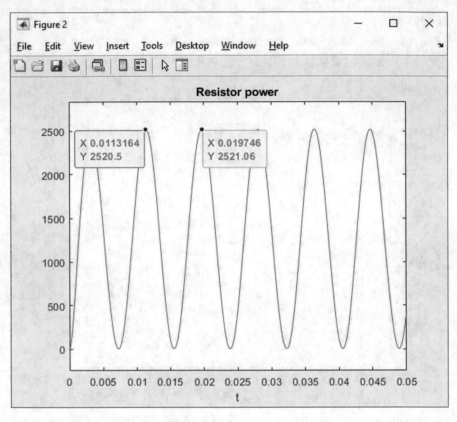

Fig. 2.168 Instantaneous power of resistor

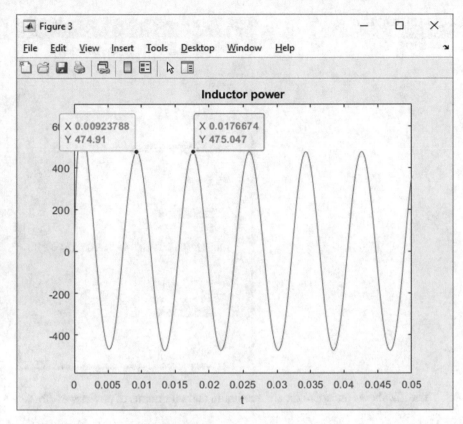

Fig. 2.169 Instantaneous power of inductor

Note that the frequency of power waveforms is two times bigger than the source V1 frequency. The cursors in Figs. 2.167, 2.168 and 2.169 are placed manually in the two consecutive maximum points. That is why the result shown in Fig. 2.170 is not exactly 120 Hz.

Fig. 2.170 MATLAB
calculations

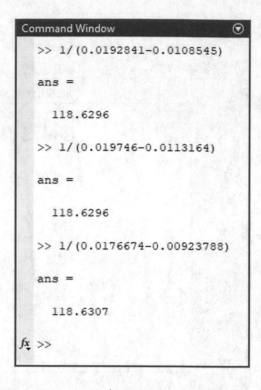

Table 2.2 shows the approximate maximum and minimums of power waveforms.

Table 2.2 Maximum and
minimums for power
waveforms

Waveforms	Min (W)	Max (W)
Total load power	2607.6	−86.3
Resistor power	0.3	2521.1
Inductor power	−475.9	475.9

The average power for resistor can be calculated by averaging the maximum and
minimum for resistor instantaneous power. According to Fig. 2.171, the average
power dissipated in the resistor is 1.26 kW.

Fig. 2.171 Average power
of resistor

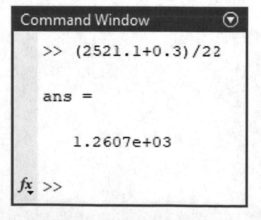

Let's verify the obtained results with Multisim. Click the interactive button (Fig. 2.172) or simulate> analyses and simulation (Fig. 2.173).

Fig. 2.172 Interactive button

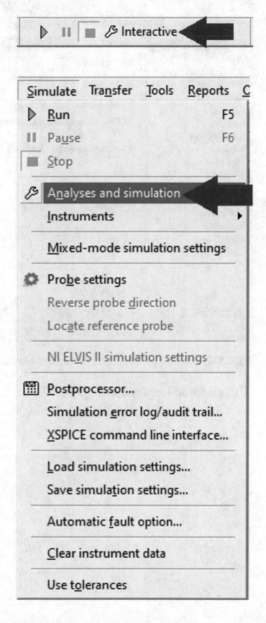

Fig. 2.173 Simulate> analyses and simulation

Select transient analysis in the appeared window (Fig. 2.174). We want to analyze the circuit for duration of 0.1 s.

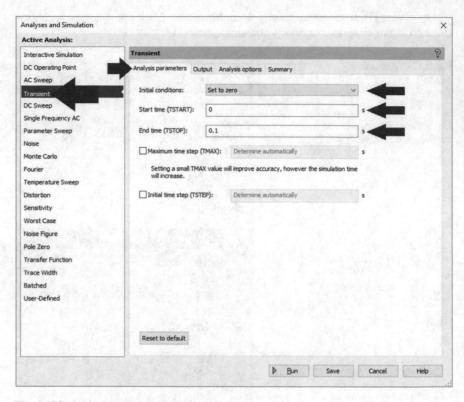

Fig. 2.174 Analyses parameters tab of transient analysis

Click the add expression (Fig. 2.175).

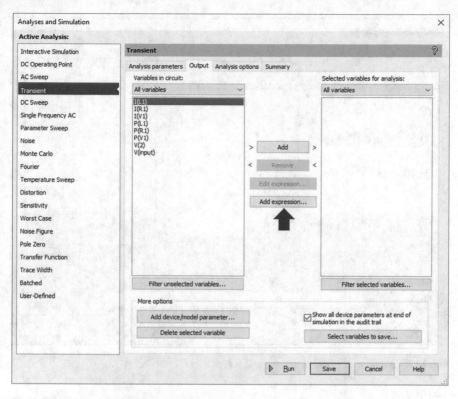

Fig. 2.175   Output tab of transient analysis

Double click the P(R1) in the appeared window. This adds the P(R1) to the expression box. P(R1) is the instantaneous power waveform resistor R1 (Fig. 2.176).

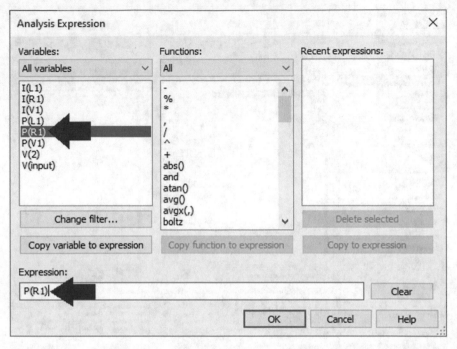

Fig. 2.176 Addition of P(R1) to the expression box

Now double click the + sign. This adds the + sign to the expression box (Fig. 2.177).

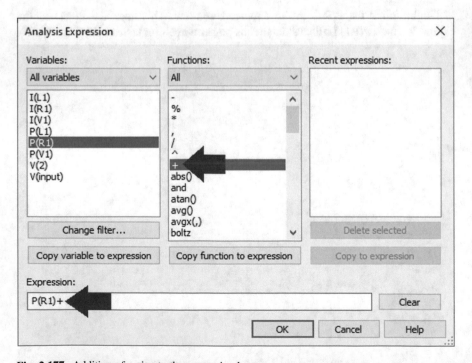

Fig. 2.177 Addition of + sign to the expression box

Now double click the P(L1). This adds the P(L1) to the expression box. P(L1) is the instantaneous power of inductor L1. The P(R1) + P(L1) in the expression box asks Multisim to draw the summation of instantaneous power of resistor R1 and inductor L1. Instead of using the procedure described above, you can simply enter the P(R1) + P(L1) to the expression box using your keyboard.

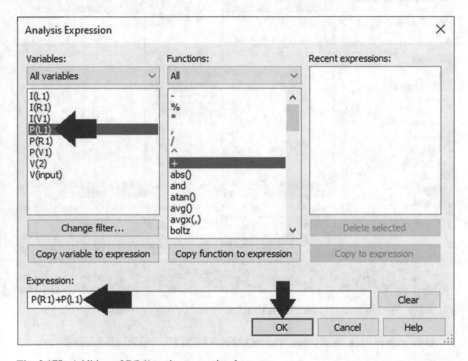

Fig. 2.178 Addition of P(L1) to the expression box

After clicking the OK button in Fig. 2.178, the expression will be added to the selected variables for analysis column of output tab. Now click the run button (Fig. 2.179).

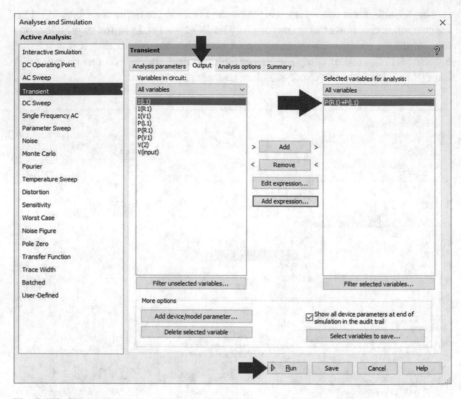

Fig. 2.179 P(R1) + P(L1) is added to the right list

After running the simulation, the result shown in Fig. 2.180 is obtained.

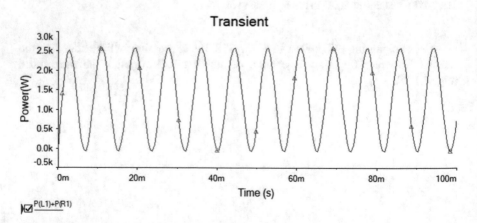

Fig. 2.180 Simulation result (summation of instantaneous power resistor and inductor)

You can zoom in the graph (Fig. 2.181) and use the cursors to measure the frequency. According to Fig. 2.182, the waveform frequency is 120 Hz and its maximum is 2.6083 kW. This number is quite close to the number in Table 2.2.

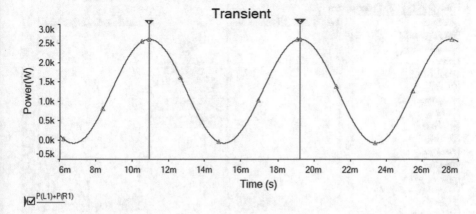

Fig. 2.181   Addition of cursors to the graph

Fig. 2.182   Coordinates of
cursors

Cursor		✕
		P(L1)+P(R1)
x1		10.8948m
y1	➤	2.6083k
x2		19.2281m
y2		2.6082k
dx		8.3333m
dy		-20.2818m
dy/dx		-2.4338
1/dx	➤	120.0005

Now we want to measure the average power dissipated in the load. Click the transient button (Fig. 2.183) to open the analysis and simulation window (Fig. 2.184). Now click the remove button (Fig. 2.184) to remove the P(R1) + P (L1) from the selected variables for analysis list (Fig. 2.185).

Fig. 2.183   Transient button

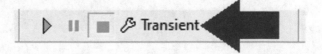

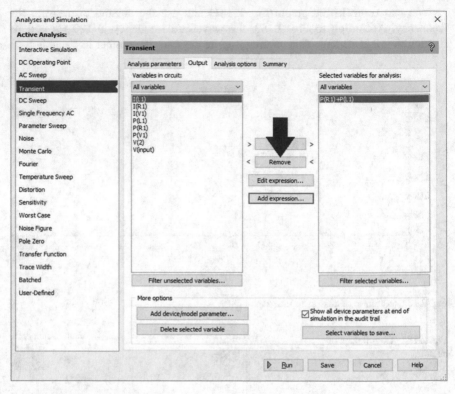

Fig. 2.184 The remove button

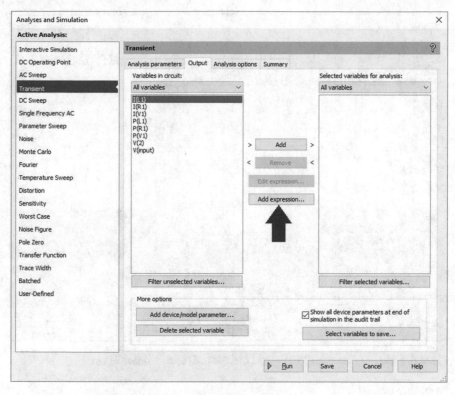

Fig. 2.185 The selected variables for analysis list is cleared

Click add expression button in Fig. 2.185. Now type the "avgx(P(R1) + P (L1),16.6667e-3)" in the expression box and click the OK button (Fig. 2.186). this expression asks Multisim to draw the moving average of summation of instantaneous power for resistor and inductor. Since the frequency is 60 Hz, the averaging interval has the length of $\frac{1}{60\ \text{Hz}} = 16.6667$ ms.

After clicking the OK button in Fig. 2.186, the entered expression will be added to selected variables for analysis list (Fig. 2.187).

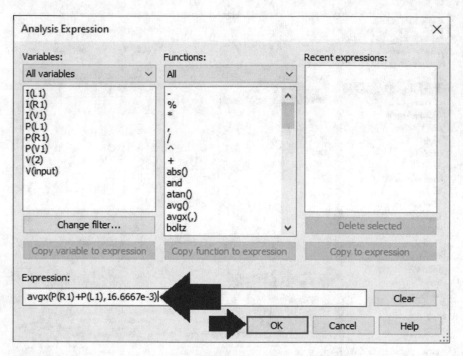

Fig. 2.186 Entering the avgx(P(R1) + P(L1),16.6667e-3) to the expression box

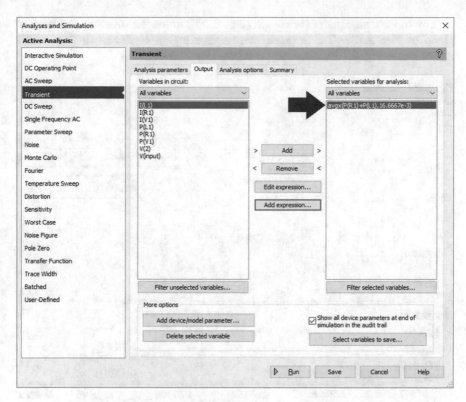

Fig. 2.187 Entered expression is added to the right list

Click the run button in Fig. 2.187 to run the simulation. The simulation result is shown in Fig. 2.188.

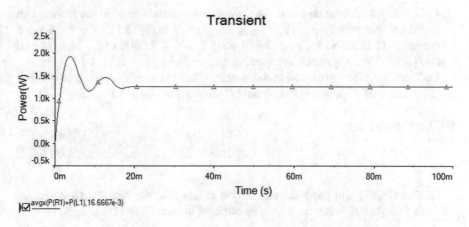

Fig. 2.188 Simulation result

You can use the zoom in area icon (Fig. 2.189) to see the steady state region of graph of Fig. 2.188. The zoomed steady state region is shown in Fig. 2.190. According to Fig. 2.190, the steady state value is 1.2598 kW.

Fig. 2.189 Zoom in area icon

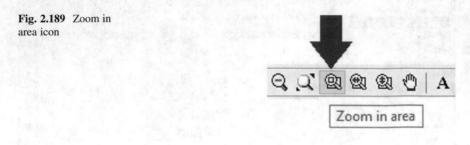

Zoom in area

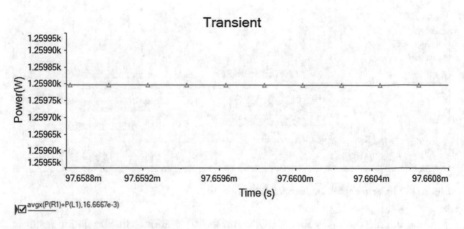

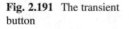

Fig. 2.190 Steady state value of Fig. 2.188

Let's verify the Multisim result. The average power dissipated in the resistor can be calculated using the $P_{AVG} = R \times I_{RMS}^2$. According to Fig. 2.157, the RMS value of current is 11.2286 A. So, $P_{AVG} = 10 \times 11.2286^2 = 1.2608$ kW. The Multisim result (1.2598 kW) is quite close to calculated number.

Let's see the power drawn from AC source. Click the transient button in order to see the waveform of power drawn from AC source (Fig. 2.191).

Fig. 2.191 The transient button

Click on P(V1) and press the add button to add it to the selected variables for analysis list. Then run the simulation by clicking the run button (Fig. 2.192).

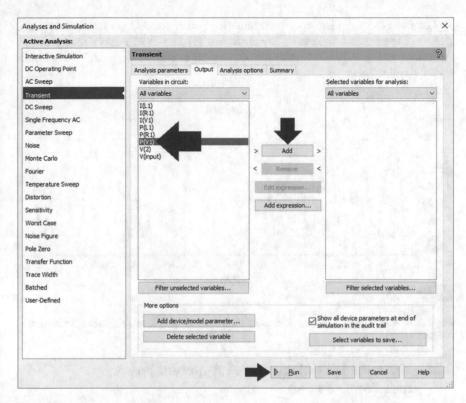

Fig. 2.192 Analyses and simulation window

The simulation result is shown in Fig. 2.193. The instantaneous power waveform for AC source is negative. This tells us that AC source supply power to the circuit. Remember from basic circuit theory that positive power means consumption of power.

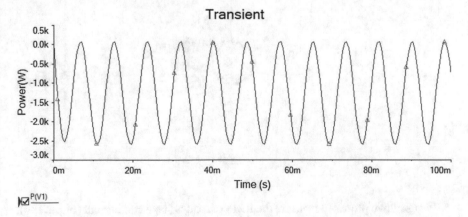

Fig. 2.193 Simulation result (instantaneous power of AC source)

Let's see the summation of instantaneous power for AC source, inductor and resistor. Figure 2.194 shows the waveform for P(L1) + P(R1) + P(V1) expression. According to Fig. 2.194, the maximum of waveform is about 30×10^{-9} W which is a very small number.

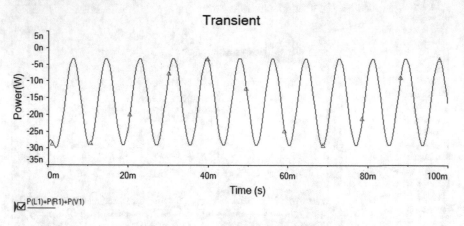

Fig. 2.194 Simulation result (summation of instantaneous power of AC source, inductor and resistor)

2.24   Example 17: Calculation of Apparent Power and Power Factor

We want to calculate the apparent power and power factor for the linear circuit shown in Fig. 2.195.

Fig. 2.195 Schematic of example 17

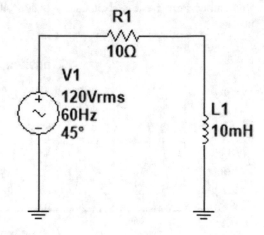

We used two probes to measure the RMS of load voltage and current (Fig. 2.196).

Fig. 2.196 Simulation result

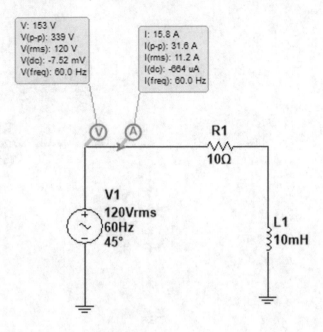

The apparent power is calculated in Fig. 2.197.

Fig. 2.197 Apparent power

Now, we need the average power drawn from the AC source. You can calculate the average power of AC source using the transient analysis (see example 16) or you can use a power probe (Fig. 2.198) to measure it. According to Fig. 2.198, the supplied average power is 1.26 kW.

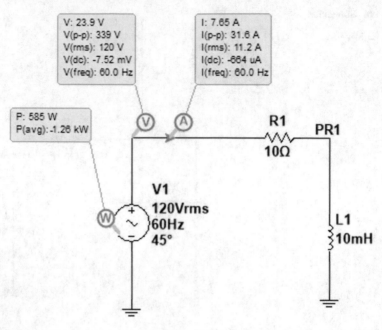

Fig. 2.198 Simulation result

The power factor can be calculated using the P.F. $= \frac{P_{average}}{P_{apparent}}$ equation. According to Fig. 2.199, the power factor is 0.9375.

Fig. 2.199 Calculation of power factor

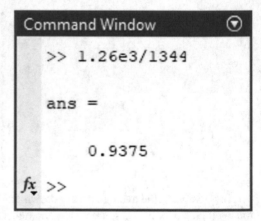

Let's verify the result obtained by Multisim. The MATLAB code shown in Fig. 2.200 shows that the obtained result is correct.

Fig. 2.200 Calculation of
power factor

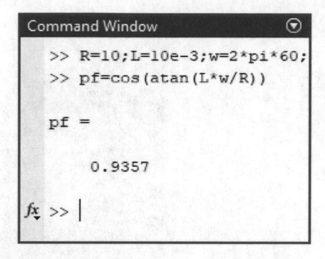

```
Command Window                                    ⊙

    >> R=10;L=10e-3;w=2*pi*60;
    >> pf=cos(atan(L*w/R))


    pf =


           0.9357

fx >> |
```

2.25   Example 18: Wattmeter Block

Wattmeter block (Figs. 2.201 and 2.202) can be used to measure the average power
and power factor of linear circuit, i.e. a circuit with voltage of $v(t) = V_m \sin (\omega t + \theta)$
and current of $i(t) = I_m \sin (\omega t + \varphi)$. You can use the wattmeter in the DC circuits as
well. In this example we want to measure the average power and power factor of
previous example.

Fig. 2.201 The wattmeter
block

Fig. 2.202 Simulate>
Instruments> Wattmeter

Assume the circuit shown in Fig. 2.203.

Fig. 2.203 Schematic of
example 18

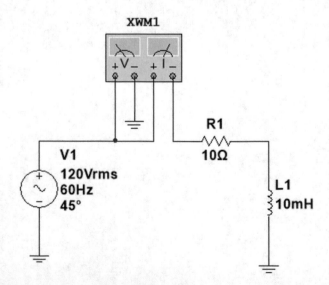

Run the simulation. The simulation result is shown in Fig. 2.204. The obtained average power and power factor are the same as previous example.

Fig. 2.204 Wattmeter reading

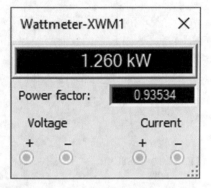

The Wattmeter block does not calculate the power factor of non-linear circuits correctly. For instance, assume the circuit shown in Fig. 2.205. In this circuit, the voltage is $v(t) = 120\sqrt{2}\sin\left(\omega t + \frac{\pi}{4}\right)$ however, the current is not in the form of i (t) = $I_m \sin(\omega t + \varphi)$. The current is half wave rectified sinusoidal waveform which contains many harmonics. Because of this, the load is not linear.

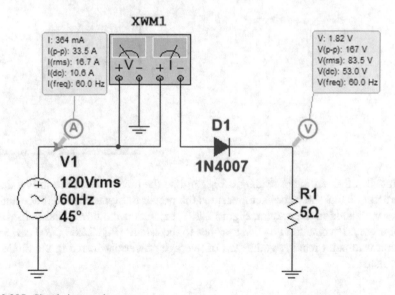

Fig. 2.205 Simulation result

Run the circuit. According to Fig. 2.206, the power factor is 0.91581. However, this is not correct. The correct power factor is calculated in Fig. 2.207.

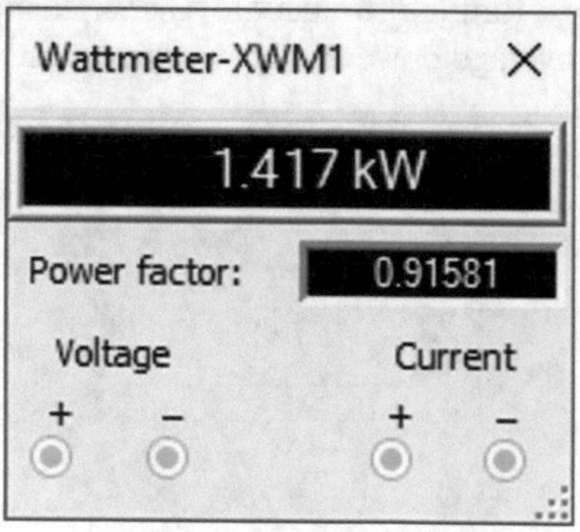

Fig. 2.206 Wattmeter
reading

Fig. 2.207 MATLAB
calculations

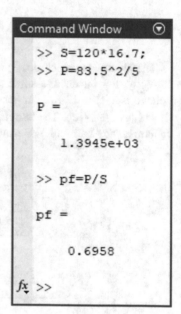

Note that the wattmeter reading is bigger than the power calculated in Fig. 2.207. The reason is that the wattmeter measures the power dissipated in the diode and the power dissipated in the resistor. Figure 2.207, calculates the power dissipated in the resistor only. If you add two power probes to the circuit (Fig. 2.208), you can easily see that wattmeter reading is the sum of average power dissipated in the diode and the resistor.

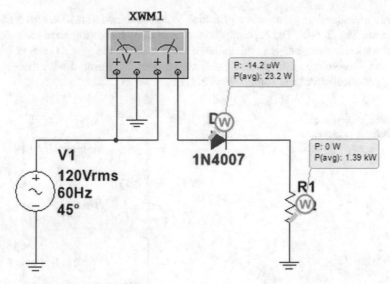

Fig. 2.208 Simulation result

2.26   Example 19: Current/Voltage Measurement with Voltage Controlled Voltage Source Block

A voltage controlled voltage source can act as a current and voltage sensor. For instance, assume that we want to see the instantaneous power waveform for the inductor shown in Fig. 2.209. We need the instantaneous current and voltage of inductor to calculate the power waveform. Remember that instantaneous power equals to the product of instantaneous current into the instantaneous voltage.

Fig. 2.209 Schematic of example 19

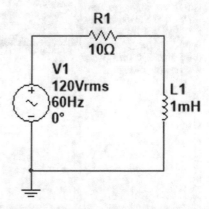

Let's measure the current waveform first. We need to add a small sense resistor to the circuit (Fig. 2.210). The current produces a small voltage drop across the resistor and if we divide that voltage to the value of sense resistor, the passing current will be obtained. The voltage drop across the sense resistor is measured with the aid of a voltage controlled voltage source (Fig. 2.211).

Fig. 2.210 Addition of sense resistor to the circuit

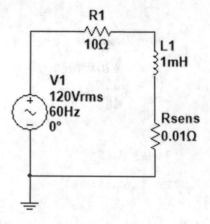

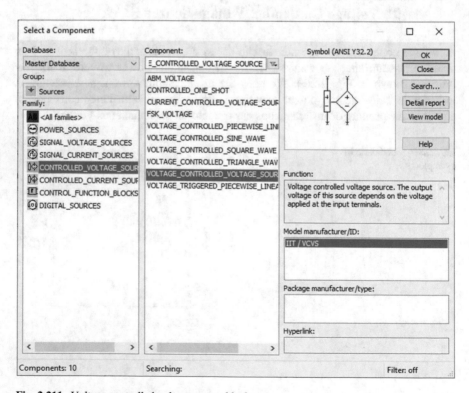

Fig. 2.211 Voltage controlled voltage source block

Add a voltage controlled voltage source in parallel to the sense resistor (Fig. 2.212).

Fig. 2.212 Addition of a voltage controlled voltage source to the circuit

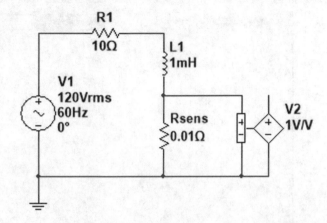

Double click the voltage controlled voltage source block and enter 1/Rsens to the voltage gain box (Fig. 2.213). In this example Rsens = 0.1 Ω so, 1/Rsens = 1/ 0.01 = 100. The voltage controlled voltage source measures the voltage across the sense resistor. According to Ohm's law, when we multiply the measurement by 1/Rsens, we obtain the current through the sense resistor which is also the circuit current.

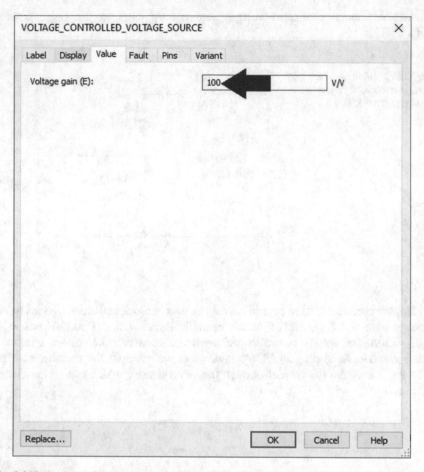

Fig. 2.213 Settings of the voltage controlled voltage source block

You can use the schematic shown in Fig. 2.214 to measure the voltage across the resistor Rsens as well. This schematic uses a voltage summer block (Fig. 2.215). Setting of the voltage summer block is shown in Fig. 2.216.

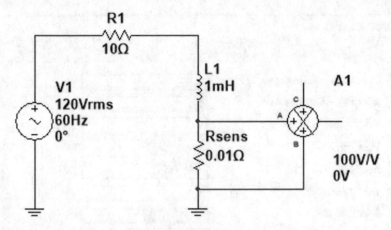

Fig. 2.214 Use of a voltage summer block to measure the voltage of the sense resistor

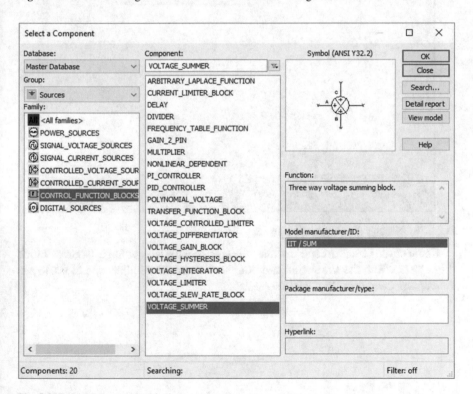

Fig. 2.215 Voltage summer block

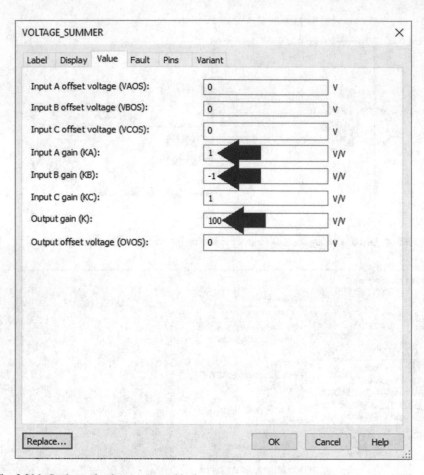

Fig. 2.216 Settings of voltage summer block

Connect an oscilloscope to the voltage controlled voltage source block (Fig. 2.217). Run the simulation and double click on the oscilloscope block to see the current waveform (Fig. 2.218).

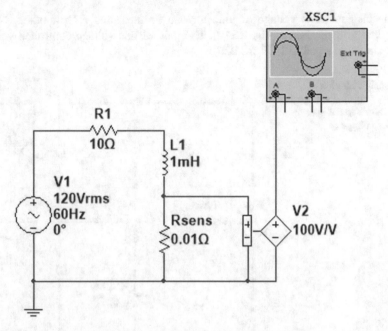

Fig. 2.217 An oscilloscope displays the output of voltage controlled voltage source block

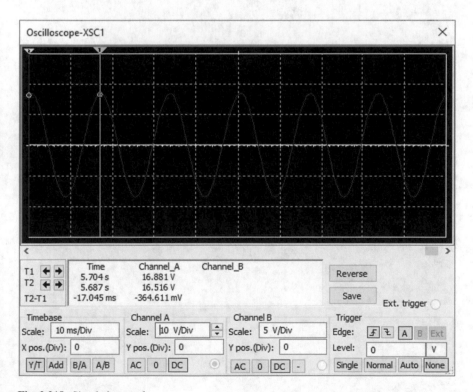

Fig. 2.218 Simulation result

The instantaneous voltage of inductor can be measured with another voltage controlled voltage source (Fig. 2.219). The gain of this voltage controlled voltage source must be equal to 1 (Fig. 2.220).

Fig. 2.219 Addition of a voltage controlled voltage source block to measure the inductor voltage

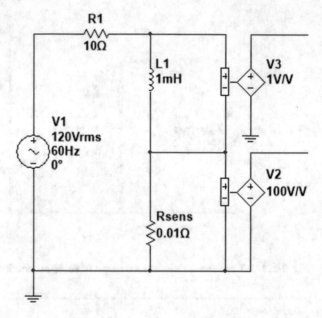

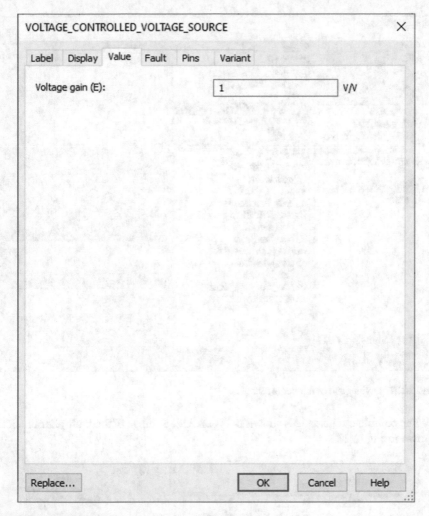

Fig. 2.220 Settings of voltage controlled voltage source V3

We need a multiplier block in order to multiply the measured voltage and current signal. The multiplier (and even divider) block can be found in the control function blocks section of sources group (Fig. 2.221).

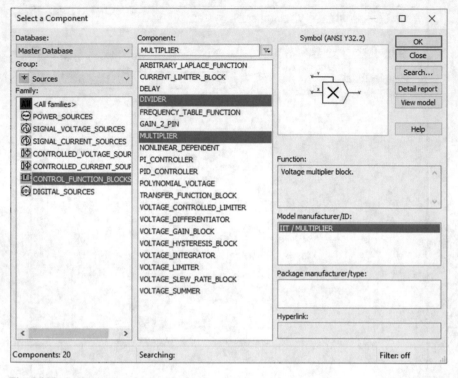

Fig. 2.221 Divider and multiplier blocks

The complete schematic is shown in Fig. 2.222. Settings of the multiplier block is shown in Fig. 2.223.

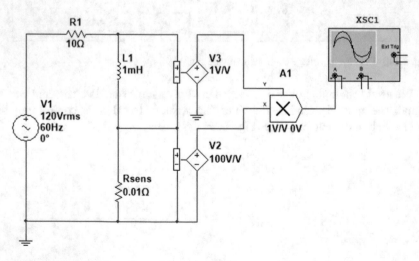

Fig. 2.222 Multiplying the voltage and current together

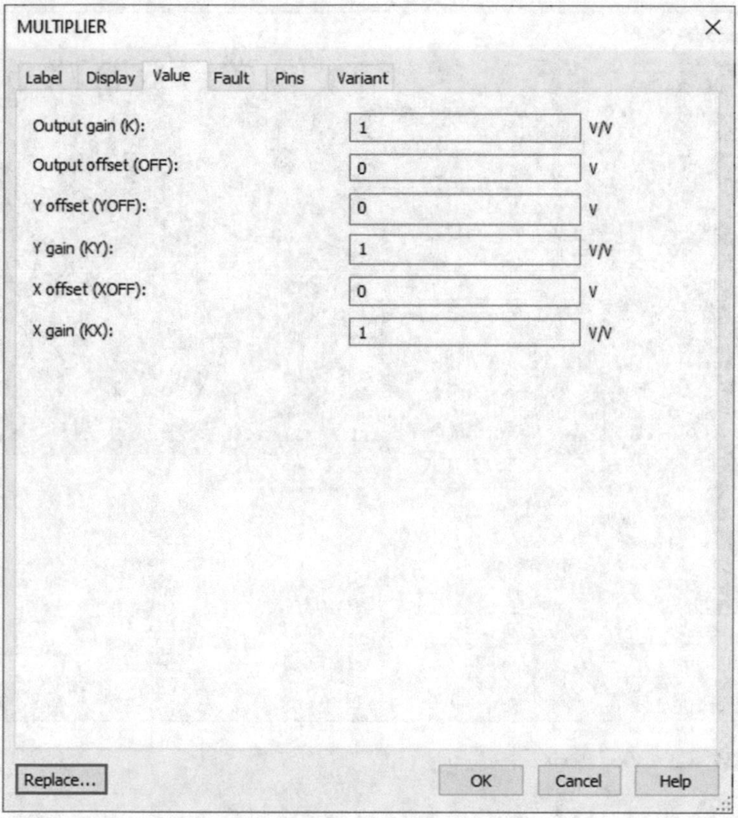

MULTIPLIER   ✕

Label   Display   Value   Fault   Pins   Variant

Output gain (K): | 1 | V/V

Output offset (OFF): | 0 | V

Y offset (YOFF): | 0 | V

Y gain (KY): | 1 | V/V

X offset (XOFF): | 0 | V

X gain (KX): | 1 | V/V

Replace...   OK   Cancel   Help

Fig. 2.223 Settings of multiplier block

Run the simulation. Double click the oscilloscope to see the power waveform. According to Fig. 2.224, the power waveform has period of 8.333 ms ($\approx$ 120 Hz). The maximum instantaneous power of the inductor is 53.726 W.

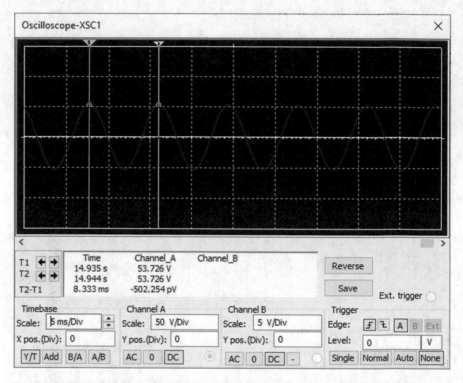

Fig. 2.224 Simulation result

Let's verify the obtained result. The MATLAB code shown in Fig. 2.225, draws the inductor instantaneous current. After running the code, the result shown in Fig. 2.226 is obtained. Period of this function is about 8.3 ms (Fig. 2.227). The maximum of waveform in Fig. 2.226 is quite close to the Multisim result (Fig. 2.224).

```
>> R=10; L=1e-3; Vmax=120*sqrt(2);w=2*pi*60;
>> syms i(t)
>> ode=L*diff(i,t)+R*i == Vmax*sin(w*t);
>> cond=i(0)==0;
>> iSol(t)=dsolve(ode,cond);
>> pL=Vmax*sin(w*t)*iSol-R*iSol^2;
>> ezplot(p,[0 43e-3]);
>> grid on
fx >> |
```

Fig. 2.225 MATLAB calculations

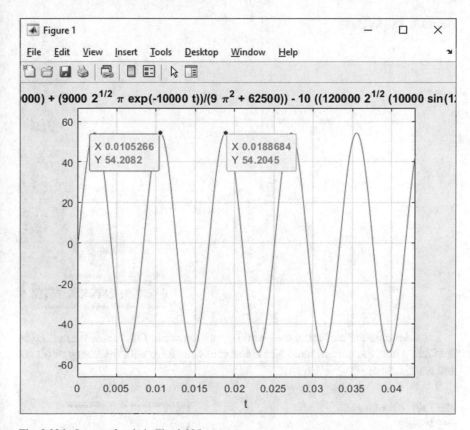

Fig. 2.226 Output of code in Fig. 2.225

Fig. 2.227 Calculation of period of wave in Fig. 2.226

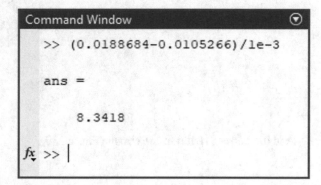

2.27   Example 20: Current Clamp Element

Multisim has an element called current clamp (Fig. 2.228). This element acts like a current sensor. You can measure the current quite easily with it.

Fig. 2.228 Current clamp

Current clamp

Let's measure the current of the simple circuit shown in Fig. 2.229. According to the Ohm's law, the current must be a sinusoidal waveform with frequency of 60 Hz and amplitude of $\frac{120\sqrt{2}}{100} = 1.697$ A.

Fig. 2.229 Schematic of example 20

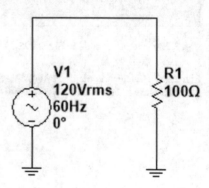

Add the current clamp to the circuit (Fig. 2.230).

Fig. 2.230 Addition of
current clamp to the
schematic of Fig. 2.229

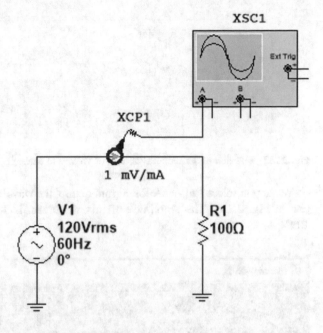

Double click the current clamp and select the desired ratio of voltage to current. When you click the section shown with arrow in Fig. 2.231, a list appears (Fig. 2.232) and you can select the desired unit. For instance, 1 mV/mA means that the output voltage of 1 mV shows 1 mA of current. In other words, there is no amplification, i.e. current is multiplied by 1 and goes to output. When you select 10 mV/mA, 1 mA of current produce 10 mV of output voltage. In other words, the current is multiplied by 10 and goes to output.

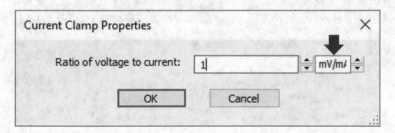

Fig. 2.231 Current clamp properties window

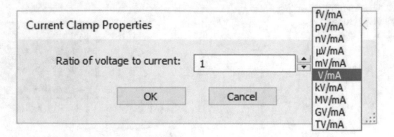

Fig. 2.232 Selection of desired unit for ratio of voltage to current box

When you select 1 mV/mA for current clamp, the waveform will be similar to the one in Fig. 2.233. The amplitude of this waveform is 1.694 V which equals to 1.694 A.

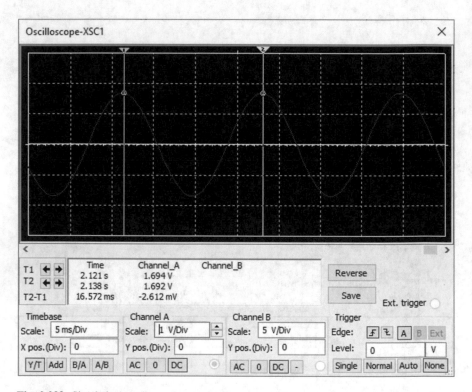

Fig. 2.233 Simulation result

2.28   Example 21: Measurement of Phase Difference with Oscilloscope

In this example, we want to measure the phase difference between point A and B in Fig. 2.234.

Fig. 2.234 Schematic of example 21

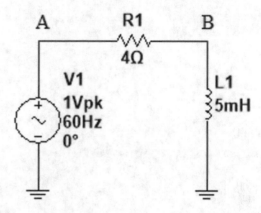

The required schematic is shown in Fig. 2.235 This schematic uses the function generator and 2 channel oscilloscope blocks (Fig. 2.236).

Fig. 2.235 Multisim equivalent of Fig. 2.234

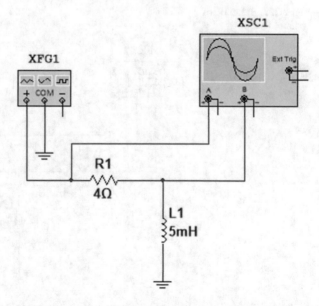

Fig. 2.236 Function
generator and oscilloscope
blocks

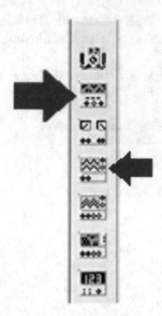

Click the wire that connects the signal generator to channel A of oscilloscope
(Fig. 2.237).

Fig. 2.237 Selection of
channel 1 wire

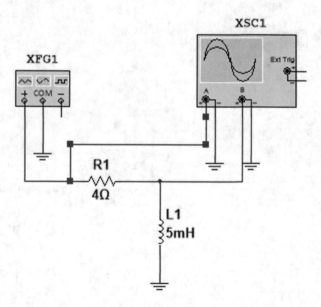

Right click on the selected wire select segment color (Fig. 2.238).

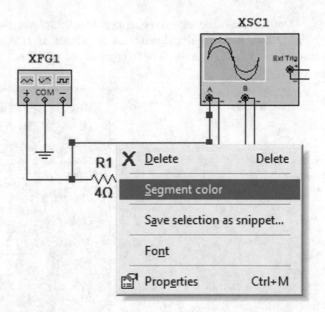

Fig. 2.238 Segment color determine the color of wire. Color of signal shown by oscilloscope is the same as the wire color

After clicking the segment color, the window shown in Fig. 2.239 appears and permits you to select the desired color for signal which will be shown on oscilloscope screen for channel A. It is recommended to use different colors for channel A and B. This help you to identify each channel rapidly and easily.

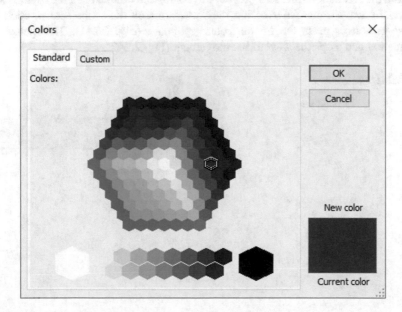

Fig. 2.239 Colors window

Double click the signal generator block and do the settings as shown in Fig. 2.240. A sinusoidal signal with amplitude of 1 V and frequency of 60 Hz, i.e. $v(t) = 1 \sin (2 \times \pi \times 60 \times t)$, is produced with these settings.

Fig. 2.240 Function generator settings

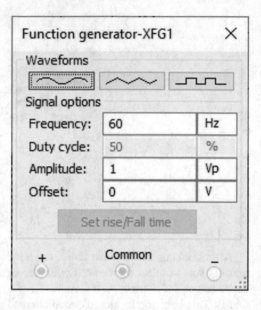

Run the simulation (press F5 on your keyboard) and double click the oscilloscope to see the waveforms. After seeing the waveforms on the scope screen, click the pause button or press the F6 on your key board (Fig. 2.241). This pause the simulation and stops the waveforms movement (Fig. 2.242).

Fig. 2.241 Pause button

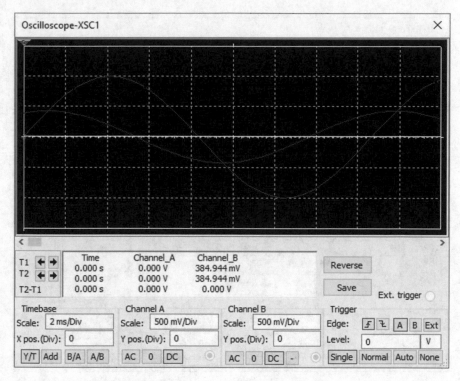

Fig. 2.242 Simulation result

Click the reverse button, if you prefer white background (Fig. 2.243).

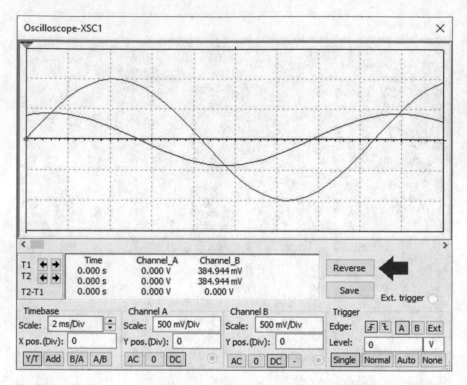

Fig. 2.243 Clicking the reverse button cause the oscilloscope background to change to white

Now use the cursors (Fig. 2.244) to measure the time difference between the two waveforms. Put the cursors as close as possible to the zero crossing point of channel A and channel B. Read the time difference from the T2-T1 row (Fig. 2.245).

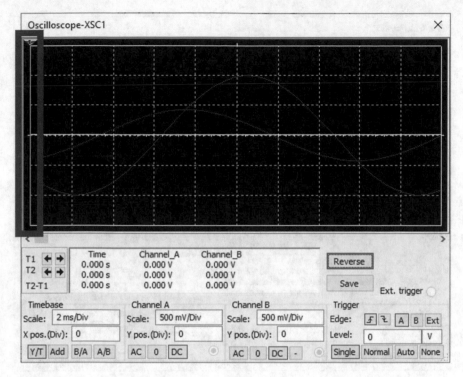

Fig. 2.244 Oscilloscope cursors

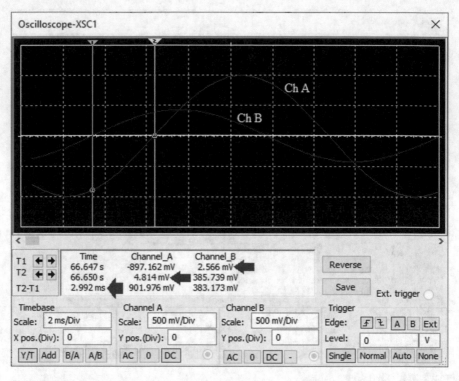

Fig. 2.245 Measuring the time difference between the two waveforms

According to Fig. 2.245, the time difference between the two waveforms is 2.992 ms. The calculations shown in Fig. 2.246 convert this time delay to angle.

Fig. 2.246 Calculation of
phase difference

Let's   check   this   result: $\frac{j\times L\times\omega}{R+j\times L\times\omega} = \frac{j\times L\times 2\pi f}{R+j\times L\times 2\pi f} = \frac{j\times 5m\times 377}{4+j\times 5m\times 377} = \frac{1.885j}{4+1.885j} =$ $0.426e^{\,j64.76°}$. This shows that the obtained result is correct.

2.29   Example 22: Measurement of Phase Difference with Lissajous Curves

You can measure the phase difference with the aid of Lissajous curves as well. In this example we measure the phase difference of previous example with the aid of Lissajous curve. Remember that the phase difference for the Lissajous curve shown in Fig. 2.247 is $\Delta\varphi = \sin^{-1}\left(\frac{B}{A}\right)$.

Fig. 2.247 A typical Lissajous curve

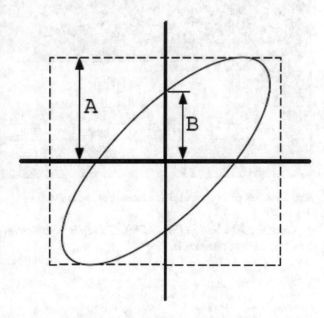

Click the B/A button (Fig. 2.248). The B/A button draws the graph of channel B as a function of channel A.

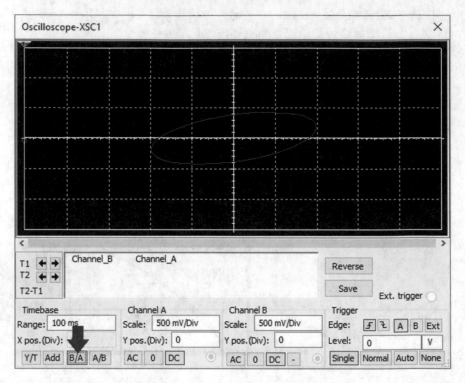

Fig. 2.248 Graph of channel B as a function of channel A

Click the A/B button (Fig. 2.249). The A/B button draws the graph of channel A as a function of channel B.

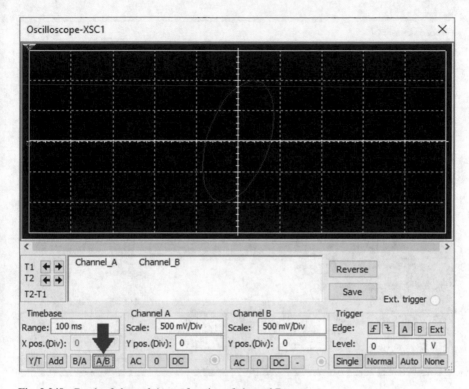

Fig. 2.249 Graph of channel A as a function of channel B

You can use either of Figs. 2.248 or 2.249 for phase difference measurement. Let's use the Fig. 2.249 to measure the values of A and B. According to Fig. 2.250, A = 998.404 mV and B = 903.636 mV.

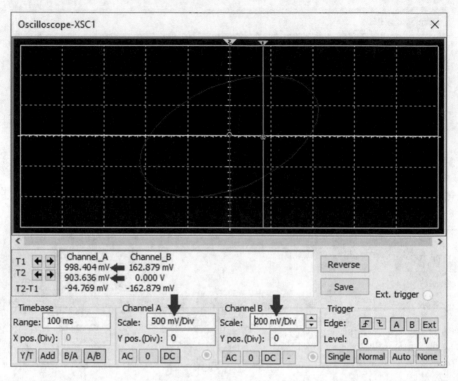

Fig. 2.250 Measurement of A and B

According to Fig. 2.251, the phase difference is 64.8341°.

Fig. 2.251 Calculation of
phase difference

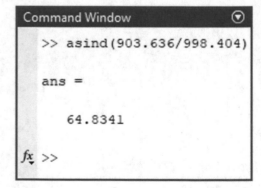

2.30   Example 23: Agilent® and Tektronix® Measurement Tools

Agilent 33120A function generator, Agilent 54622D digital oscilloscope, Agilent 34401A digital multimeter and Tektronix TDS 2024 digital oscilloscope (Fig. 2.252) are simulated in Multisim. This permits you to experiment with these tools (without being worry to damage the device) in the software environment. Such a capability is a great educational tool for students.

Fig. 2.252 Agilent and Tektronix measurement devices

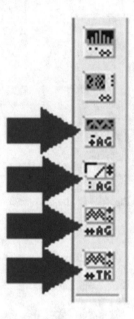

Let's measure the phase difference of example 22 with the aid of these tools. The required schematic is shown in Fig. 2.253.

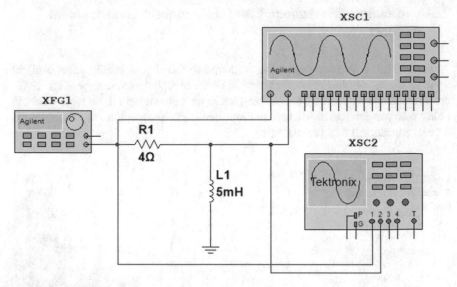

Fig. 2.253   Multisim schematic to measure the phase difference

Run the simulation (Fig. 2.254).

Fig. 2.254   The run button

Double click the function generator block. The window shown in Fig. 2.255 appears.

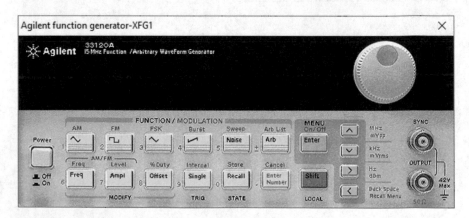

Fig. 2.255   The Agilent function generator window

Click the power button to turn on the device. Use the Freq, Ampl and the rotating knob to obtain 60 Hz and 1 Vp (Fig. 2.256).

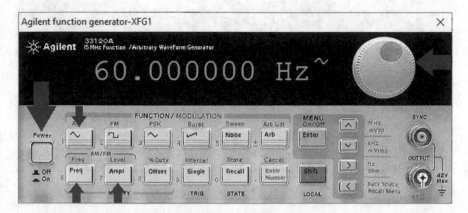

Fig. 2.256   Production of 60 Hz sinusoidal voltage with peak of 1 V

After setting the output of signal generator to 1 Vp and 60 Hz, double click the Tektronix oscilloscope. After double clicking, the window shown in Fig. 2.257 appears. Click the power button to turn on the device.

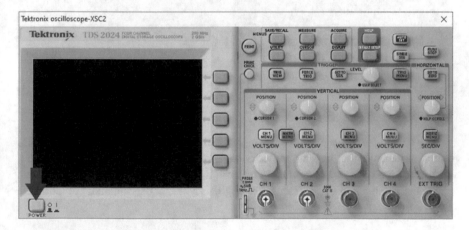

Fig. 2.257   Tektronix oscilloscope window

Use the cursor button to measure the time difference between the two waveforms. You can take a look to Tektronix TDS 2024 user manual to learn the details of using cursors. You can obtain the use manual by searching for "Tektronix TDS 2024 user manual" in Google.

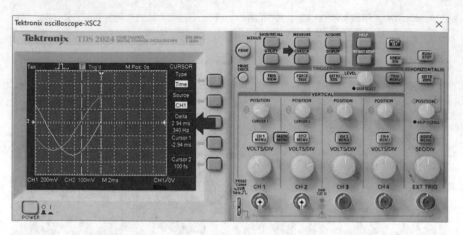

Fig. 2.258   Measurement of time difference between the two waveforms

According to Fig. 2.258, the time difference is 2.94 ms. According to Fig. 2.259, the phase difference is 63.5043°.

Fig. 2.259   Calculation of
phase difference

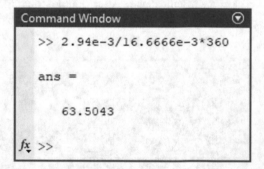

You can do the measurement with the Agilent oscilloscope as well. Use the cursor button to measure the time difference between the two waveforms. You can take a look to Agilent 54622D user manual to learn the details of using cursors. You can obtain the use manual by searching for "Agilent 54622D user manual" in Google.

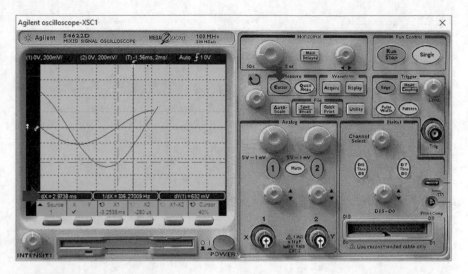

Fig. 2.260 Agilent oscilloscope window

According to Fig. 2.260, the time difference is 2.9738 ms. According to Fig. 2.261, the phase difference is 64.1523°.

Fig. 2.261 Calculation of phase difference

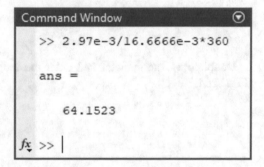

2.31   Example 24: Transients of an RLC Circuit

In this example we want to simulate the transient of the circuit shown in Fig. 2.262. The initial condition of circuit is iL = 0.1 A and vC = 2 V.

Fig. 2.262 Schematic of
example 24

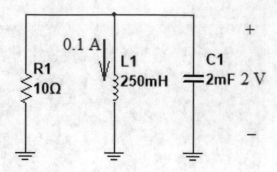

Draw the schematic of circuit (Fig. 2.263).

Fig. 2.263 Multisim
schematic of Fig. 2.262

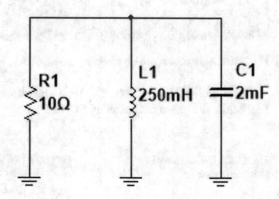

Double click the inductor and enter the inductor value and its initial condition
(Fig.2.264).

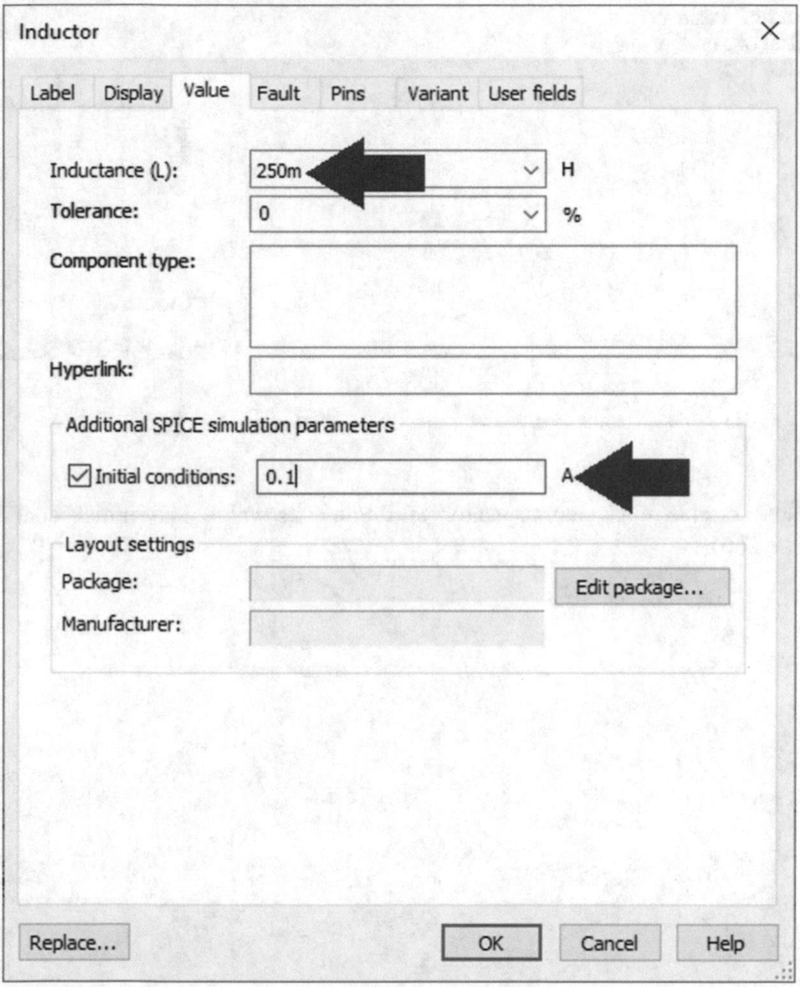

Fig. 2.264 Entering the inductor value and its initial current

The terminal which is connected to mouse pointer during the placement of component on the schematic is considered to be positive and the current which enters to that terminal is assumed to be positive (Fig. 2.265). So, when you enter +0.1 to the initial condition box, you mean that 0.1 A is entered to the terminal which is connected to the mouse pointer.

Fig. 2.265 Positive
terminal of capacitor and
inductor

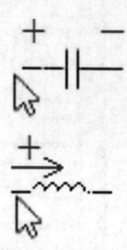

Now double click the capacitor and enter its value and initial condition
(Fig. 2.266).

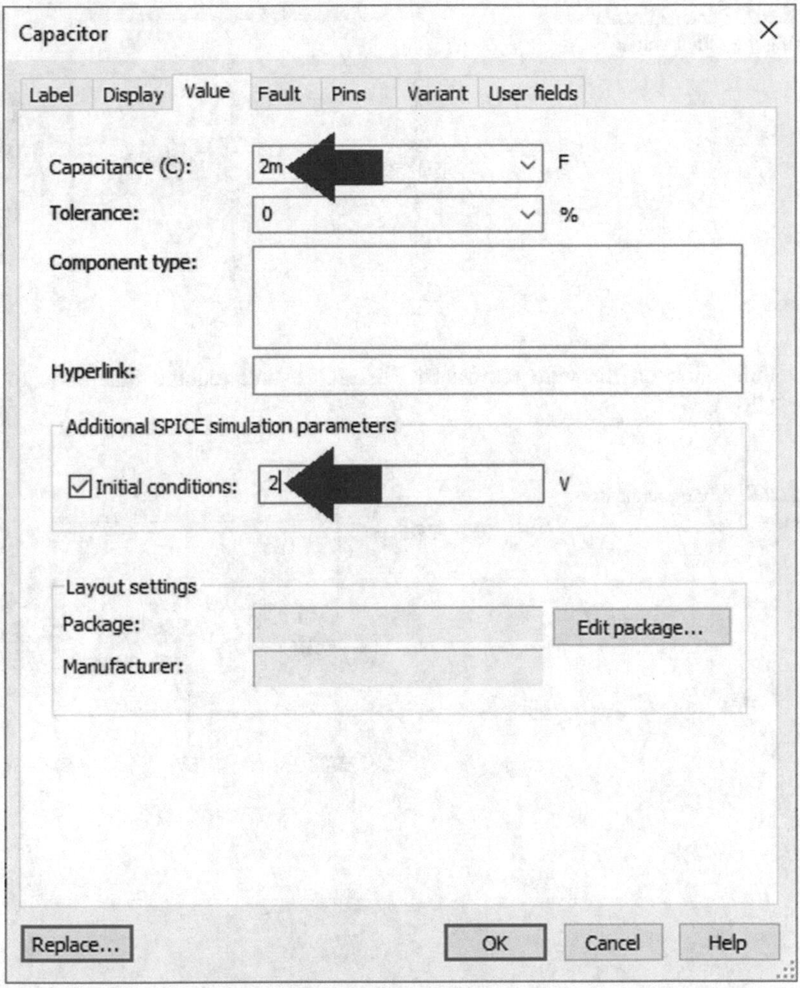

Fig. 2.266 Entering the capacitor value and its initial voltage

Now, the schematic must look like Fig. 2.267.

Fig. 2.267 Schematic after
defining the initial values

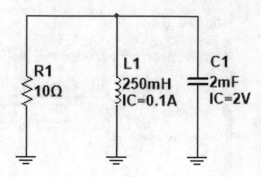

Double click on the wire shown in Fig. 2.268 and rename that node to Vo
(Fig. 2.269).

Fig. 2.268 Giving name to
the node

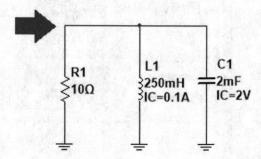

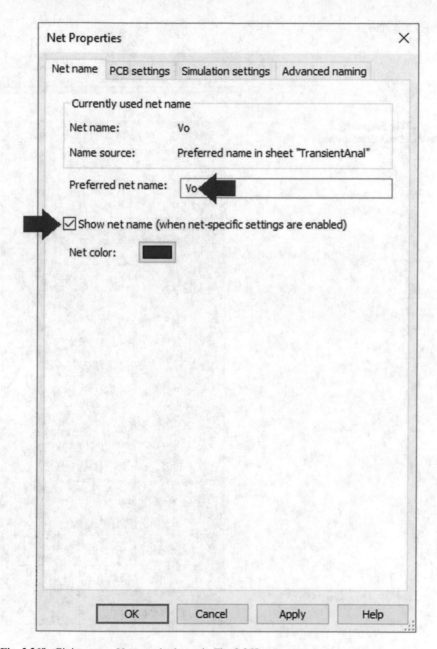

Fig. 2.269 Giving name Vo to node shown in Fig. 2.268

Click the interactive (Fig. 2.270) or simulate> analyses and simulation (Fig. 2.272). The window shown in Fig. 2.271 will appear.

Fig. 2.270 The interactive
button

Fig. 2.271 Simulate>
Analyses and simulation

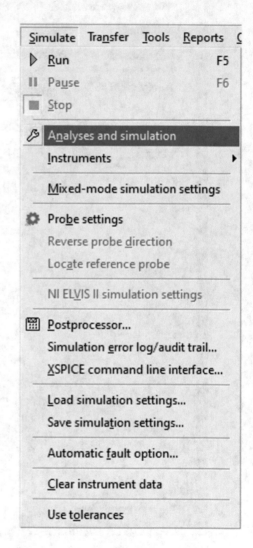

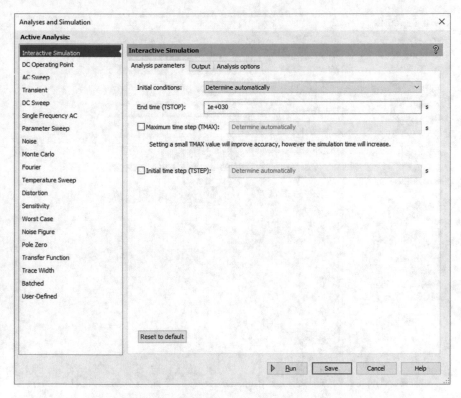

Fig. 2.272 Analyses and simulation window

Go to the transient section (Fig. 2.273).

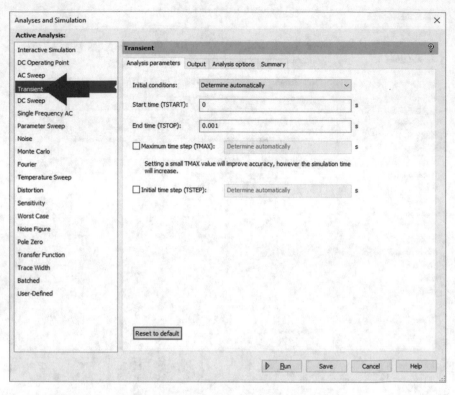

Fig. 2.273   Transient analysis section of analyses and simulation window

Do the setting similar to Fig. 2.274. We want to simulate the circuit behavior of 0.5 s interval. Ensure that user defined is selected for initial condition box.

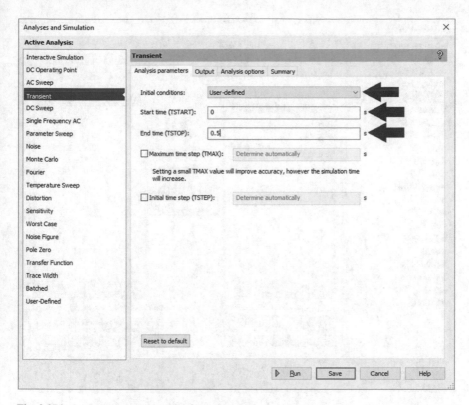

Fig. 2.274 Analyses parameters tab of transient analysis

Now go to the output tab and select the variable which you want to see its graph. Click the V(vo) from the left list and click the add button to add it to the right list (Fig. 2.275). The V(vo) tells the Multisim that we want to see the voltage of node vo. The I(C1), I(L1) and I(R1) represents the current of capacitor C1, inductor L1 and resistor R1, respectively. The current is assumed to be positive if it enters to the positive terminal. The positive terminal is the terminal which is connected to the mouse pointer during the placement of the element on the schematic (Fig. 2.276). P (C1), P(L1) and P(R1) shows the power waveforms of capacitor C1, inductor L1 and resistor R1, respectively. The power waveform of these components are calculated using the $(V_{+terminal} - V_{other\ terminal}) \times I_{entered\ to\ the\ +\ terminal}$ equation.

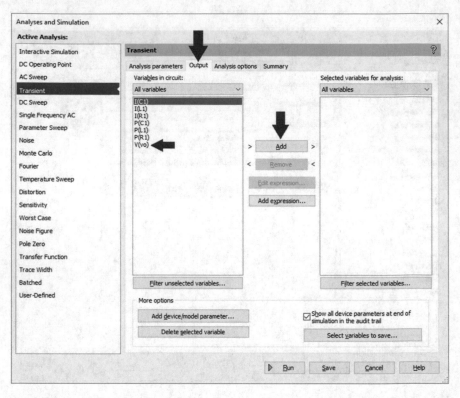

Fig. 2.275 Output tab of transient analysis

Fig. 2.276 Positive
terminal of capacitor,
inductor and resistor

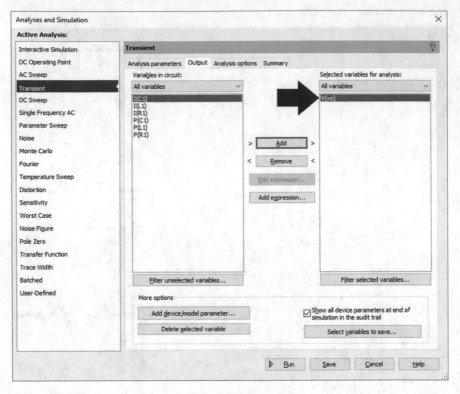

Fig. 2.277 Addition of node Vo voltage to the right list

Click the run button in Fig. 2.277. The result is shown in Fig. 2.278.

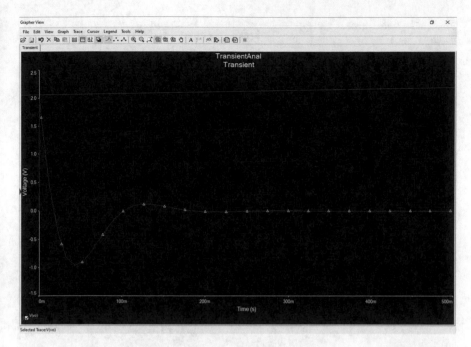

Fig. 2.278 Simulation result

You can set the width and color of traces with the aid of trace menu (Figs 2.279 and 2.280).

Fig. 2.279 Trace> trace
width

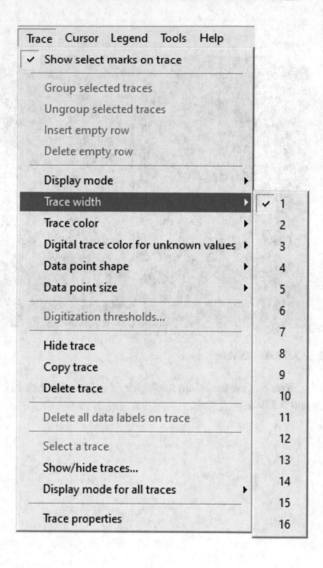

Fig. 2.280 Trace> trace color

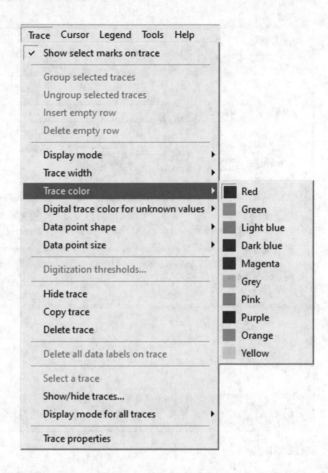

Double click on the graph. The window shown in Fig. 2.281 is appeared. You can enter the desired title for the graph in the title box.

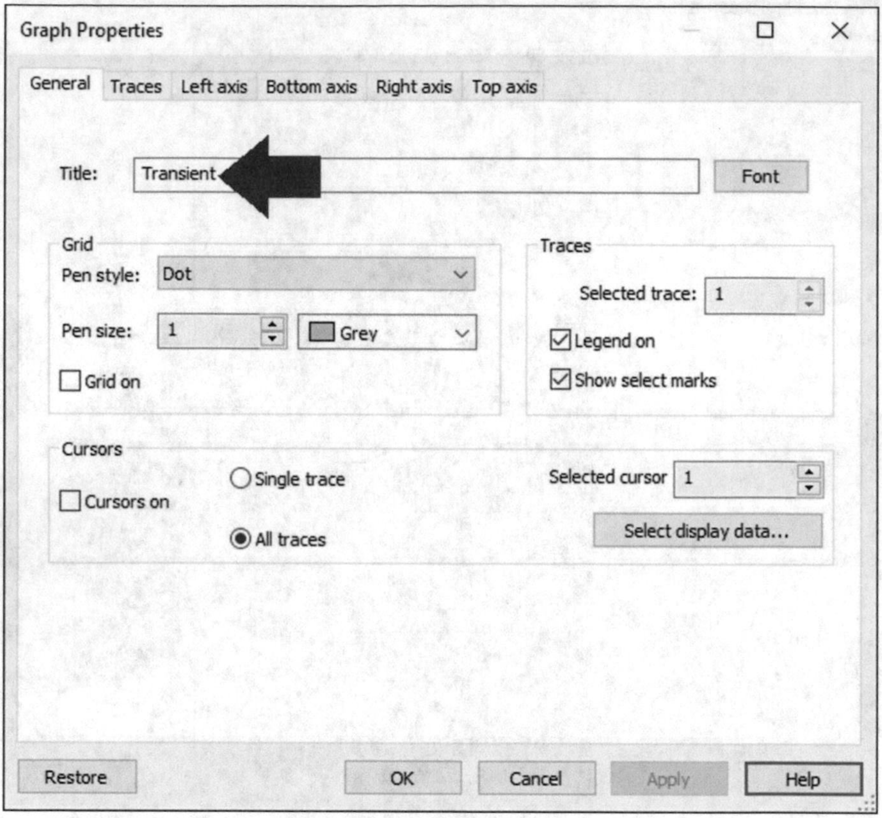

Fig. 2.281 Entering the title

Sometime you select more than one variable to be shown on the screen. In this case, you can hide the unwanted ones by going to the trace tab and uncheck the show box for that graph (Fig. 2.282).

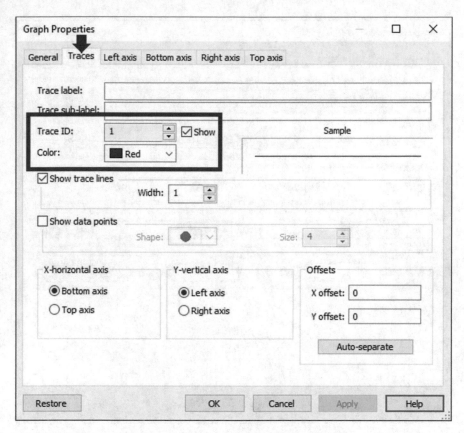

Fig. 2.282   Hiding the unwanted waveform by unchecking the show box

The label of y axis and its range is controllable with the aid of left axis tab (Fig. 2.283).

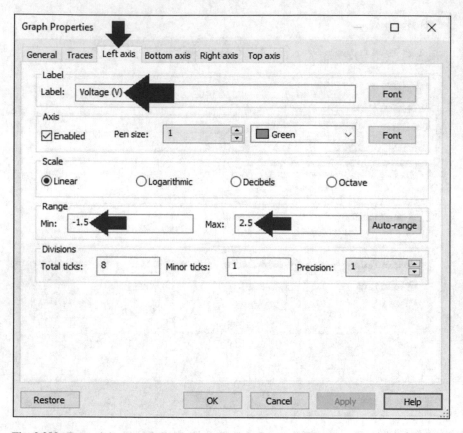

Fig. 2.283 Determining the label and range of values for x axis

Go to the bottom axis tab. The label of x axis and its range is controllable with the aid of this tab (Fig. 2.284).

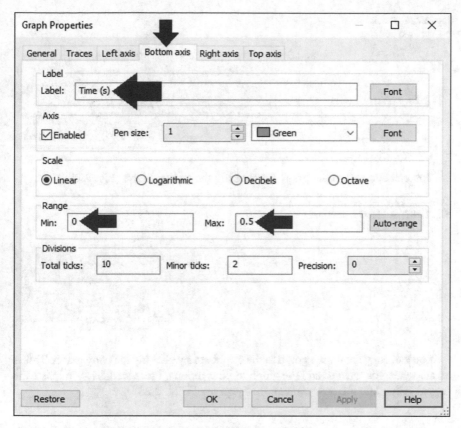

Fig. 2.284 Determining the label and range of values for y axis

You can add grid to the graph by clicking the show grid icon (Fig. 2.285).

Fig. 2.285 Show grid
button

By default, the color of background is black. You can change the background color to white by clicking the black background icon (Fig. 2.286).

Fig. 2.286 Black
background button

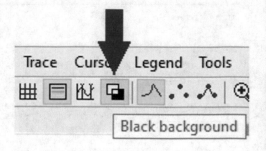

You can add text to the graph by clicking the add text icon (Fig. 2.287).

Fig. 2.287 Add text button

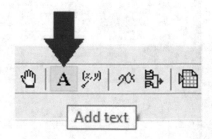

You can copy the drawn graph to the clipboard by clicking the copy graph. This is useful when you want to add the graph to your reports. Press the Ctrl+V to paste the copied graph into the environment that you want (Fig. 2.288).

Fig. 2.288 Edit> Copy
graph

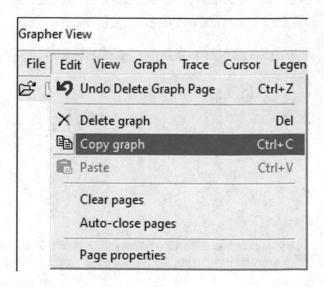

You can add cursors to the graph by clicking the show cursor icon (Fig. 2.289).

Fig. 2.289 Show cursors
button

You need to left click on the cursor and hold down the mouse left button to move the cursors to the desired location. The coordinate of point, the difference between the two cursor and frequency can be calculated easily with the aid of cursors (Fig. 2.290).

Fig. 2.290 The cursor
window shows the
coordinate of cursors

Cursor	V(vo)
x1	136.0264m
y1	116.0046m
x2	159.9340m
y2	56.8636m
dx	23.9077m
dy	−59.1410m
dy/dx	−2.4737
1/dx	41.8276

Sometime it is difficult to bring the cursor to a specific point using the mouse cursor. You can use the cursor menu for these cases (Fig. 2.291).

Fig. 2.291 Useful function
of cursor menu

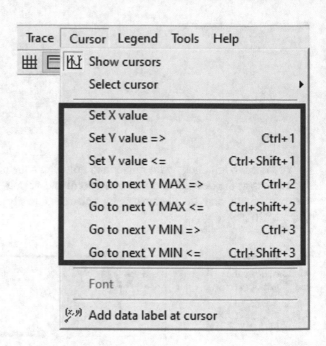

For instance, assume that you want to put the cursor 1 to t = 26.793 ms. In order
to do this, click on the cursor 1 to select it. The click the cursor> set x value and enter
26.793 m to the appeared box. After clicking the OK button in Fig. 2.292, the cursor
1 will go to t = 26.793 ms (Fig. 2.293).

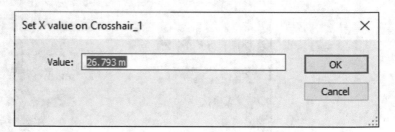

Fig. 2.292 Set X value on crosshair window

Fig. 2.293 Cursor 1 went to the point that entered in Fig. 2.292

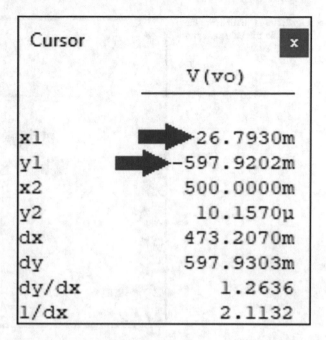

As another example assume that you want to find the time instant which the graph equals to −0.6. To do this, select one of the cursors (click on one of the cursors), then click the cursors> set y value =>. Then enter -600 m or − 0.6 to the window appeared and click the OK button (Fig. 2.294). The set y value => searches the time instants after the initial position of cursor and finds the time instant which its value equals to -600 m (Fig. 2.295).

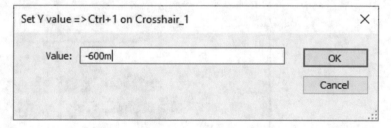

Fig. 2.294 Set Y value window

Fig. 2.295 Cursor 1 y value
is equal to the value entered
in Fig. 2.294

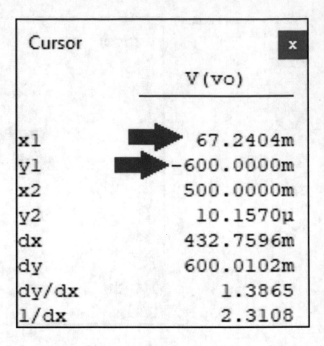

You can use y value <= to search the time instants before the initial position of
the selected cursor (Fig. 2.296).

Fig. 2.296 Cursor 1 y value
is equal to the value entered
in Fig. 2.294

Cursor   **x**

	V(vo)
x1	26.8395m
y1	-600.0000m
x2	500.0000m
y2	10.1570μ
dx	473.1605m
dy	600.0102m
dy/dx	1.2681
1/dx	2.1134

2.32   Example 25: Decreasing the Time Step of Simulation

Sometimes the simulation result is not smooth enough. In this cases you need to decrease the time step of simulation to obtain a smoother graph, however the simulation takes more time to be done. For instance, assume that we want to increase the smoothness of result of previous example. In order to do this, click the transient button (Fig. 2.297).

Fig. 2.297 Transient button

Check the maximum time step box. The default value is 1e-5 which means 10^{-5} s. Change the default value to a smaller value (i.e. 1e-6) and click the run button (Fig. 2.298).

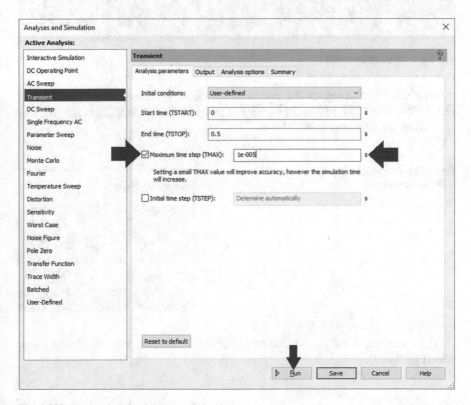

Fig. 2.298 Decreasing the time step of simulation

The result for maximum time step of 1e-6 ($=10^{-6}$) is shown in Fig. 2.299. This figure is more smooth in comparison to Fig. 2.278.

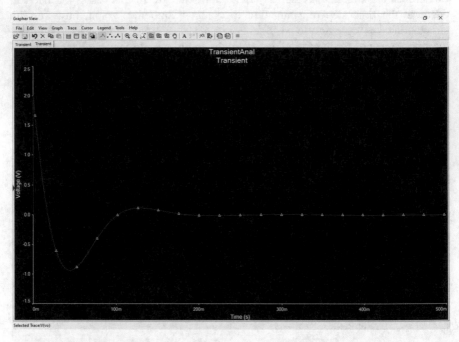

Fig. 2.299   Simulation result

Note that Multisim saves the results of different simulations. The Table 2.1 of Fig. 2.300 is the result of simulation for time step 1e-5 and Table 2.2 is the result of simulation for time step 1e-6.

Fig. 2.300   Results of different simulations

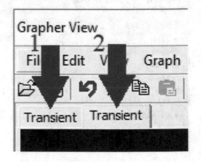

You can remove any of the tabs by clicking on it and clicking the delete graph icon (Fig. 2.301).

Fig. 2.301 Removing a tab
from the grapher view
window

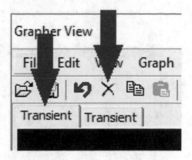

2.33 Example 26: Exporting from Grapher View to MS Excel®

You can export the result of analysis from Grapher View (the environment which shows the graph of simulations) to MS Excel quite easily. In this example we want to export the result of previous example to Excel. In order to do this, click the tools> export to Excel (Fig. 2.302) or export to excel icon (Fig. 2.303).

Fig. 2.302 Tools> export
to Excel

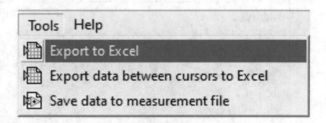

Fig. 2.303 Export to Excel
button

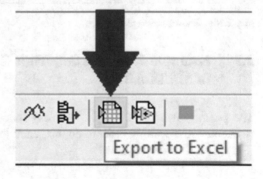

Now, select the trace that you want to be exported to the Excel and click OK button (Fig. 2.304).

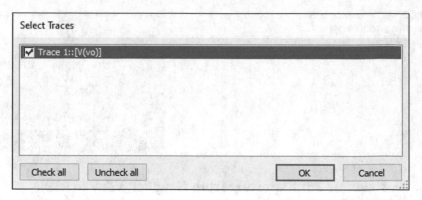

Fig. 2.304 Select traces window

The window shown in Fig. 2.305 will appear. Click the OK button. After clicking the OK button, the Excel will be opened and it contains the data for waveform(s) that you selected.

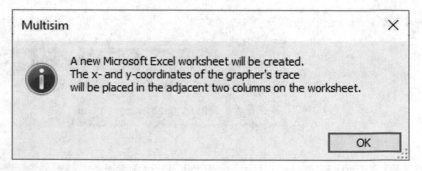

Fig. 2.305 Shown message

2.34 Example 27: Exporting from Grapher View to MATLAB®

The graph drawn in the Graph View environment can be exported to MATLAB environment as well. In order to do this, click the file> save as (Fig. 2.306).

Fig. 2.306 File> save as

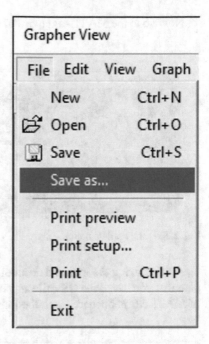

Now select the CSV format and save the file in the desired path (Fig. 2.307).

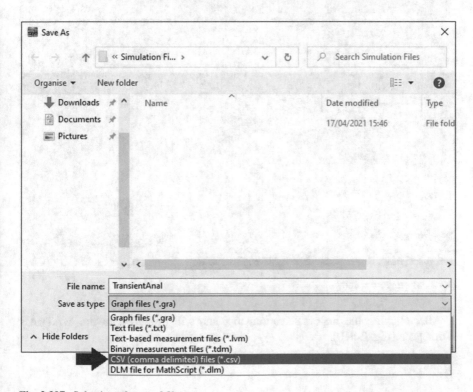

Fig. 2.307 Selection of type of file

Now go to the MATLAB environment and click the import data icon (Fig. 2.308).

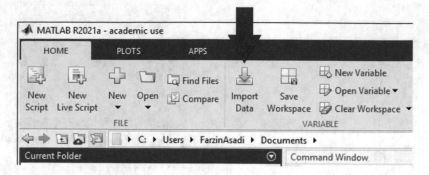

Fig. 2.308 Import data button

After clicking the import data icon, the import window will be opened. Go to the path that you saved the CSV file and select it. After few seconds, the import window of MATLAB is opened. Click the import selection (Fig. 2.309).

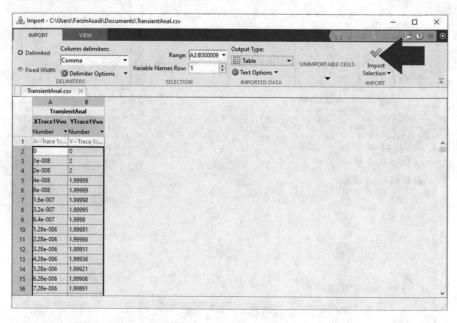

Fig. 2.309 Import window

After clicking the import selection, the data will be added to the MATLAB Workspace (Fig. 2.310).

Fig. 2.310 Data of CSV file
is exported to MATLAB
environment

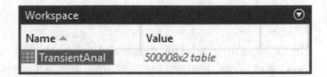

Use the commands shown in Fig. 2.311 to read and draw the plot of imported
data. Result is shown in Fig. 2.312. This is the same as Fig. 2.299.

```
Command Window                                        ▼
    >> t=table2array(TransientAnal(:,1));
    >> Vo=table2array(TransientAnal(:,2));
    >> plot(t,Vo)
fx >> |
```

Fig. 2.311 Drawing the graph of imported CSV file

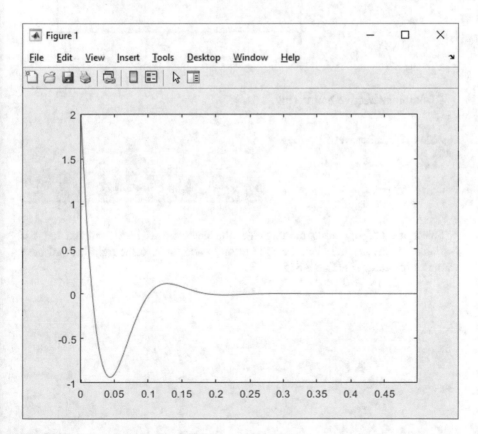

Fig. 2.312 Output of code in Fig. 2.311

2.35   Example 28: DC Operating Point

The DC operating point analysis calculates the desired quantity for the circuit in steady state. Remember that in steady state, the inductors behave as short circuit and the capacitors behave as open circuit.

Assume that we want to obtain the steady state voltage of nodes for the schematic shown in Fig. 2.313. The numbers behind the wires show the name of the nodes.

Fig. 2.313 Schematic of example 28

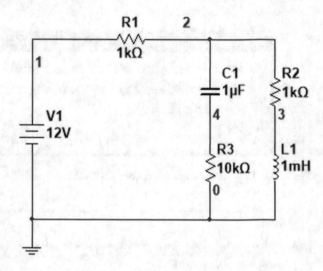

Click the interactive button (Fig. 2.314).

Fig. 2.314 The interactive button

Select the DC operating point. Select the desired variables from the left list (variables in circuit) and click the add button to add them to the right list (selected variables for analysis) (Fig. 2.315).

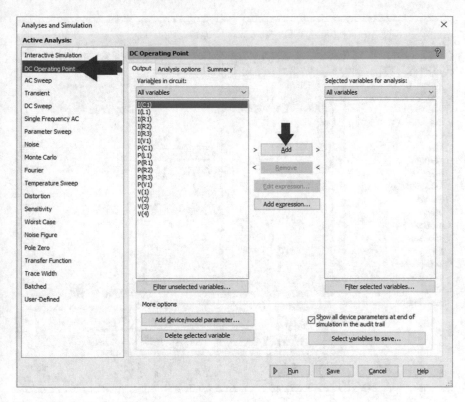

Fig. 2.315 Analysis and simulation window

In this example we want to calculate the steady state node voltages. So, we added V(1), V(2), V(3) and V(4) to the right list. Click the run button to run the simulation (Fig. 2.316).

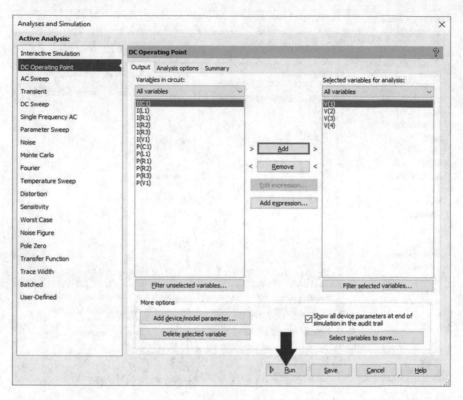

Fig. 2.316 Output tab of DC operating point

The simulation result is shown in Fig. 2.317.

Fig. 2.317 Simulation
result

	Variable	Operating point value
1	V(1)	12.00000
2	V(2)	6.00000
3	V(3)	0.00000e+000
4	V(4)	0.00000e+000

Let's verify the obtained result. The equivalent circuit for steady state is shown in Fig. 2.318. According to the schematic, voltage of node 1 is 12 V and voltage of node 2 is $\frac{1k}{1k+1k} \times 12 = 6$ V. Voltage of node 3 and 4 equals to zero. This results verify the Multisim result.

Fig. 2.318 DC equivalent of Fig. 2.313

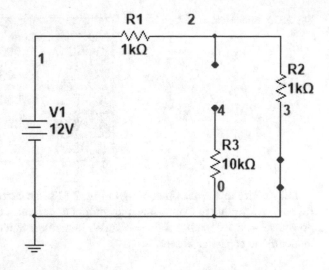

You can calculate the steady state power for components by adding the P(C1), P (L1), P(R1), P(R2) and P(R3) to the right list as well (Fig. 2.319). After running the simulation, the result shown in Fig. 2.320 is obtained.

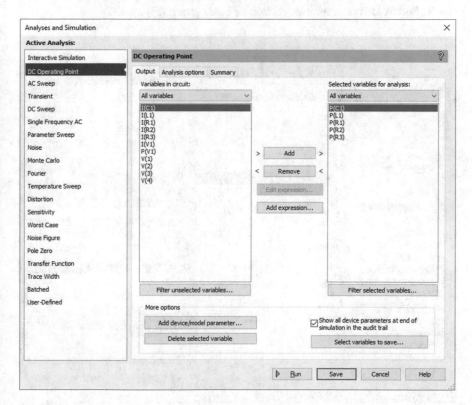

Fig. 2.319 Output tab of DC operating point

Fig. 2.320 Simulation
result

	Variable	Operating point value
1	P(C1)	0.00000e+000
2	P(L1)	0.00000e+000
3	P(R1)	36.00000 m
4	P(R2)	36.00000 m
5	P(R3)	0.00000e+000

Let's verify this result. According to Fig. 2.318, the current pass through R1 and R2 only. According to Ohm's law, the amount of current is 6 mA. So, the dissipated power is $P = R \times I^2 = 1k \times 36\mu = 36$ mW. Remember that the steady state power of inductor and capacitor is zero.

2.36   Example 29: Potentiometer Block

Potentiometer is one of the components that is commonly used in circuits. This block can be find in the potentiometer section of basic group (Fig. 2.321).

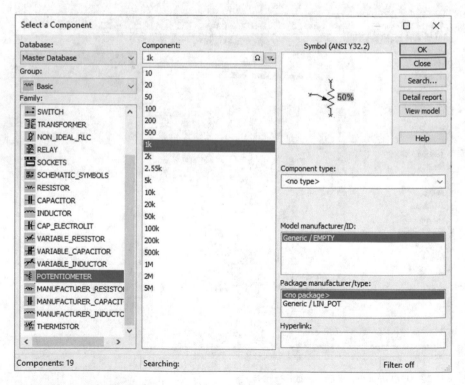

Fig. 2.321 Potentiometer block

Consider the schematic shown in Fig. 2.322. The value of potentiometer can be changed by moving the slider behind the block.

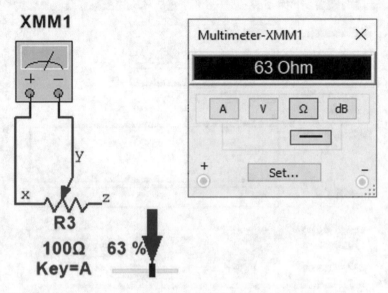

Fig. 2.322 Value of potentiometer can be changed by keyboard or the slider behind it

You can change the value of potentiometer by the keyboard as well. Double click the potentiometer block and use the value tab to assign the desired key to it (Fig. 2.323). For instance, in Fig. 2.322, the potentiometer is controlled by key A. You can press the A key of keyboard to increase the resistance between y and x (and decrease the resistance between the y and z). You can decrease the resistance between the y and x (and increase the resistance between the y and z) by pressing the Shift+A keys of the keyboard.

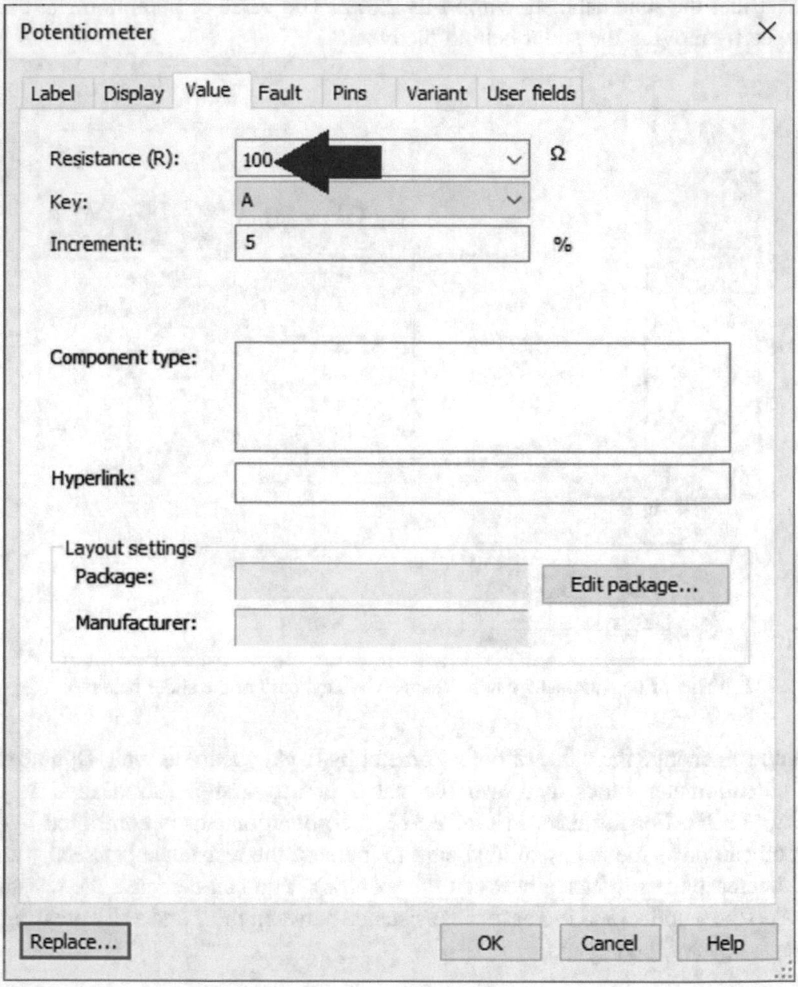

Fig. 2.323 Settings of the potentiometer

The schematic shown in Fig. 2.324 shows an amplifier with variable gain. The gain of amplifier is controlled by changing the potentiometer value.

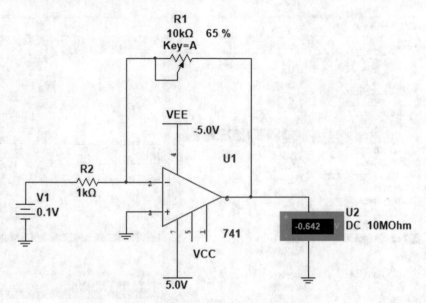

Fig. 2.324 An inverting amplifier with variable gain amplifier

Place of the op amp 741 and VCC-VEE elements are shown in Figs. 2.325 and 2.326.

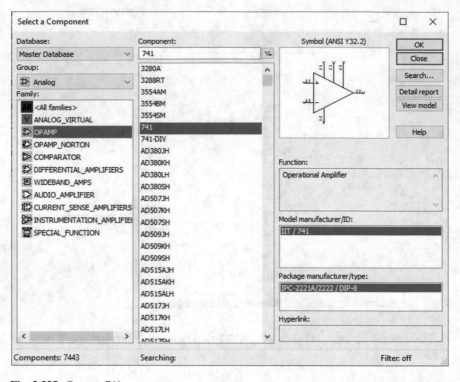

Fig. 2.325 Op amp 741

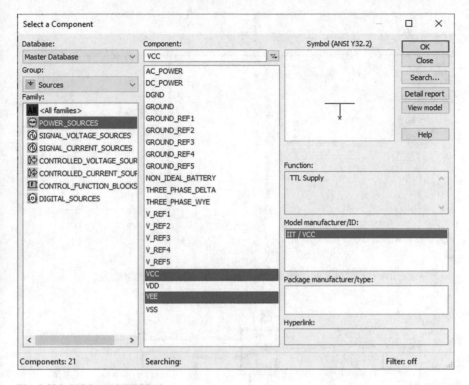

Fig. 2.326 VCC and VEE blocks

2.37 Example 30: Simulation of Circuits Containing a Switch

In this example we study a circuit which contains a switch. In the schematic shown in Fig. 2.327, the switch is in the state 1 for period of 10 ms and after that it goes to state 2 and stay there. Initial voltage of capacitor and initial current of inductor are assumed to be zero.

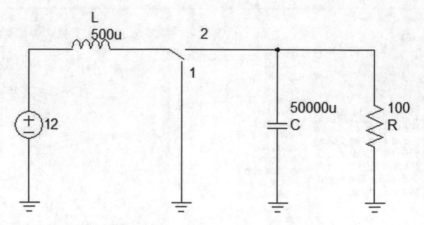

Fig. 2.327 Schematic of example 30

We use the schematic shown in Fig. 2.328 to simulate the circuit.

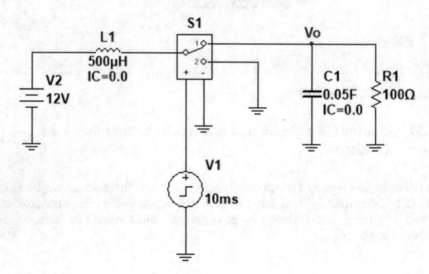

Fig. 2.328 Multisim equivalent of Fig. 2.327

Place of voltage controlled SPDT and step voltage blocks are shown in Figs. 2.329 and 2.330.

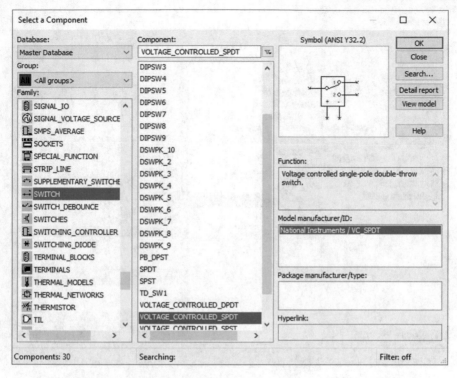

Fig. 2.329   Voltage controlled SPDT block

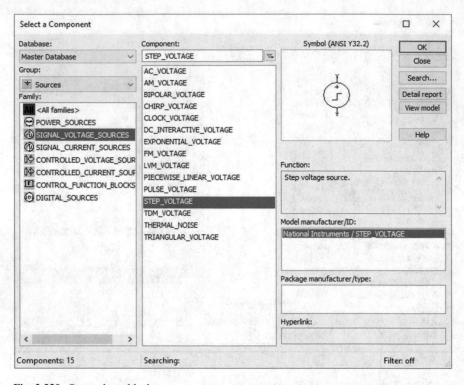

Fig. 2.330 Step voltage block

Double click the voltage controlled SPDT block and do the settings as shown in Fig. 2.331.

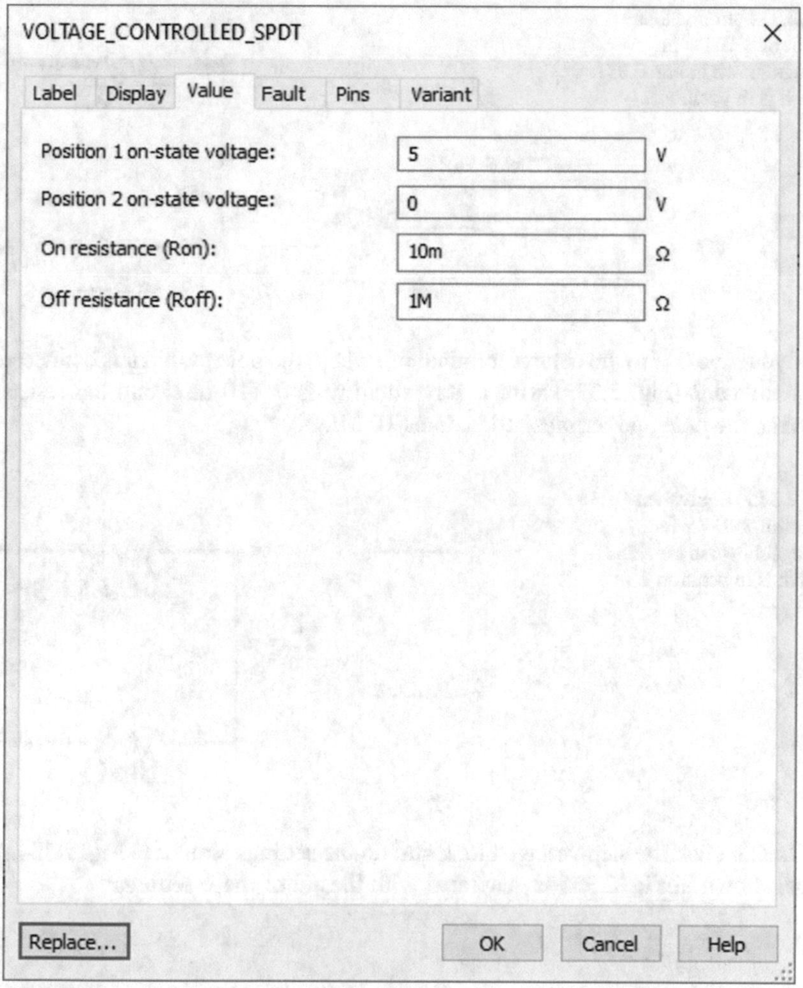

Fig. 2.331 Settings of voltage controlled SPDT block

According to Fig. 2.331, if you give 5 V to the control terminal of switch, the pole terminal (terminal 0 in Fig. 2.332) is connected to the terminal 1 with a very small resistor (10 mΩ) and the resistance between the pole and terminal 2 becomes 10 MΩ.

Fig. 2.332 Equivalent circuit of SPDT with settings shown in Fig. 2.331. Switch is in position 1

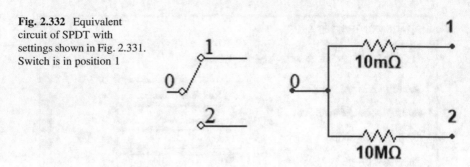

if you give 0 V to the control terminal of switch, the pole terminal is connected to the terminal 2 (Fig. 2.333) with a very small resistor (10 mΩ) and the resistance between the pole and terminal 1 becomes 10 MΩ.

Fig. 2.333 Equivalent circuit of SPDT with settings shown in Fig. 2.331. Switch is in position 2

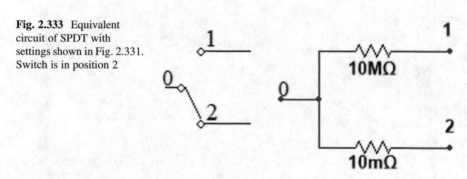

Double click the step voltage block and do the settings similar to Fig. 2.334. The signal shown in Fig. 2.335 is generated with the aid of these settings.

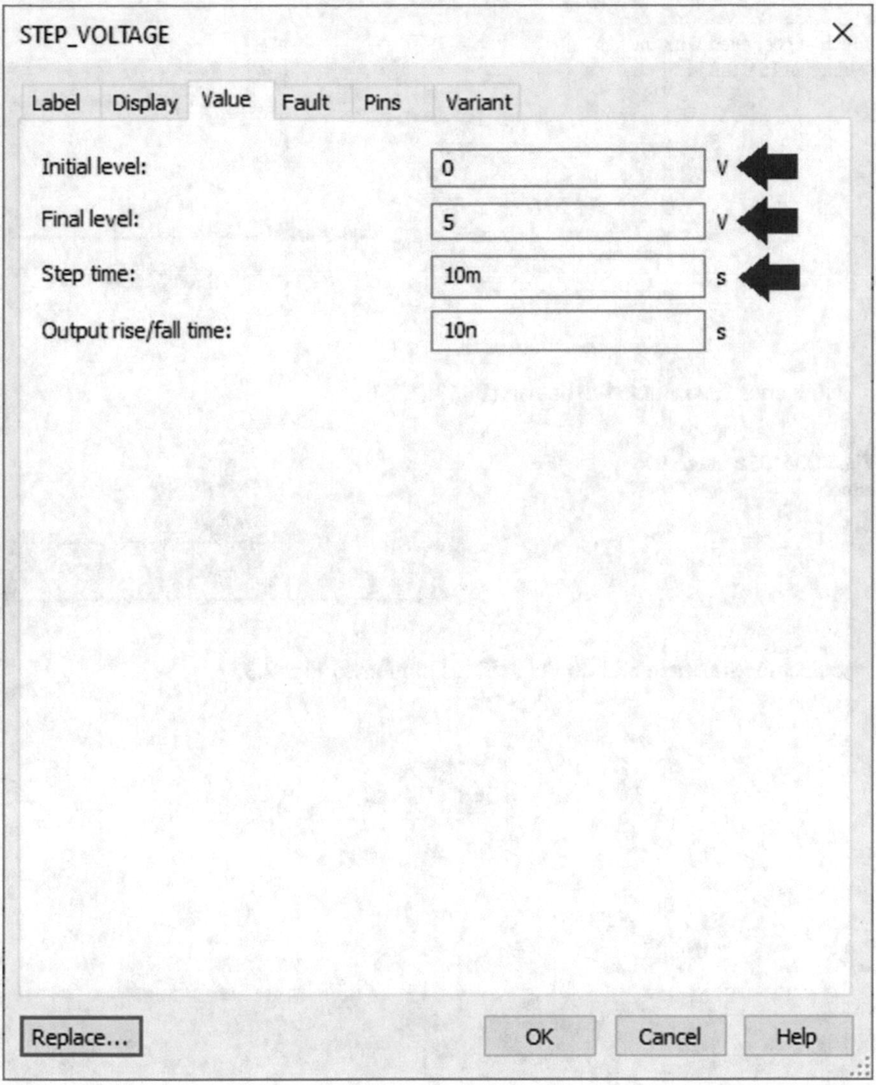

Fig. 2.334 Settings of step voltage block

Fig. 2.335 Waveform
which is generated with the
settings of Fig. 2.334

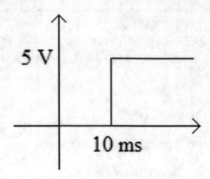

Now click the interactive button (Fig. 2.336).

Fig. 2.336 The interactive
button

Select the transient and do the settings similar to Fig. 2.337.

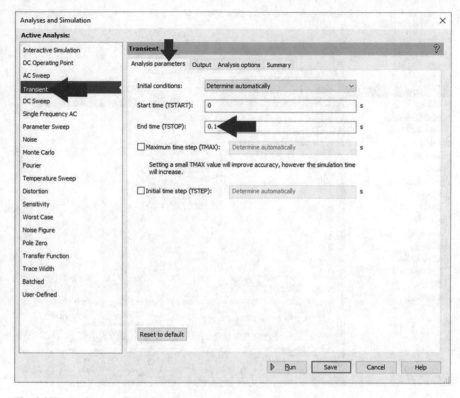

Fig. 2.337 Analyses and simulation window

Go to the output tab and add V(vo) to the right list and click the run button (Fig. 2.338).

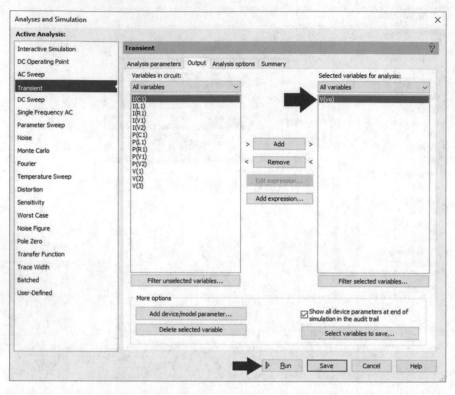

Fig. 2.338 Output tab of transient analysis

The simulation result is shown in Fig. 2.339. The voltage is zero in the first 10 ms, since the switch isolates the inductor and source from the node Vo.

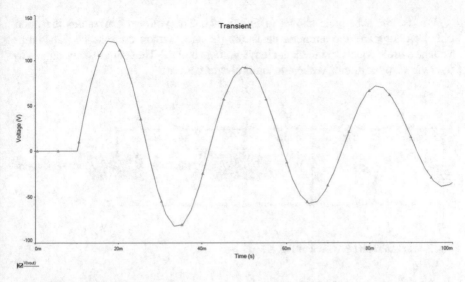

Fig. 2.339 Simulation result

2.38   Example 31: Thevenin and Norton Equivalent Circuits I

In this example, we want to obtain the Thevenin equivalent circuit seen from points a and b (Fig. 2.340).

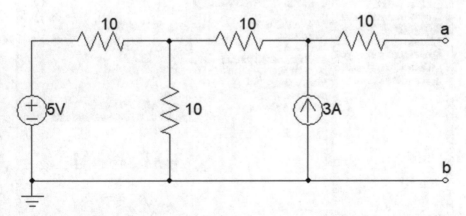

Fig. 2.340 Schematic of example 31

We use the schematic shown in Fig. 2.341. The voltmeter measures the open circuit voltage and the ammeter measures the short circuit current. The Thevenin voltage source equals to the open circuit voltage and the Thevenin resistor equals to the ratio of open circuit voltage to short circuit current.

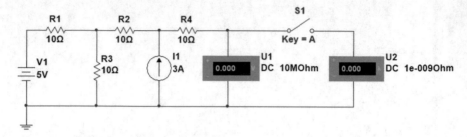

Fig. 2.341 Multisim equivalent of Fig. 2.340

Place of DC current source and switch are shown in Figs. 2.342 and 2.343.

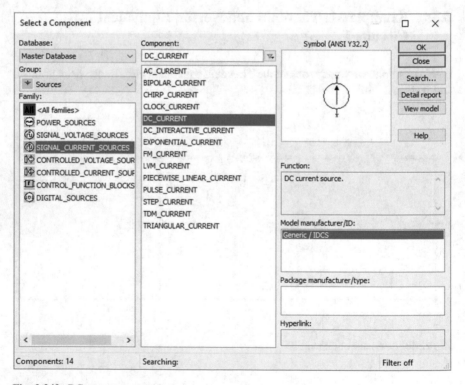

Fig. 2.342 DC current source block

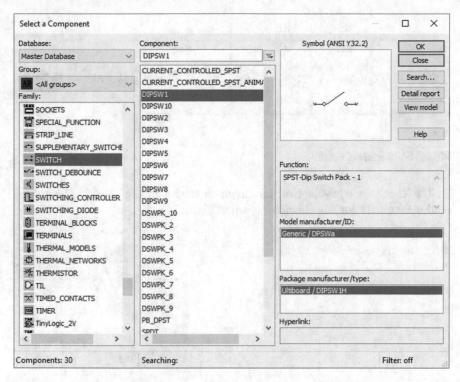

Fig. 2.343 DIPSW1 switch block

Run the simulation. The voltmeter reads 47.5 V(Fig. 2.344). This is the Thevenin voltage.

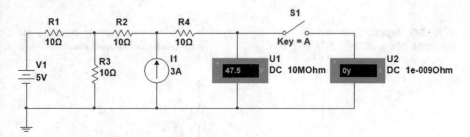

Fig. 2.344 Simulation result

Close the switch. The ammeter measures 1.9 A (Fig. 3.345). So the Thevenin resistor is 47.5/1.9 = 25 Ω.

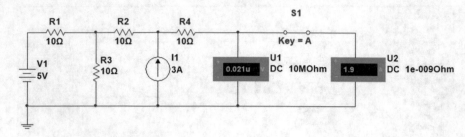

Fig. 2.345 Simulation result

The Thevenin equivalent of the circuit is shown in Fig. 2.346. The Norton equivalent circuit for this circuit is shown in Fig. 2.347.

Fig. 2.346 Thevenin
equivalent of Fig. 2.340

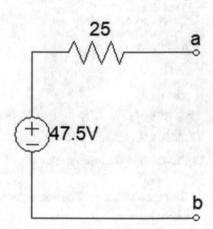

Fig. 2.347 Norton
equivalent of Fig. 2.346

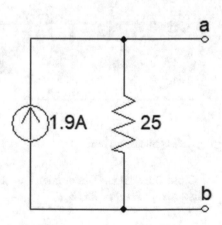

Let's verify the obtained result. The circuit is redrawn in Fig. 2.348 for easy reference. We need to kill the independent sources (i.e. replace the current sources with open circuit and voltage sources with short circuit) and measure the resistance seen from a and b. According to Fig. 2.349, the seen resistance is 25 Ω.

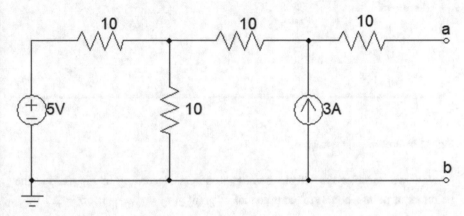

Fig. 2.348 Schematic of example 31

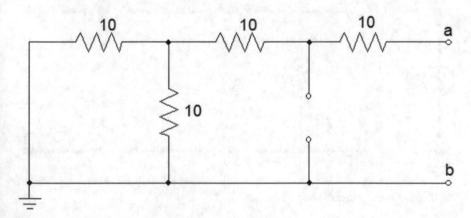

Fig. 2.349 Independent sources are killed

Now, we need to find the open circuit voltage. We use the super position principle. Let's consider the effect of current source first. According to Fig. 2.350, the potential of point a equals to the potential of point x. $v_x = \left(10 + \frac{10 \times 10}{10 + 10}\right) \times 3 = 45$ V.

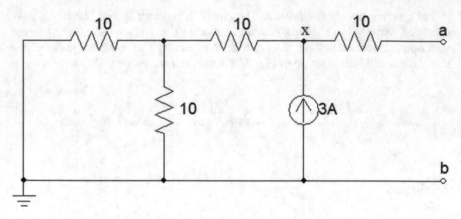

Fig. 2.350 Voltage source is killed

Now we consider the effect of voltage source. According to Fig. 2.351, the potential of point a equals to the potential of point y. $v_y = \left(\frac{10}{10+10}\right) \times 5 = 2.5$ V.

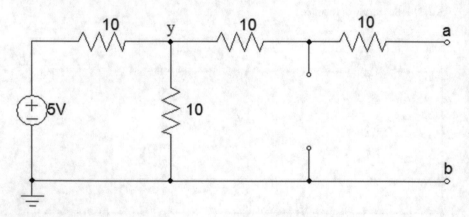

Fig. 2.351 Current source is killed

So, the open loop voltage difference between point a and b is $45 + 2.5 = 47.5$ V.

2.39 Example 32: Thevenin and Norton Equivalent Circuits II

The previous example is composed by independent sources only. In this example we study a circuit which contains the dependent sources. In this example we want to obtain the Thevenin equivalent circuit seen from points a and b for the circuit shown in Fig. 2.352.

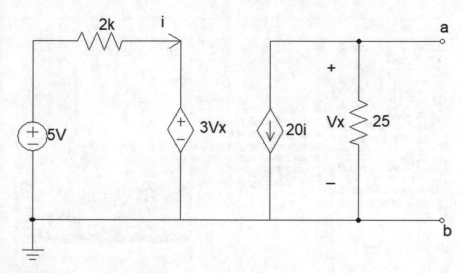

Fig. 2.352   Schematic of example

The Multisim schematic of this circuit is shown in Fig. 2.353. The schematic uses voltage controlled voltage source and current controlled current source. The place of these blocks are shown in Figs. 2.354 and 2.355.

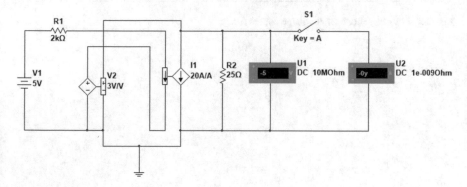

Fig. 2.353   Multisim equivalent of Fig. 2.352

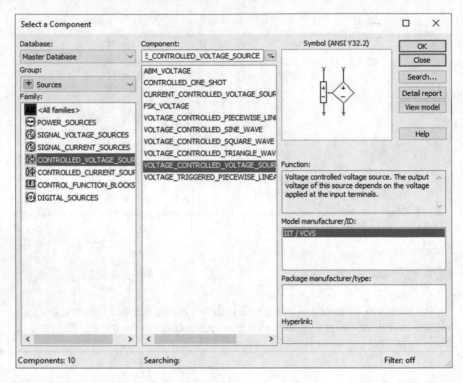

Fig. 2.354   Voltage controlled voltage source block

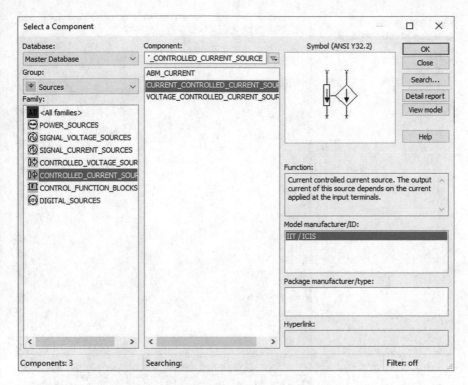

Fig. 2.355 Current controlled current source block

Double click the voltage controlled voltage source and enter 3 to the voltage gain box (Fig. 2.356).

Fig. 2.356 Settings of voltage controlled voltage source block

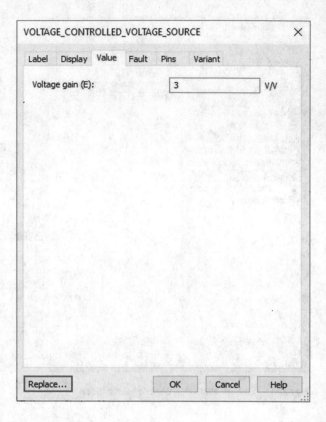

Double click the current controlled current source and enter 20 to the current gain box (Fig. 2.357).

Fig. 2.357 Settings of current controlled current source block

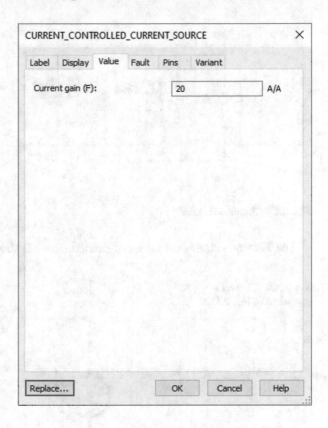

Run the simulation. The open circuit voltage is -5 V (Fig. 2.358). So, the Thevenin voltage source is -5 V.

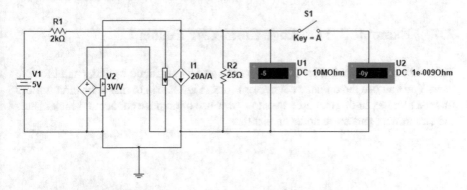

Fig. 2.358 Simulation result

Close the switch. The ammeter measure -0.05 A (Fig. 2.359). So, the Thevenin resistance is $-5/-0.05 = 100\ \Omega$.

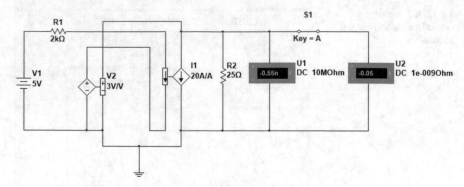

Fig. 2.359 Simulation result

The Thevenin equivalent circuit of studied circuit is shown in Fig. 2.360.

Fig. 2.360 Thevenin
equivalent of Fig. 2.352

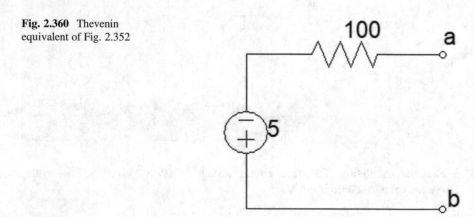

2.40   Example 33: On Page Connector Element

The schematic shown in Fig. 2.353 is messy and it is difficult to understand for the user. You can use the on page connector block (Fig. 2.361) to make the circuit more understandable. In this example we show how use of on page connector block makes the circuit neat and more understandable.

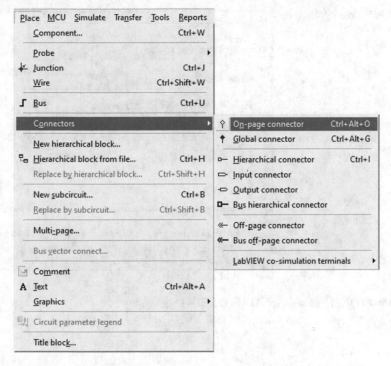

Fig. 2.361 On page connector block

Let's remove the connection from Fig. 2.353. Result is shown in Fig. 2.362.

Fig. 2.362 Removing the
wires of Fig. 2.353

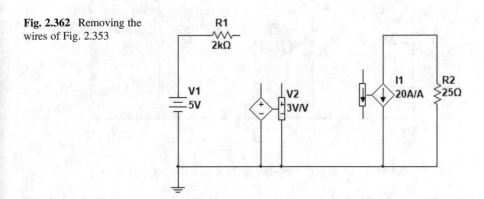

Now click the place> connector> on page connector or press the Ctrl+Alt+O. The
window show in Fig. 2.363 is appeared. Enter i1 to the connector name box and click
the OK button.

Fig. 2.363 Entering the connector name

Connect the i1 to resistor R1 (Fig. 2.364).

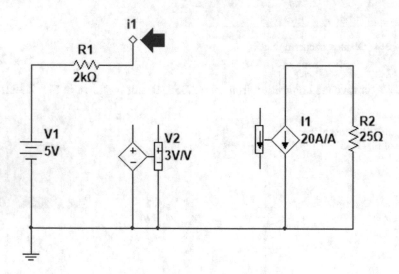

Fig. 2.364 Connecting the on page connector to the resistor R1

Now add another on page connector to the circuit and name it o1. Connect the o1 to the schematic as shown in Fig. 2.365.

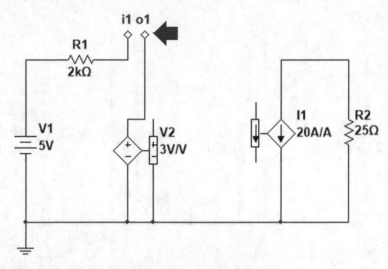

Fig. 2.365 Connecting the on page connector to V2

Add another on page connector to the circuit and connect it to the current controlled current source as shown in Fig. 2.366. Multisim considers the two on page connector block with the same name as connected. So, according to Fig. 2.366, the right terminal of the resistor is connected to the upper terminal of current controlled current source.

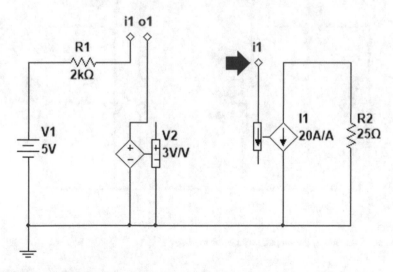

Fig. 2.366 Connecting the on page connector to I1

Connect an on page connector block to the lower terminal of current controlled current source (Fig. 2.367).

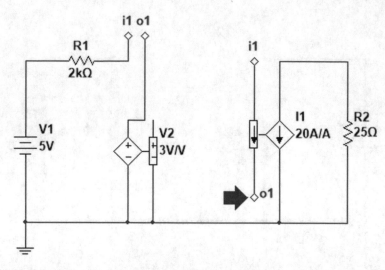

Fig. 2.367 Connecting the on page connector to I1

Connect two more on page connector block to the circuit and name them a. Connect these two blocks to the circuit as shown in Fig. 2.368.

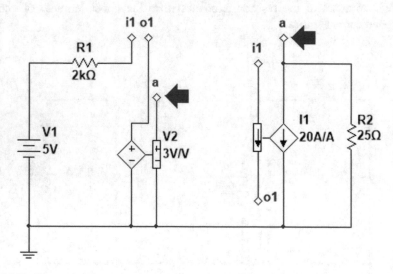

Fig. 2.368 Connecting the on page connector to I1

The final schematic is shown in Fig. 2.369. The open circuit voltage is −5 V.

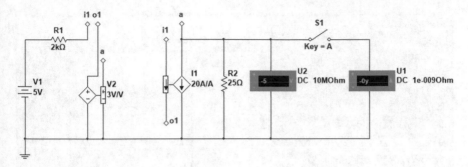

Fig. 2.369 Simulation result

The short circuit current is -0.05 A (Fig. 2.370). These results are the same as example 39. However the schematic is less messy in comparison to Fig. 2.353.

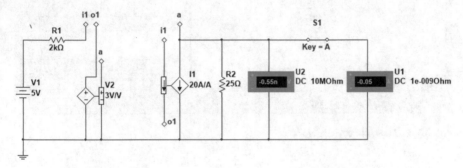

Fig. 2.370 Simulation result

2.41   Example 34: Three Phase Circuits and Capacitive Compensation

Simulation of three phase circuits is very easy in Multisim. Multisim has three phase sources. Place of these blocks are shown in Figs. 2.371 and 2.372.

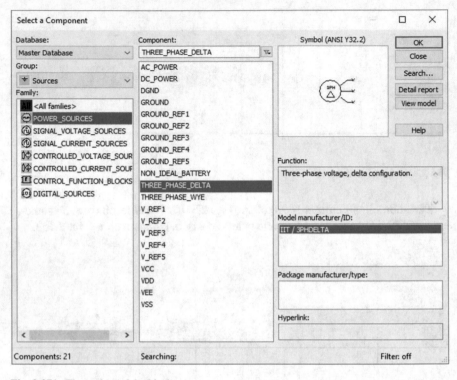

Fig. 2.371 Three phase delta block

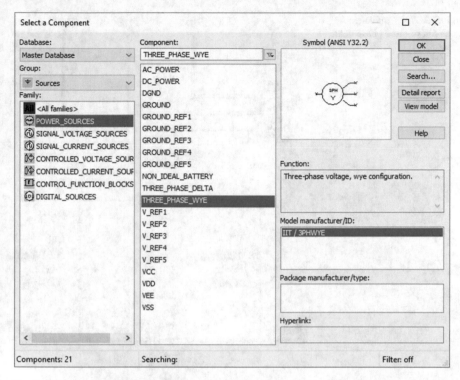

Fig. 2.372 Three phase wye block

The equivalent circuit of these sources are show in Figs. 2.373 and 2.374.

Fig. 2.373 Equivalent circuit of three phase delta source

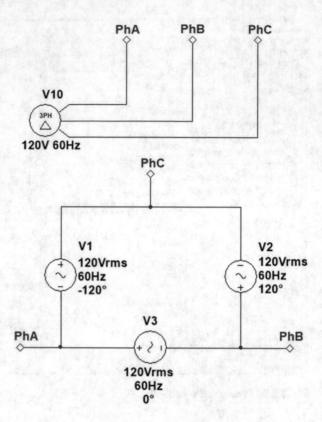

Fig. 2.374 Equivalent circuit of three phase wye source

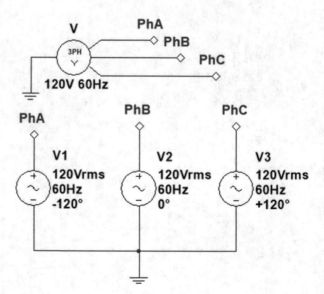

If you like to make your own delta source, you need to put some small resistor in series to your sources (Fig. 2.375). These resistors are added to solve the convergence problem of circuit and avoids the circuit matrix to become singular. Setting of the voltage source V1, V2 and V3 are shown in Figs. 2.376, 2.377 and 2.378, respectively.

Fig. 2.375 Addition of small resistors to the sources

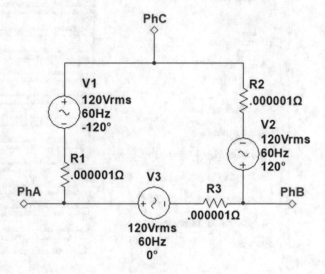

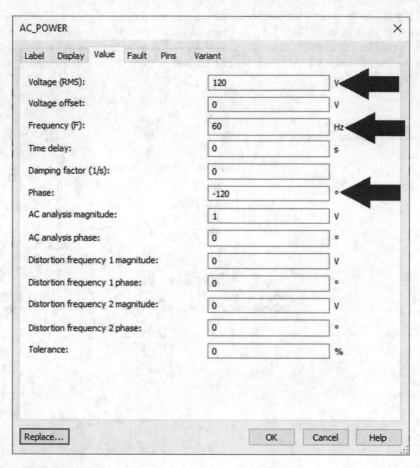

Fig. 2.376 Settings of source V1

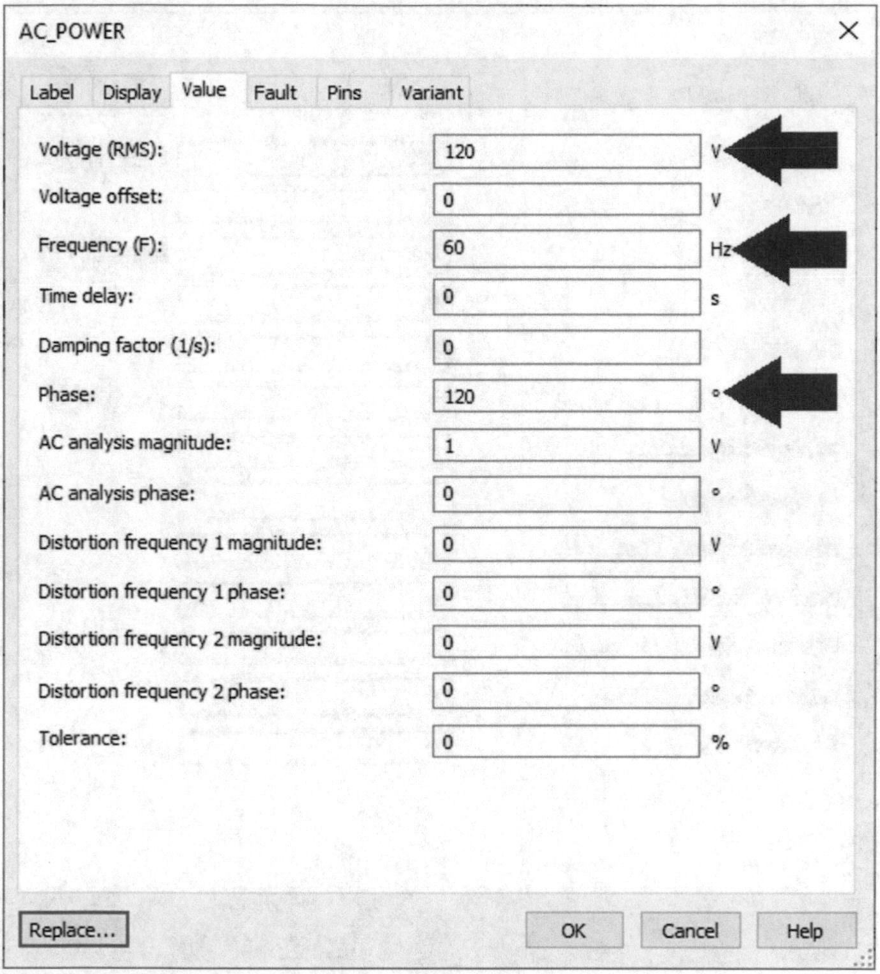

Fig. 2.377 Settings of source V2

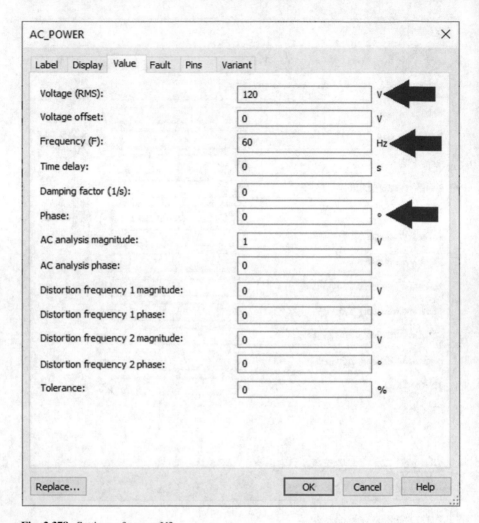

Fig. 2.378 Settings of source V3

Let's see a sample simulation. Assume the schematic shown in Fig. 2.379. We want to measure the average power and current of the circuit.

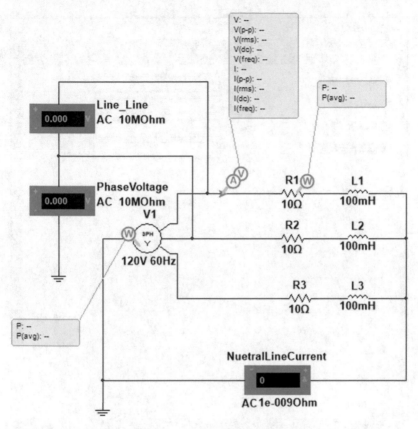

Fig. 2.379 Schematic of example 34

Settings of AC source V1 is shown in Fig. 2.380. The line-neutral voltage is 120 Vrms.

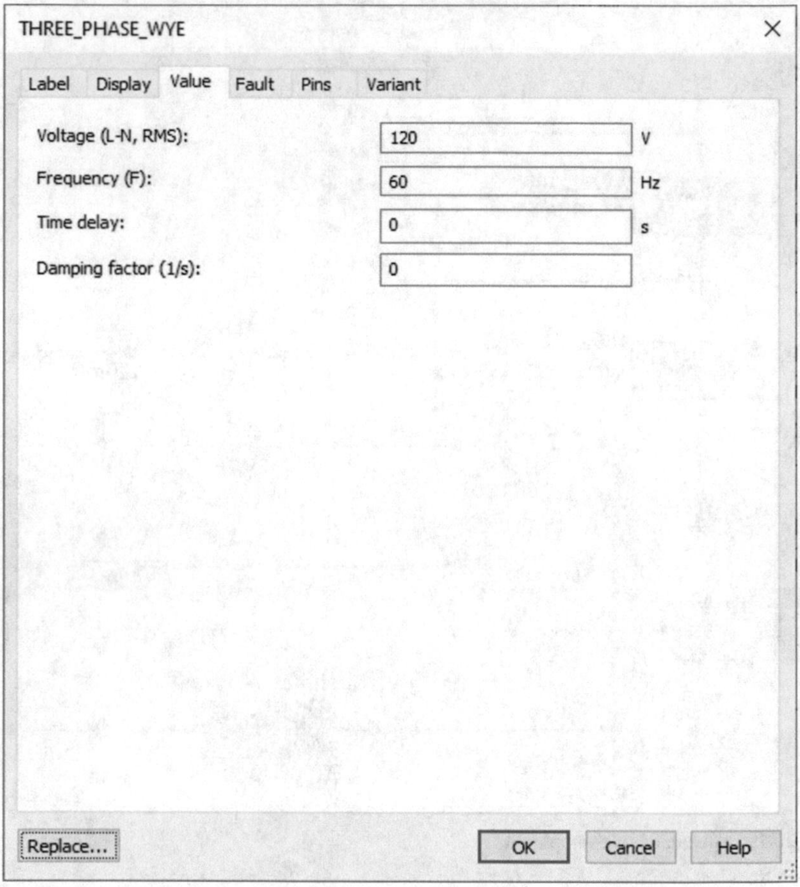

THREE_PHASE_WYE   ✕

Label   Display   Value   Fault   Pins   Variant

Voltage (L-N, RMS):   120   V

Frequency (F):   60   Hz

Time delay: 0   s

Damping factor (1/s):   0

Replace...   OK   Cancel   Help

Fig. 2.380 Three phase wye source V1 settings

Run the circuit. The simulation result is shown in Fig. 2.381. Note that the current is the neutral line is very small. This is expected since the load is balanced.

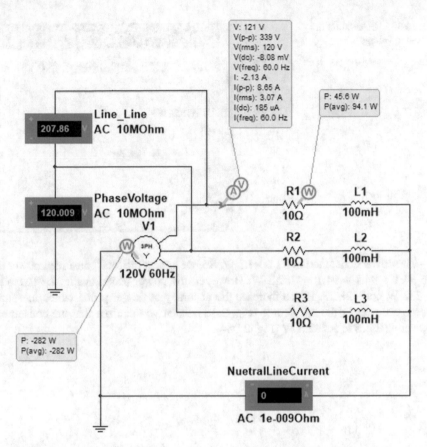

Fig. 2.381 Simulation result

The line-line voltage is 207.86 V. Let's verify this result (Fig. 2.382).

Fig. 2.382 Calculation of
line-line voltage

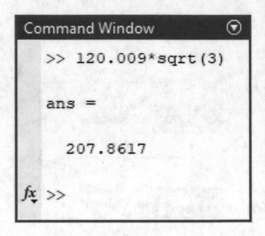

The power consumed in R1 is 94.1 W. So, we expect the total consumed power to be 94.1 + 94.1 + 94.1 = 282.3 W. However, the power probe in Fig. 2.381 reads −282 W not −282.3. If you increase the accuracy of power probe which measure average power of the AC source (Fig. 2.383), then you can see that the consumed power changes to −282.3 W (Fig. 2.384).

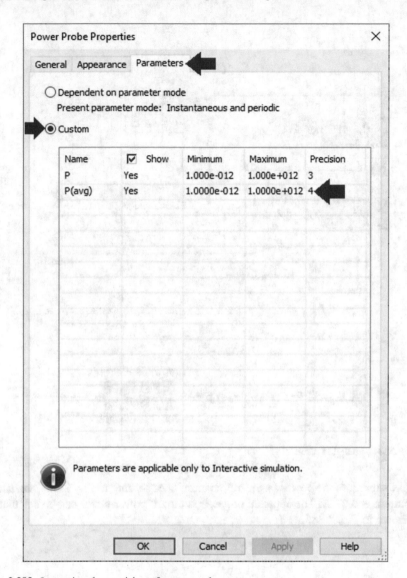

Fig. 2.383 Increasing the precision of power probe

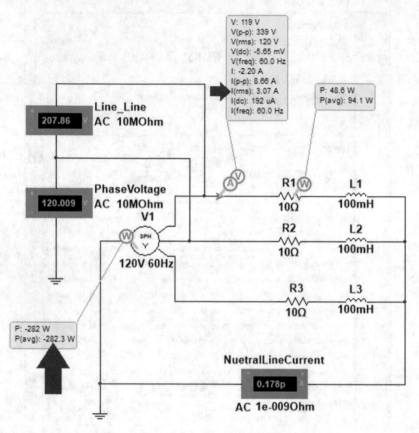

Fig. 2.384 Simulation result

Let's measure the power factor of this circuit. According to Fig. 2.384, the RMS of current is 3.07 A. The apparent power and circuit power factor can be calculated as shown in Fig. 2.385.

Fig. 2.385 Calculation of power factor

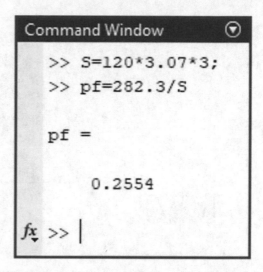

According to Fig. 2.385, the power factor is 0.2554 which is quite low. Let's add some capacitor to the circuit in order to increase the power factor. Calculation of required capacitor is shown in Fig. 2.386.

```
Command Window
>> P=94.1;Vrms_phase=120;Irms_phase=3.07;S=Vrms_phase*Irms_phase;f=60;w=2*pi*f;
>> Q=sqrt(S^2-P^2);
>> C=Q/(w*Vrms_phase^2)

C =

   6.5611e-05

fx >>
```

Fig. 2.386 MATLAB calculations

Figure 2.387 shows the compensated circuit.

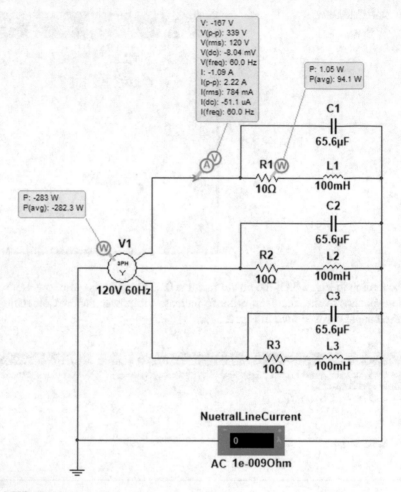

Fig. 2.387 Simulation result

Let's measure the circuit power factor. The calculations shown in Fig. 2.388 shows that the power factor is unity.

Fig. 2.388 Calculation of power factor

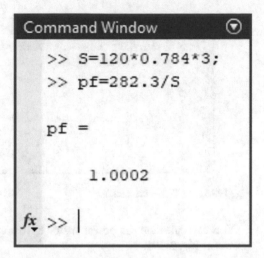

Let's study another example. Assume the schematic shown in Fig. 2.389. The AC source is a delta connected source.

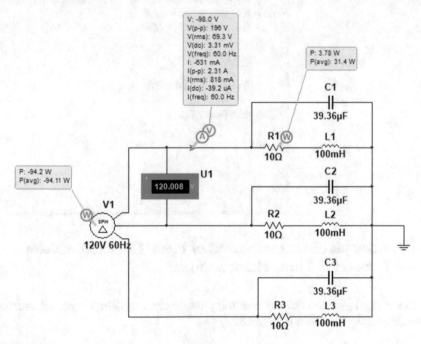

Fig. 2.389 Simulation result

The power factor is calculated in Fig. 2.390.

```
Command Window                                              ⊙
    >> P=31.4;Vline=69.3;Iline=.818;S=Vline*Iline;
    >> pf=P/(Vline*Iline)

   pf =

        0.5539

fx >>
```

Fig. 2.390 MATLAB calculations

You can calculate the power factor by using the angle of load. This method is shown in Fig. 2.391.

Fig. 2.391 MATLAB calculations

```
Command Window                                              ⊙
    >> R=10;L=0.1;C=39.36e-6;;f=60;w=2*pi*f;
    >> X1=j*L*w;Xc=-j/w/C;
    >> Z=(R+X1)*Xc/((R+X1)+Xc)

   Z =

     46.2638 +69.9815i

    >> cos(angle(Z))

   ans =

        0.5515

fx >> |
```

2.42 Example 35: Measurement of Phase Current for Delta Connected Three Phase Sources

In this example we want to measure the phase current of delta connected sources. Assume the circuit shown in Fig. 2.392.

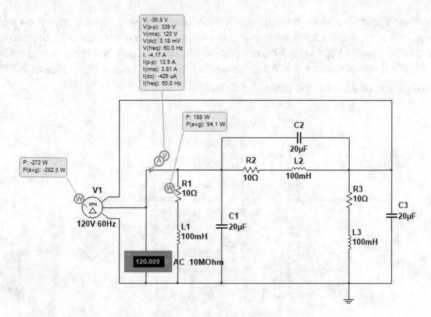

Fig. 2.392 Simulation result

We can't measure the phase current when we use the Multisim ready to use three phase source. The reason is that we have no access to the sources which formed the block. So, we need to make our three phase source (Fig. 2.393).

Fig. 2.393 Addition of small resistor in series to the sources

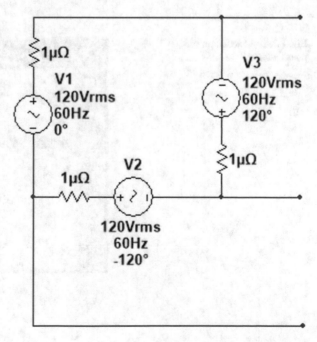

Now replace the Multisim delta connected three phase source with what shown in Fig. 2.393. Now you have access to the sources and you can measure the phase current. According to Fig. 2.394, the phase current is 2.20 A and the line current is 3.81 A. Remember that for delta connected sources, $I_{phase} = \frac{I_{line}}{\sqrt{3}}$. Figure 2.395 verifies this relation for the measured values.

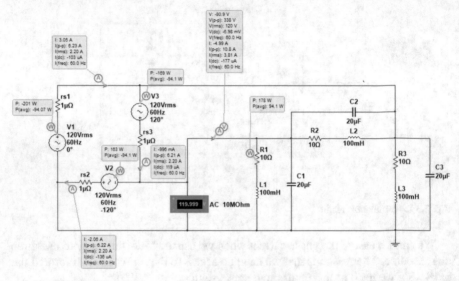

Fig. 2.394 Simulation results

Fig. 2.395 Calculation of phase current

2.43   Example 36: Maximum Power Transfer

In this example we want to calculate the value of load resistor such that it takes the maximum power from the source. The circuit is shown in Fig. 2.396. The R1 and L1 show the internal impedance of the source.

Fig. 2.396 Schematic of example 36

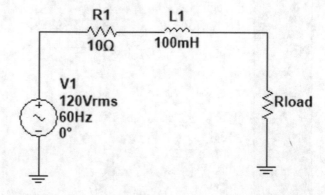

Let's use MATLAB to find the value of resistor which consumes the maximum power. The code shown in Fig. 2.397 draws the consumed power as a function of the load resistor. The result of this code is shown in Fig. 2.398. You can use the cursor to find the maximum of the curve. According to Fig. 2.398, the load which consumes the maximum power is about 39.03 Ω.

Fig. 2.397 MATLAB calculations

```
Command Window                                    ⊙
>> syms x
>> R1=10;L=100e-3;f=60;w=2*pi*f;
>> XL=j*L*w;
>> I=120*exp(j*0)/(R1+x+XL);
>> P=x*abs(I)^2

P =

(14400*x)/abs(x + pi*12i + 10)^2

>> ezplot(P,[0 100])
fx >>
```

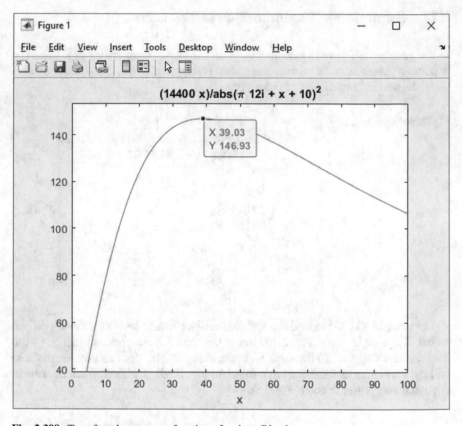

Fig. 2.398 Transferred power as a function of resistor Rload

We need to use the parameter sweep analysis to solve this problem. Parameter sweep analysis changes something and permits us to see the effect of change on the desired output variable. Click the interactive button to start a parameter sweep analysis (Fig. 2.399).

Fig. 2.399 Interactive
button

Go to the parameter sweep section in the opened window and do the settings as shown in Fig. 2.400. With these settings, the resistor is changed from 0 to 100 Ω with 2.04082 Ω steps. If you increase the number of points (decreasing the step size), the accuracy of simulation increases however the simulation takes more time to be completed.

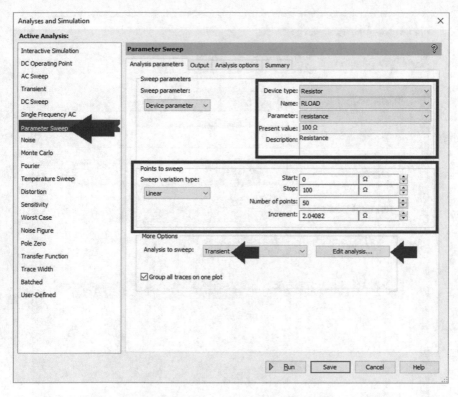

Fig. 2.400 Analysis parameters tab of parameter sweep analysis

In the next steps we need to determine the output variable which we want to see the effect of changes on it. Click the edit analysis button in the Fig. 2.400 and enter 0.1 to the end time box (Fig. 2.401).

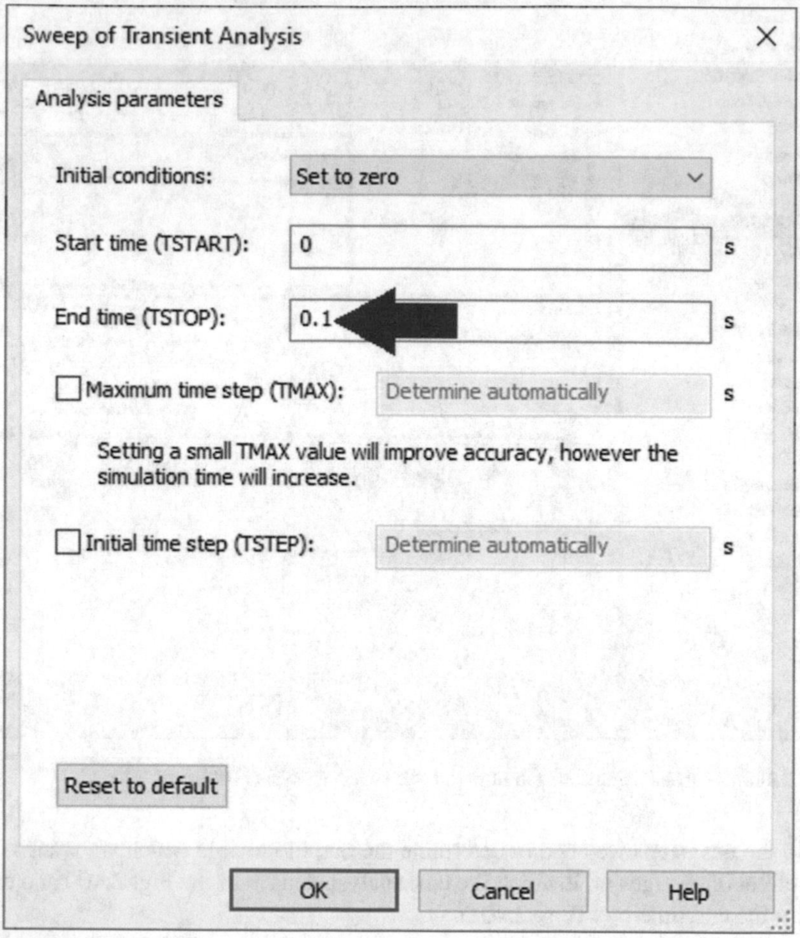

Fig. 2.401 Determining the end time of simulation

Go to the output tab and click the add expression button (Fig. 2.402).

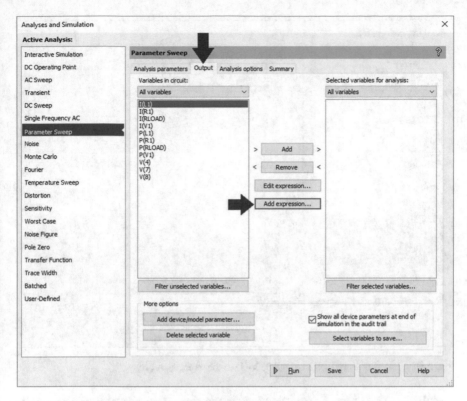

Fig. 2.402 Output tab of parameter sweep analysis

Enter the avgx(P(RLOAD),8.333e-3) to the expression box. This expression calculates the moving average of resistor RLOAD power with frequency of 60 Hz.

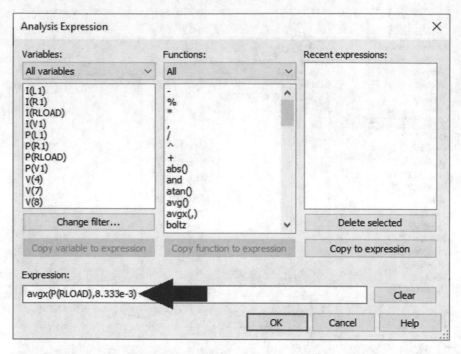

Fig. 2.403 Entering avgx(P(RLOAD),8.333e-3) to the expression box

Click the OK button in Fig. 2.403. Now the entered expression must be seen in the right side selected variables for the analysis list. Click the run button to start the simulation (Fig. 2.404).

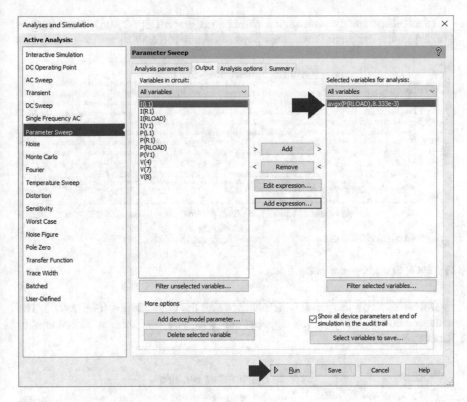

Fig. 2.404 Output tab of parameter sweep analysis

The simulation result is shown in Fig. 2.405.

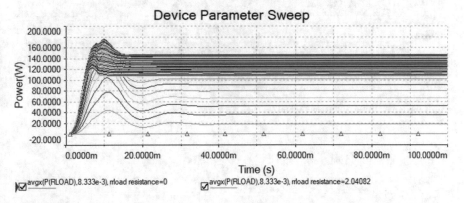

Fig. 2.405 Simulation result

Zoom in the steady state region of obtained graph (Fig. 2.406). According to Fig. 2.406, the maximum dissipated power is about 146.6 W.

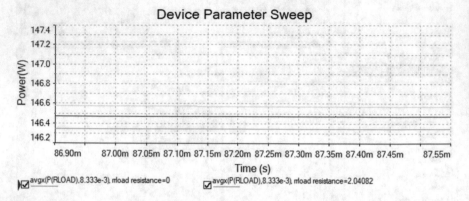

Fig. 2.406 Zoomed version of Fig. 2.405

Double click on the graph which is above all the other graphs (Fig. 2.407). The window shown in Fig. 2.408 appears and tells you that this curve belongs to Rload = 38.775 Ω.

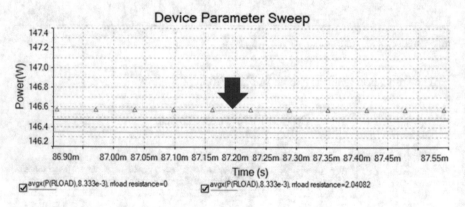

Fig. 2.407 Maximum curve is clicked

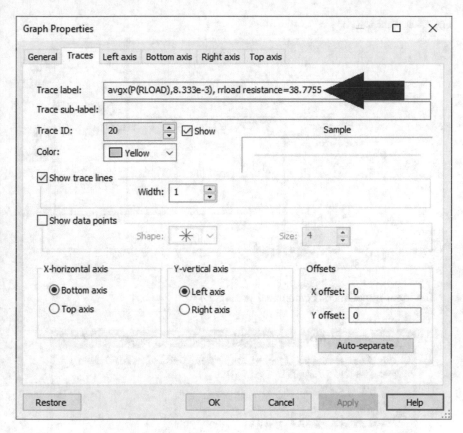

Fig. 2.408 Reading the value of Rload

2.44 Example 37: Transformer

In this example we want to measure the impedance seen from a transformer. Assume
the load shown in Fig. 2.409.

Fig. 2.409 Schematic of
example 37

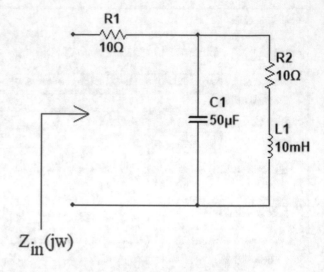

Fig. 2.409 Schematic of example 37

The input impedance is calculated in Fig. 2.410.

```
Command Window

>> f=60;w=2*pi*f;R1=10;C1=50e-6;R2=10;L1=10e-3;
>> Xc=-j/(w*C1);
>> XL=j*w*L1;
>> Z2=R2+XL;
>> Z=R1+Xc*Z2/(Xc+Z2)

Z =

   21.1302 + 1.7998i

fx >>
```

Fig. 2.410 MATLAB calculations

Let's connect this load to the secondary of a transformer and measure the impedance seen from the primary. Place of transformer block is shown in Fig. 2.411. Note that nPmS shows a transformer with n primary coils and m secondary coils.

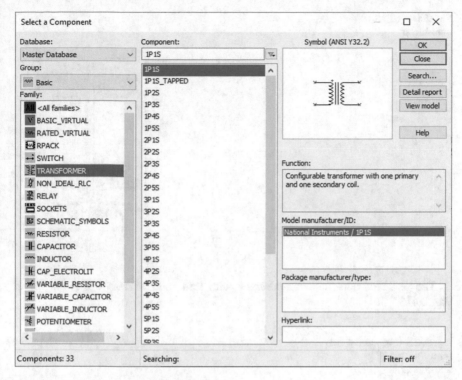

Fig. 2.411   different types of transformers

Assume the 2P2S transformer shown in Fig. 2.412. The primary side coils (2) are numbered from the top down starting from the corner with the primary side designator (1). The coils on the secondary side (4) are numbered from the top down on the opposite side of the component's symbol.

The numbers on the bottom of the symbol (3) indicate the number of coil turns in each of the coils. In the example shown in Fig. 2.412, primary coil 1 has ten turns, primary coil 2 has five turns, secondary coil 1 (Coil 3 in Fig. 2.412) has one turn, and secondary coil 2 (Coil 4 in Fig. 2.412) has two turns.

Fig. 2.412 2P2S
transformer

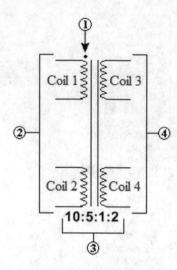

The place of dots for a transformer with turn ratio of 10:5:1:2 is shown in Fig. 2.413.

Fig. 2.413 Location of dots
for a 10:5:1:2 2P2S
transformer

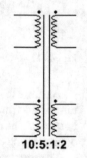

You can change the place of dots by entering a negative turn ratio. For instance, the place of dots for a transformer with turn ratio of 10:-5:-1:2 is shown in Fig. 2.414.

Fig. 2.414 Location of dots
for a 10:-5:-1:2 2P2S
transformer

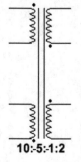

Assume the schematic shown in Fig. 2.415.

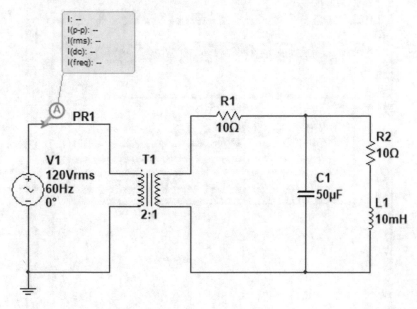

Fig. 2.415 Multisim schematic of example 37

Double click the transformer and do the settings as shown in Fig. 2.416.

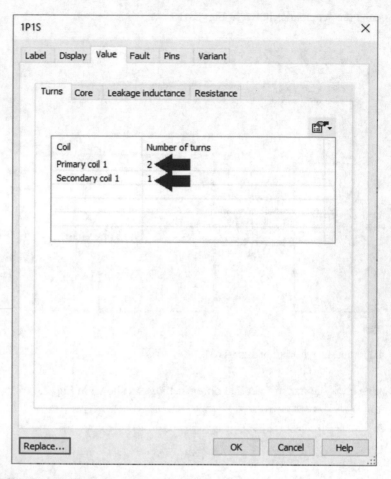

Fig. 2.416 Settings of transformer T1 in Fig. 2.415

Run the simulation. The RMS of current is 1.41 A (Fig. 2.417).

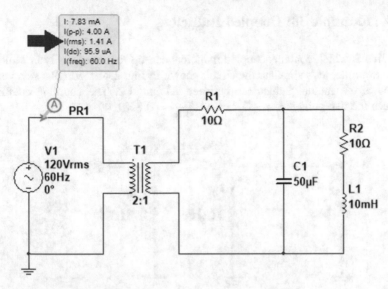

Fig. 2.417 Simulation result

Let's verify the result given by Multisim. The calculation shown in Fig. 2.418 shows that the Multisim result is correct.

Fig. 2.418 MATLAB calculations

```
Command Window                                    ⊙
  >> n=2;Z=21.1302 + 1.7998i;
  >> I=120/(n^2*Z)


  I =


     1.4095 - 0.1201i


  >> abs(I)


  ans =


     1.4146

fx >>
```

2.45 Example 38: Coupled Inductors

The circuits which contains coupled inductors can be simulated easily in Multisim. As an example, let's simulate the circuit shown in Fig. 2.419. Vin is a step voltage and M is the mutual inductance between L1 and L2. The coupling coefficient between the two coils is $k = \frac{M}{\sqrt{L_1 L_2}} = \frac{0.9m}{\sqrt{1m \times 1.1m}} = 0.8581$.

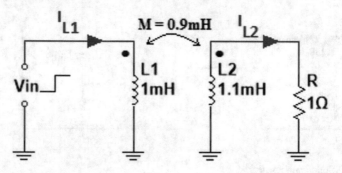

Fig. 2.419 Schematic of example 38

From basic circuit theory,

$$\begin{cases} L_1 \dfrac{di_{L1}}{dt} - M \dfrac{di_{L2}}{dt} = V_{in}(t) \\[2mm] Ri_{L2} + L_2 \dfrac{di_{L2}}{dt} - M \dfrac{di_{L1}}{dt} = 0 \end{cases}$$

Take the Laplace transform of both side:

$$\begin{bmatrix} L_1 s & -Ms \\ -Ms & R + L_2 s \end{bmatrix} \times \begin{bmatrix} I_{L1}(s) \\ I_{L2}(s) \end{bmatrix} = \begin{bmatrix} V_{in}(s) \\ 0 \end{bmatrix}$$

So,

$$\begin{bmatrix} I_{L1}(s) \\ I_{L2}(s) \end{bmatrix} = \begin{bmatrix} L_1 s & -Ms \\ -Ms & R + L_2 s \end{bmatrix}^{-1} \times \begin{bmatrix} V_{in}(s) \\ 0 \end{bmatrix}$$

$V_{in}(s) = \frac{1}{s}$, so,

$$\begin{bmatrix} I_{L1}(s) \\ I_{L2}(s) \end{bmatrix} = \begin{bmatrix} \dfrac{(11s + 10000) \times 10000}{s^2 \times (29s + 100000)} \\[4mm] \dfrac{90000}{s(29s + 100000)} \end{bmatrix}$$

You can use the commands shown in Fig. 2.420 in order to see the time domain graph of I_{L1} and I_{L2}. Graph of I_{L1} and I_{L2} are shown in Figs. 2.421 and 2.422, respectively.

```
Command Window                                              ⊙
    >> s=tf('s');
    >> I1=(11*s+10000)*10000/s/(29*s+100000);
    >> I2=90000/(29*s+100000);
    >> step(I1,[0:0.06/100:0.06]), grid on
    >> step(I2), grid on
fx >>
```

Fig. 2.420 Drawing the step response of Fig. 2.419

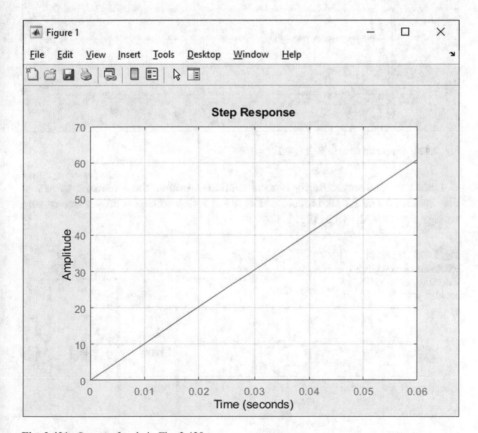

Fig. 2.421 Output of code in Fig. 2.420

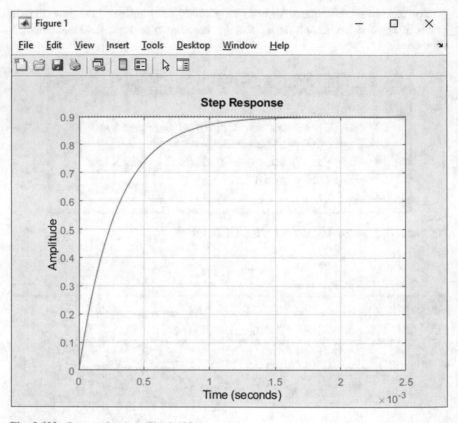

Fig. 2.422 Output of code in Fig. 2.420

Let's use Multisim to verify our calculations. Assume the schematic shown in Fig. 2.423. L1 and L2 are inductors. The V1 is a step voltage source. Place of step voltage source is shown in Fig. 2.424.

Fig. 2.423 Multisim schematic of example 38. Inductors are not coupled yet

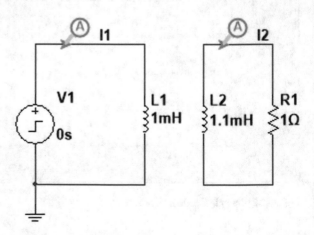

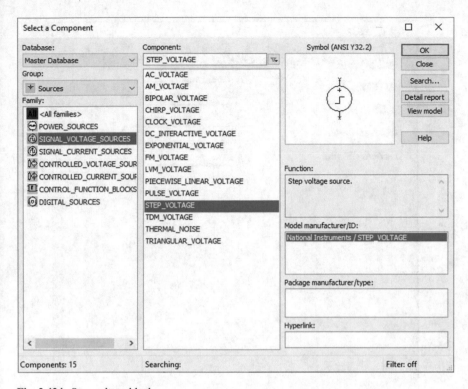

Fig. 2.424 Step voltage block

Double click on the step voltage source and do the settings as shown in Fig. 2.425. These settings produce a voltage that jumps from 0 V to 1 V at t = 0 s.

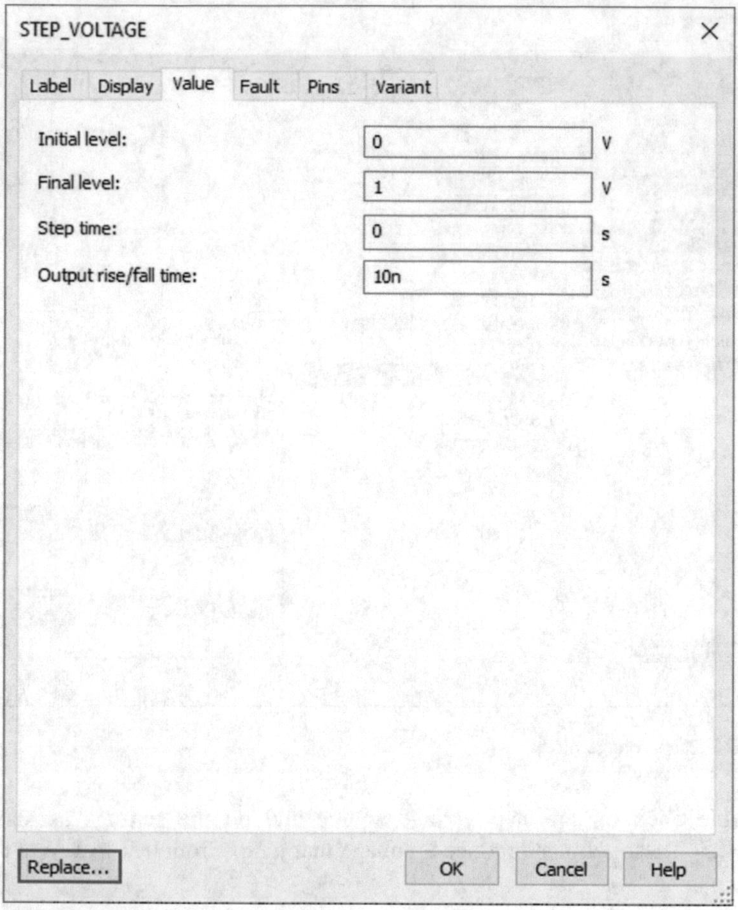

Fig. 2.425 Step voltage block settings

Now select and add a inductor coupling block (Fig. 2.426) to the schematic (Fig. 2.427).

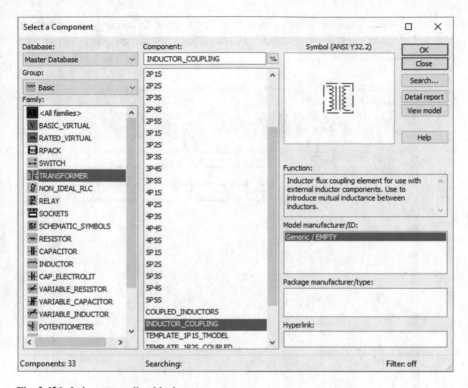

Fig. 2.426 Inductor coupling block

Fig. 2.427 Addition of inductor coupling block to the schematic

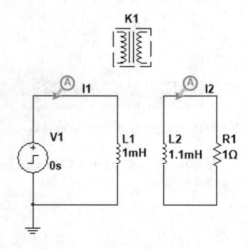

Double click on the inductor coupling block and do settings as shown in Fig. 2.428, then click the OK button. These settings set the coupling between L1 and L2 equal to 0.8581.

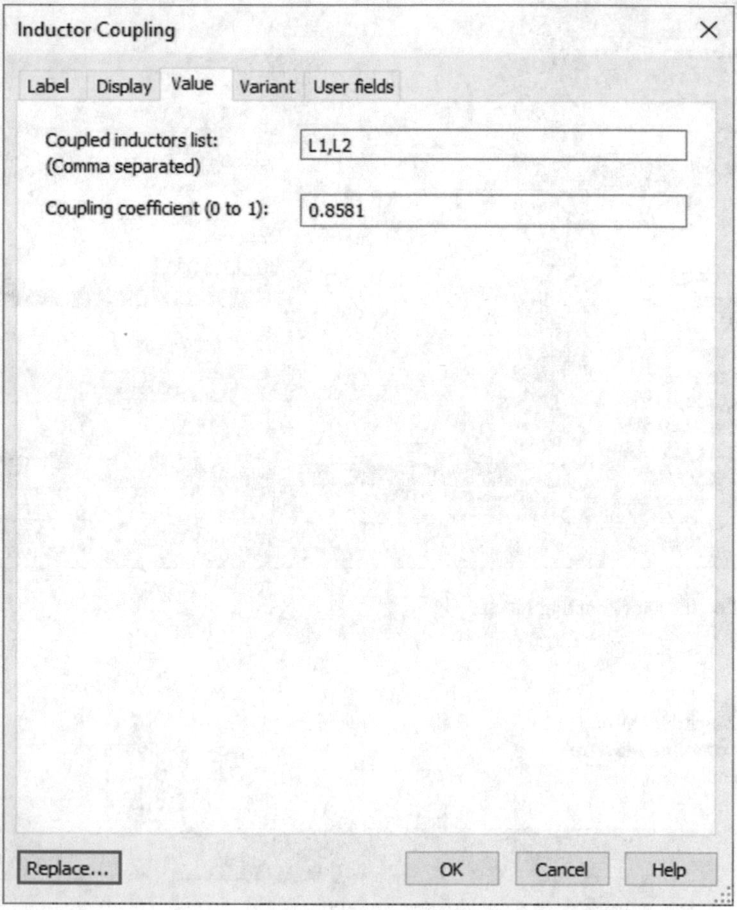

Inductor Coupling					✕
Label	Display	Value	Variant	User fields	

Coupled inductors list: L1,L2
(Comma separated)

Coupling coefficient (0 to 1):   0.8581

Replace...   OK   Cancel   Help

Fig. 2.428 Settings of inductor coupling block

Now, the schematic must look like what shown in Fig. 2.429. It is ready to apply a transient analysis.

Fig. 2.429 Inductor L1 and
L2 are coupled

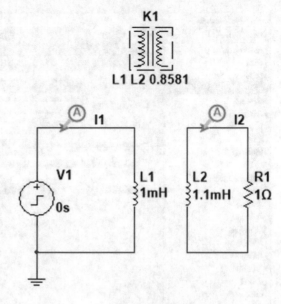

Click the interactive button (Fig. 2.430).

Fig. 2.430 Interactive
button

Do the settings as shown in Figs. 2.431 and 2.432 and run the simulation. The
result is shown in Fig. 2.433. This figure shows the current for inductor L1 and is the
same the graph shown in Fig. 2.421.

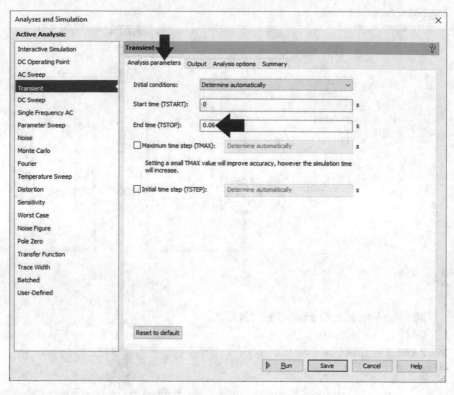

Fig. 2.431 Analysis parameters tab of transient analysis

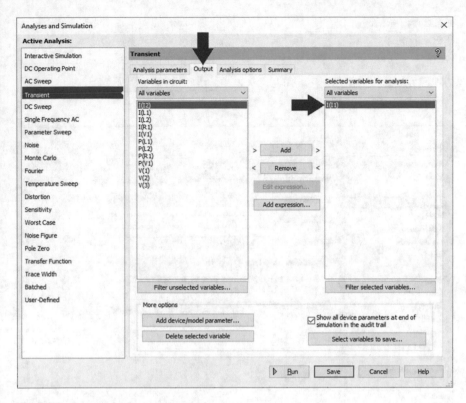

Fig. 2.432 Output tab of transient analysis

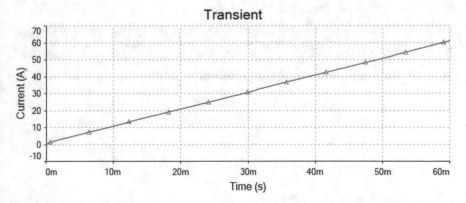

Fig. 2.433 Simulation result

Click the interactive button again (Fig. 2.434) and do the transient analysis settings as shown in Figs. 2.435 and 2.436.

Fig. 2.434 Interactive
block

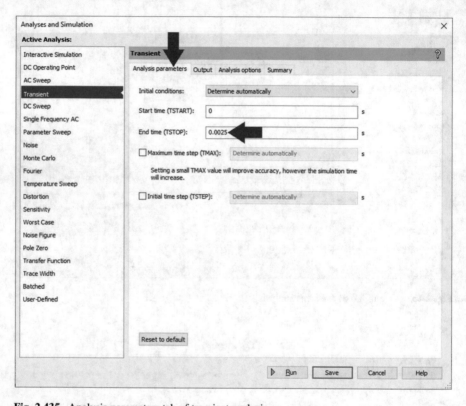

Fig. 2.435 Analysis parameters tab of transient analysis

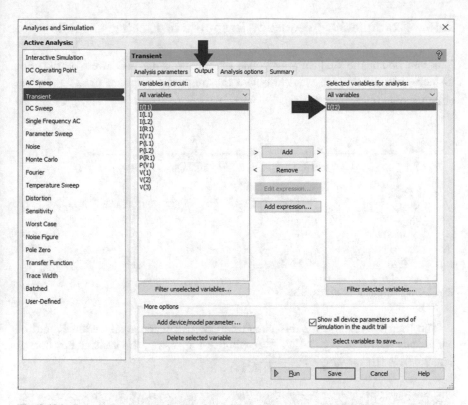

Fig. 2.436 Output tab of transient analysis

Run the simulation. The result is shown in Fig. 2.437. This figure shows the current for inductor L2 and is the same the graph shown in Fig. 2.422.

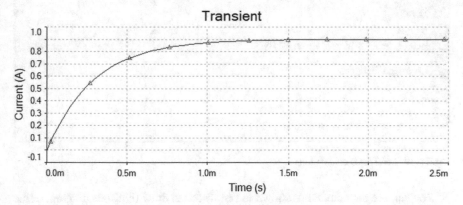

Fig. 2.437 Simulation result

2.46   Example 39: Frequency Response of Electric Circuits

You can obtain the frequency response of circuits using AC sweep easily. For instance, assume that we want to obtain the frequency response of $\frac{V_{out}(s)}{V_1(s)}$ for the circuit shown in Fig. 2.438.

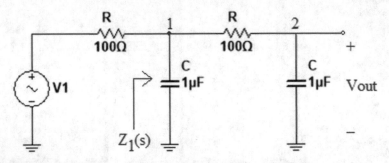

Fig. 2.438 Schematic of example 39

The transfer function for this circuit can be calculated easily with the aid of commands shown in Fig. 2.439. According to Fig. 2.439, the transfer function of Fig. 2.438 is $H(s) = \frac{V_{out}(s)}{V_1(s)} = \frac{1}{(RC)^2 s^2 + 3RCs + 1}$.

```
Command Window                                                              ⊙
 >> syms s C R
 >> Zc=1/s/C;
 >> Z1=Zc*(R+Zc)/(Zc+R+Zc);
 >> V1=Z1/(Z1+R);
 >> V2=Zc/(Zc+R)*V1

 V2 =

 1/(C^2*s^2*(R + 2/(C*s))*(R + (R + 1/(C*s))/(C*s*(R + 2/(C*s)))))

 >> simplify(V2)

 ans =

 1/(C^2*R^2*s^2 + 3*C*R*s + 1)
```

Fig. 2.439 MATLAB calculations

You can use the commands shown in Fig. 2.440 to draw the Bode diagram of the obtained transfer function. The Bode plot of the transfer function is shown in Fig. 2.441. Note that the horizontal axis unit is Rad/s, it is NOT Hertz.

Fig. 2.440 Drawing the bode diagram of transfer function H

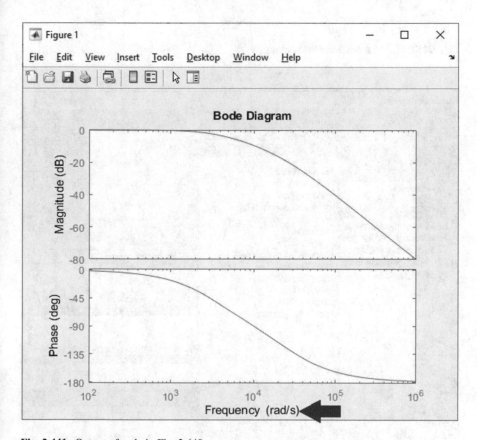

```
Command Window                                    ⊙
    >> R=100;C=1e-6;
    >> H=tf(1,[(R*C)^2 3*R*C 1]);
    >> bode(H)
ƒx >>
```

Fig. 2.441 Output of code in Fig. 2.440

Let's obtain the frequency response of the circuit ($\frac{V_{out}(s)}{V_1(s)}$) with Multisim. Assume the schematic shown in Fig. 2.442. The input node is called input and the output node is called output. V1 is a signal voltage source (Fig. 2.443) and its settings are shown in Fig. 2.444. AC analysis magnitude and AC analysis phase (Fig. 2.444) are important in AC sweep analysis. The other settings have no effect on the AC sweep analysis result. Ensure to set the mentioned boxes as shown in Fig. 2.444.

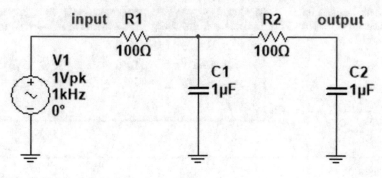

Fig. 2.442 Multisim schematic of example 39

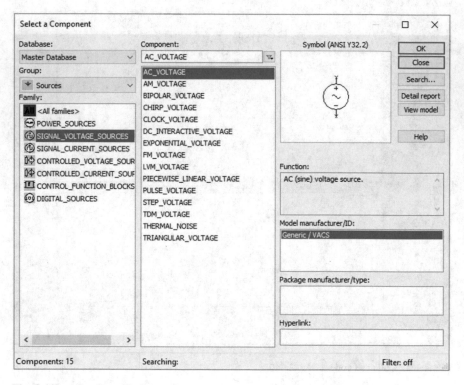

Fig. 2.443 AC voltage block

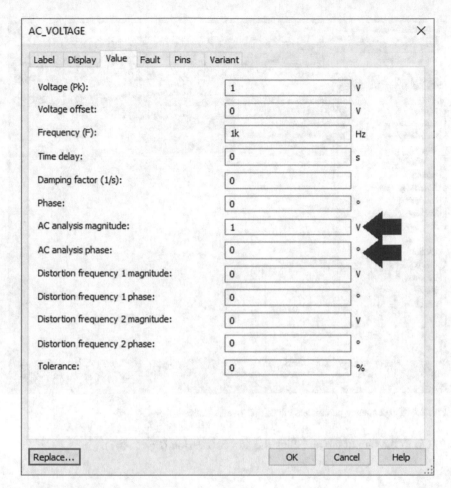

Fig. 2.444   Settings of AC source V1

Click the interactive button (Fig. 2.445).

Fig. 2.445 Interactive
button

Go to the AC sweep section and do the settings as shown in Fig. 2.446. We want to obtain the frequency response of the circuit for the 100 Hz-1 kHz interval. So, start frequency and stop frequency boxes are filled with 100 Hz and 1 kHz, respectively.

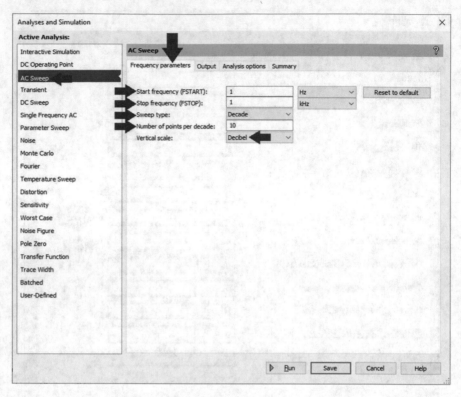

Fig. 2.446 Frequency parameters tab of AC sweep analysis

Go to the output tab and click the add expression (Fig. 2.447).

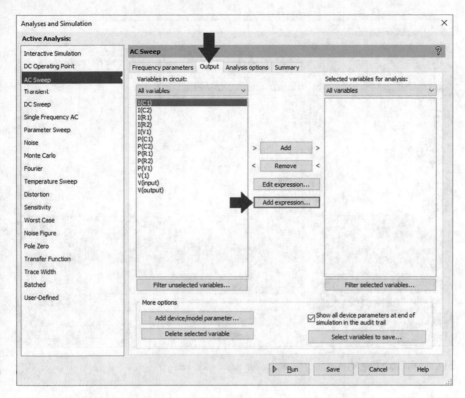

Fig. 2.447 Output tab of AC sweep analysis

Enter the expression shown in Fig. 2.448 and click the OK button. You can simply enter V(output) instead of V(output)/V(input), since the input amplitude and phase equals to 1 and 0, respectively. Remember that $\frac{V_{output}}{1e^{j0}} = \frac{V_{output}}{1} = V_{output}$.

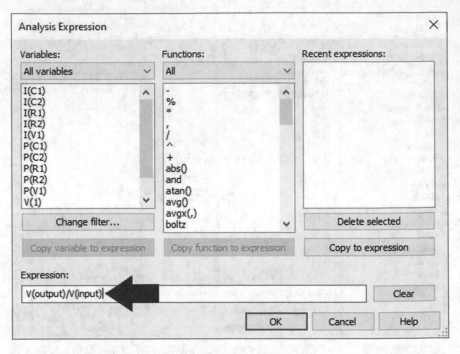

Fig. 2.448 Entering V(output)/V(input) to the expression box

The expression entered in the Fig. 2.448 appears in the selected variables for analysis list (Fig. 2.449). Click the run to start the simulation.

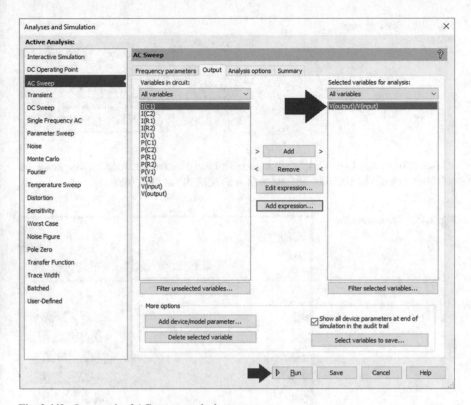

Fig. 2.449 Output tab of AC sweep analysis

The simulation result is shown in Fig. 2.450. Use the cursors (Fig. 2.451) to read different points of the graph. You can read the phase graph as well by clicking on it and then clicking the show cursor icon.

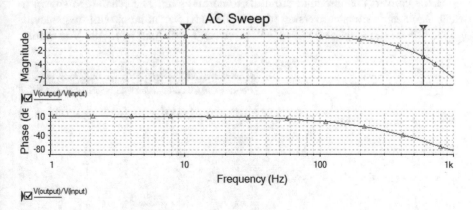

Fig. 2.450 Simulation result

Fig. 2.451 Show cursors button

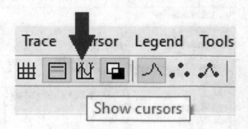

Show cursors

According to Fig. 2.452, the amplitude of the frequency response is −0.0014214 dB at 10.7534 Hz and it is −3.0132 dB at 592.7488 Hz.

Fig. 2.452 Cursors reading

Cursor	V(output)/V(input)
x1	10.7534
y1	−1.4214m
x2	592.7488
y2	−3.0132
dx	581.9954
dy	−3.0118
dy/dx	−5.1750m
1/dx	1.7182m

Let's convert the amplitude from dB to normal gain. The calculation shown in Fig. 2.453 shows this conversion for 592.7488 Hz. So, an input with frequency of 592.7488 and an amplitude of say 7 V, produces $7 \times 0.7069 = 4.9483$ V at output.

Fig. 2.453 Conversion from dB to normal gain

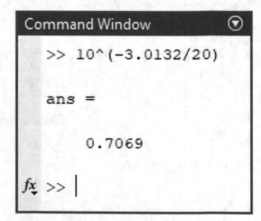

Command Window

```
>> 10^(-3.0132/20)

ans =

    0.7069

fx >> |
```

Let's verify the Multisim result. We use the graph of Fig. 2.454. Amplitude of this graph at $3.74 \times 10^3 \frac{\text{Rad}}{\text{s}}$ equals to -3.01 dB. Let's convert the $3.74 \times 10^3 \frac{\text{Rad}}{\text{s}}$ into Hertz. According to Fig. 2.455, $3.74 \times 10^3 \frac{\text{Rad}}{\text{s}}$ equals to 595.2395 Hz, which is quite close to Multisim result.

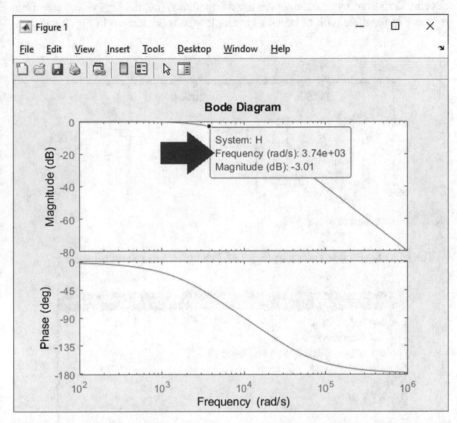

Fig. 2.454 -3 dB point of graph is located at $3.74 \times 10^3 \frac{\text{Rad}}{\text{s}}$

Fig. 2.455 Conversion of $3.74 \times 10^3 \frac{\text{Rad}}{\text{s}}$ to Hz

2.47 Example 40: Measurement of Input/Output Impedance as a Function of Frequency

In this example we show the method of extraction of input/output impedance of circuits. Let's start by calculating the input impedance (i.e. the impedance seen from source V1) for the circuit of previous example which is redrawn in Fig. 2.456.

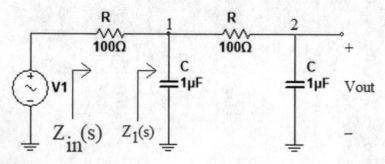

Fig. 2.456 Schematic of example 40

The MATLAB code shown in Fig. 2.457 calculates the input impedance.

```
Command Window

>> syms s C R
>> Zc=1/s/C;
>> Z1=Zc*(R+Zc)/(Zc+R+Zc);
>> Zin=R+Z1

Zin =

R + (R + 1/(C*s))/(C*s*(R + 2/(C*s)))

>> simplify(Zin)

ans =

(C^2*R^2*s^2 + 3*C*R*s + 1)/(C*s*(C*R*s + 2))

fx >> |
```

Fig. 2.457 MATLAB calculations

The value of this impedance for R = 100 Ω and C = 1 μF is calculated in Fig. 2.458. After running the code, the Bode diagram shown in Fig. 2.459 is obtained.

```
Command Window                                          ⊙
   >> R=100;C=1e-6;f1=100;f2=1e3;
   >> s=tf('s');
   >> Zin=R+(R + 1/(C*s))/(C*s*(R+2/(C*s)))

   Zin =

     1e-14 s^3 + 3e-10 s^2 + 1e-06 s
     -------------------------------
           1e-16 s^3 + 2e-12 s^2

   Continuous-time transfer function.

   >> w=logspace(log10(2*pi*f1),log10(2*pi*f2),250);
   >> bode(Zin,w)
fx >> |
```

Fig. 2.458 Drawing the bode of Zin

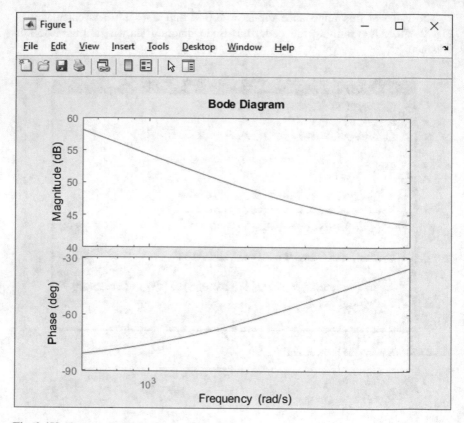

Fig. 2.459 Output of code in Fig. 2.458

Let's draw the frequency response of input impedance with Multisim. Click the interactive button (Fig. 2.460).

Fig. 2.460 Interactive
button

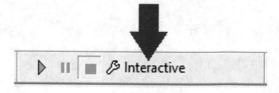

Go to ac sweep section and do the settings similar to Fig. 2.461.

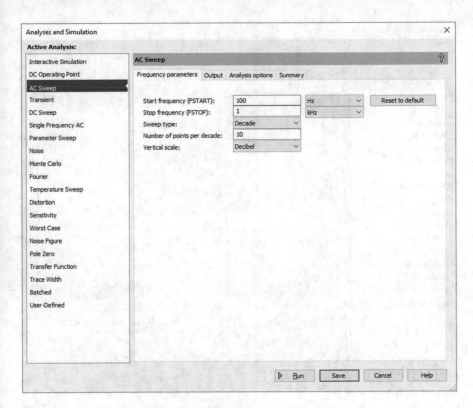

Fig. 2.461 Frequency parameters tab of AC sweep

Go to the output tab and click the add expression button (Fig. 2.462).

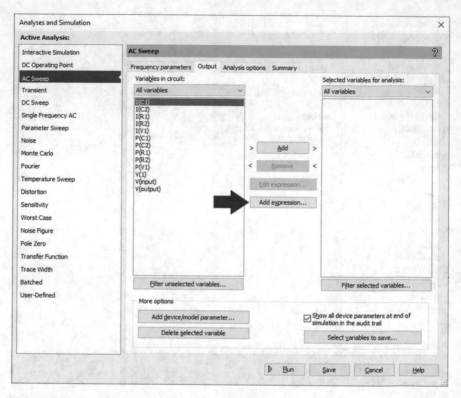

Fig. 2.462 Output tab of AC sweep

Enter V(input)/-I(V1) to the expression box (Fig. 2.463). Remember that input impedance is the ratio of input voltage to input current. The I(V1) is the current that enters to source V1 (Fig. 2.464). In order to obtain the current that goes out from V1, we need to multiply it by a negative sign.

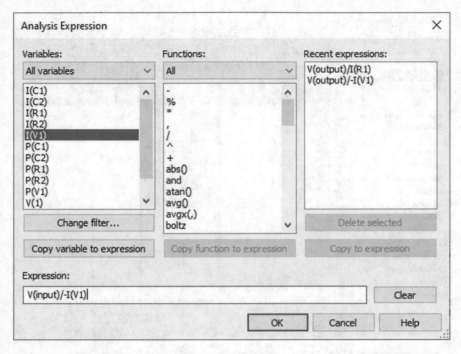

Fig. 2.463 Entering V(input)/-I(V1) to the expression box

Fig. 2.464 I(V1) is the current that enters to the positive terminal of V1

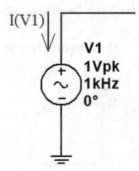

After clicking the OK button in Fig. 2.463, the V(input)/-I(V1) is added to the right list. Click the run button to do the simulation (Fig. 2.465).

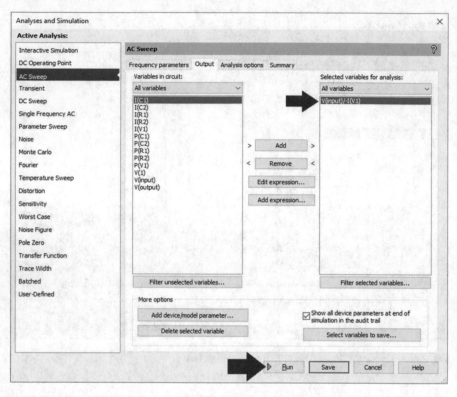

Fig. 2.465   Output tab of AC sweep analysis

The frequency response for V(input)/-I(V1) is shown in Fig. 2.466.

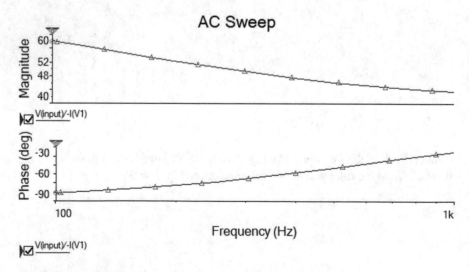

Fig. 2.466   Simulation result

You can use cursors to read the impedance for different frequencies. For instance, the impedance for 300 Hz (Fig. 2.467), is 49.4236 dB (Fig. 2.468) and its phase angle is $-64.9265°$ (Fig. 2.469). $10^{\frac{49.4236}{20}} = 295.9239$, so the input impedance at 300 Hz is $295.9239e^{j(-64.9265°)}$.

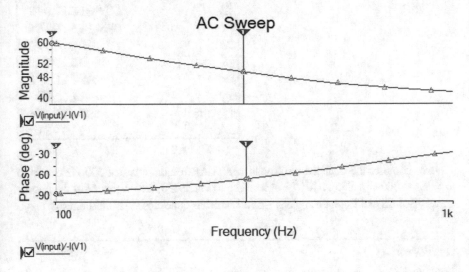

Fig. 2.467 Simulation result

Fig. 2.468 Cursor 1 of magnitude plot

Cursor	✕
	V(input)/−I(V1)
x1	➤ 299.6772
y1	➤ 49.4236
x2	100.0000
y2	58.1300
dx	−199.6772
dy	8.7064
dy/dx	−43.6023m
1/dx	−5.0081m

Fig. 2.469 Cursor 1 of phase plot

Cursor	V(input)/-I(V1)
x1	300.1972
y1	-64.9265
x2	100.0000
y2	-81.0833
dx	-200.1972
dy	-16.1568
dy/dx	80.7045m
1/dx	-4.9951m

Let's compare the obtained result with MATLAB result. Note that 300 Hz equals to $\omega = 2\pi \times 300 = 1.885 \times 10^3 \frac{Rad}{s}$. According to Fig. 2.470, the magnitude of the impedance is 49.4 dB and its phase angle is $-65.1°$ which is quite close to Multisim result.

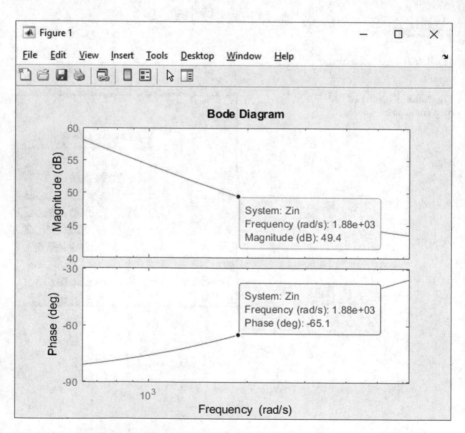

Fig. 2.470 Magnitude and phase of $1.885 \times 10^3 \frac{Rad}{s}$

You can use the linear axis for the vertical axis if you like to see the value of impedance is Ohm. To do this, select linear for vertical axis (Fig. 2.471).

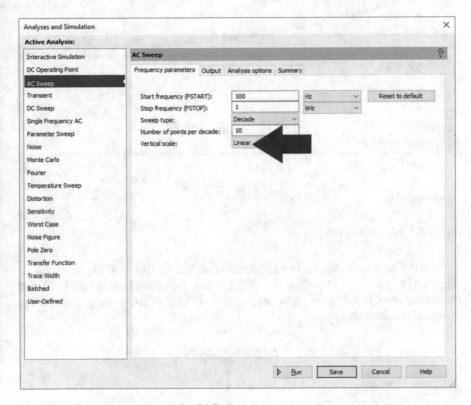

Fig. 2.471 Frequency parameters tab of AC sweep

After running the simulation, the result shown in Fig. 2.472 is obtained.

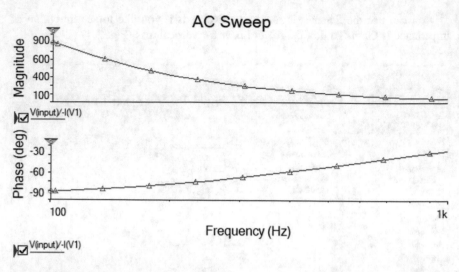

Fig. 2.472 Simulation result

Use the cursors to read the input impedance at 300 Hz (Fig. 2.473). According to Fig. 2.474, the input impedance at 300 Hz has magnitude of 296.5207 Ω and according to Fig. 2.475, its phase angle is −64.9265°. This is quite close to the result of previous analysis.

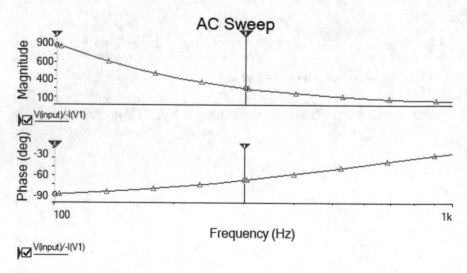

Fig. 2.473 Simulation result

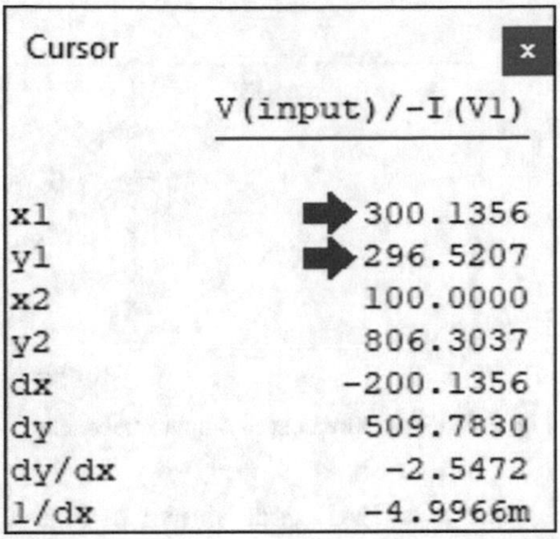

Fig. 2.474 Cursor 1 of magnitude plot

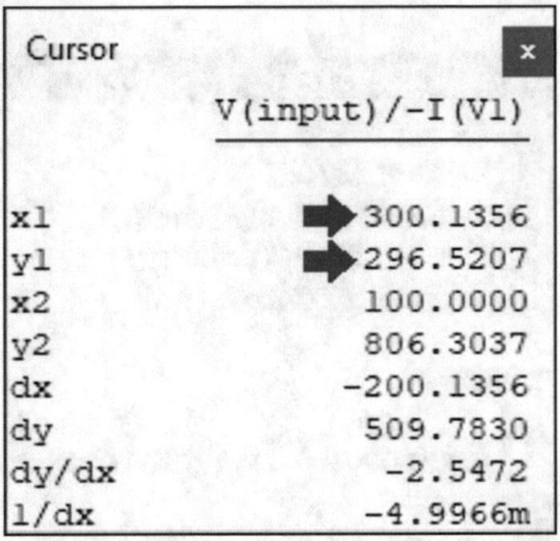

Cursor	x
	V(input)/-I(V1)
x1	➡ 300.1356
y1	➡ 296.5207
x2	100.0000
y2	806.3037
dx	-200.1356
dy	509.7830
dy/dx	-2.5472
1/dx	-4.9966m

Fig. 2.475 Cursor 1 of phase plot

Cursor	x
	V(input)/-I(V1)
x1	➡ 300.1356
y1	➡ 296.5207
x2	100.0000
y2	806.3037
dx	-200.1356
dy	509.7830
dy/dx	-2.5472
1/dx	-4.9966m

The output impedance of the circuit is defined as shown in Fig. 2.476. The output impedance can be calculated quite easily similar to input impedance. Just connect a voltage source to the output of the circuit and drawing the ratio of voltage to its current.

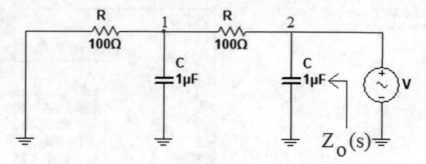

Fig. 2.476 Measurement of output impedance

The MATLAB code shown in Fig. 2.477 draws the Bode diagram of output impedance. The output of this code is shown in Fig. 2.478.

```
Command Window                                                    ⊙
  >> syms s C R
  >> Zc=1/s/C;
  >> Z2=R+(Zc*R)/(Zc+R);
  >> Zo=Zc*Z2/(Zc+Z2);
  >> simplify(Zo)

ans =

(R*(C*R*s + 2))/(C^2*R^2*s^2 + 3*C*R*s + 1)

  >> R=100;C=1e-6;s=tf('s');
  >> w=logspace(log10(2*pi*f1),log10(2*pi*f2),250);
  >> bode((R*(C*R*s + 2))/(C^2*R^2*s^2 + 3*C*R*s + 1),w)
fx >> |
```

Fig. 2.477 Drawing the bode diagram of Zo

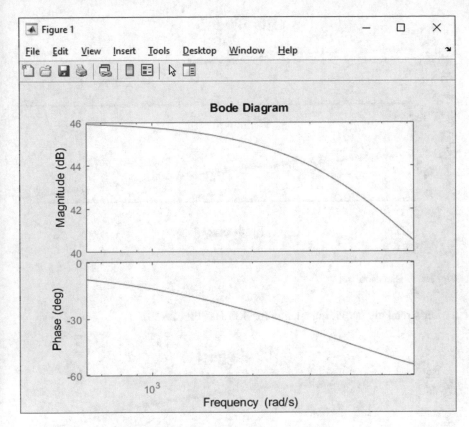

Fig. 2.478 Output of code in Fig. 2.477

The output impedance of the circuit is calculated with Multisim and is shown in Fig. 2.479. The vertical axis is linear and show the magnitude of output impedance is Ohms.

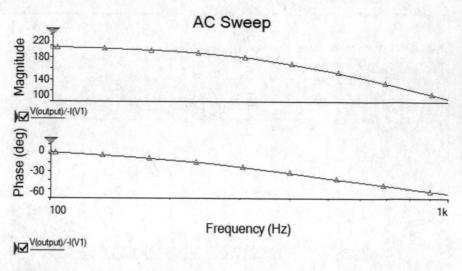

Fig. 2.479 Simulation result

Let's read the output impedance for 300 Hz (Fig. 2.480).

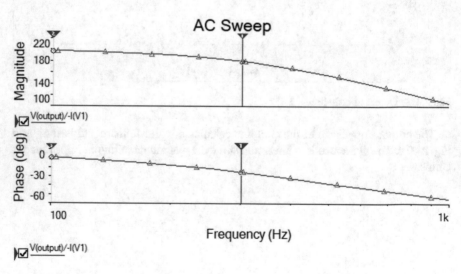

Fig. 2.480 Simulation result

According to Figs. 2.481 and 2.482 the output impedance at 300 Hz is $179.4456e^{-j25.0735°}$ Ω.

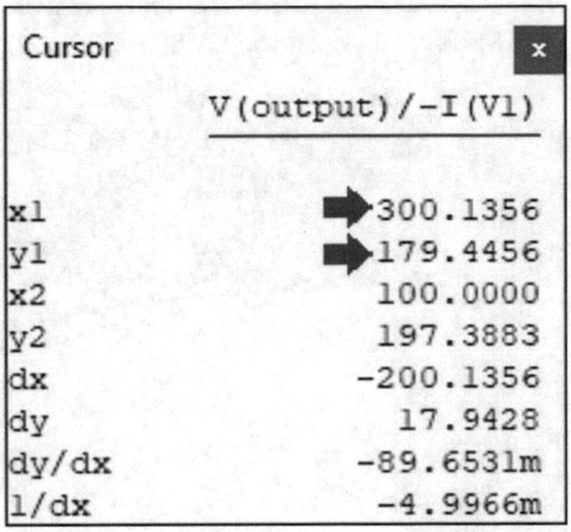

Fig. 2.481 Cursor 1 of
magnitude plot

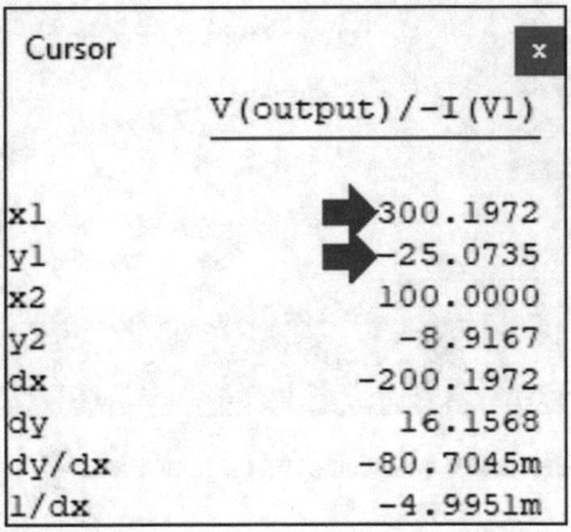

Cursor	x
	V(output)/-I(V1)
x1	300.1356
y1	179.4456
x2	100.0000
y2	197.3883
dx	-200.1356
dy	17.9428
dy/dx	-89.6531m
1/dx	-4.9966m

Fig. 2.482 Cursor 1 of
phase plot

Cursor	x
	V(output)/-I(V1)
x1	300.1972
y1	-25.0735
x2	100.0000
y2	-8.9167
dx	-200.1972
dy	16.1568
dy/dx	-80.7045m
1/dx	-4.9951m

Let's verify the obtained result. Put the cursor at $2\pi \times 300$ Hz $= 1.88\text{k} \frac{\text{Rad}}{\text{s}}$. The result is shown in Fig. 2.483. According to Fig. 2.483, the phase angle of output impedance is $-25°$. MATLAB draws the magnitude in decibel. The decibel is converted to Ohm in Fig. 2.484. So, MATLAB calculates the output impedance at 300 Hz as $179.8871e^{-j25°}$ which is quite close to Multisim result.

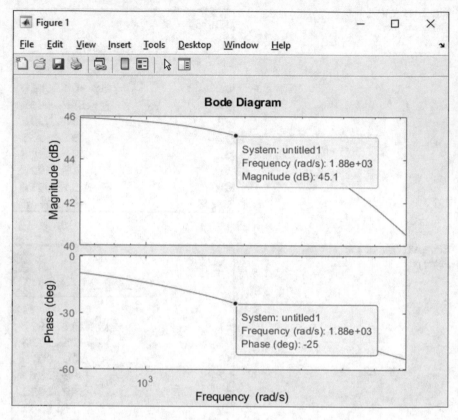

Fig. 2.483 Magnitude and phase of 1.88k $\frac{Rad}{s}$

Fig. 2.484 Conversion of
dB to Ohm

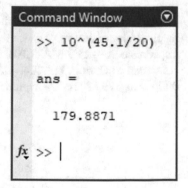

2.48   Example 41: Bode Plotter

Use of AC sweep in obtaining the frequency response of a circuit is studied in the previous example. In this example we study the another method to obtain the frequency response of a circuit. This method uses the Bode plotter (Fig. 2.485).

Fig. 2.485 Bode plotter

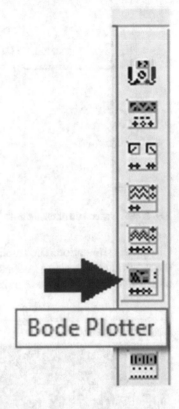

Add a Bode plotter to the schematic of previous example (Fig. 2.486).

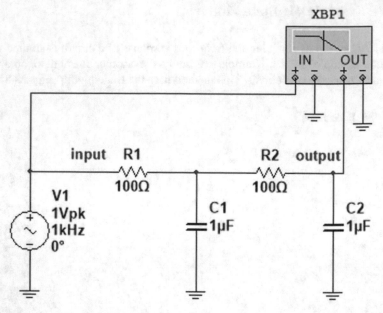

Fig. 2.486 Schematic of example 41

Run the simulation. The amplitude graph and phase graph can be seen by clicking the magnitude and phase buttons (Figs. 2.487 and 2.488). According to Figs. 2.487 and 2.488, the magnitude at 592.78 Hz is −2.989 dB and phase is −52.375°.

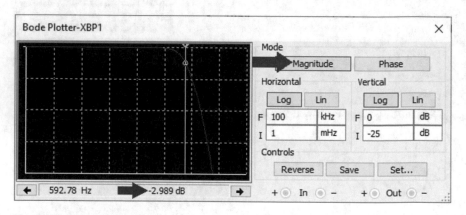

Fig. 2.487 Simulation result

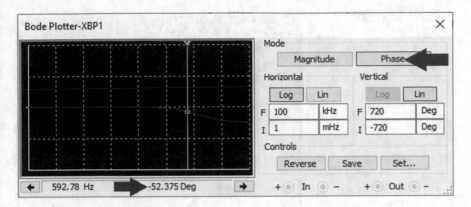

Fig. 2.488 Simulation result

Let's check the obtained result. Figure 2.489 shows that the results are correct.

Fig. 2.489 MATLAB
calculations

```
Command Window                                              ⊙

  >> R=100; C=1e-6;
  >> H=tf(1,[(R*C)^2 3*R*C 1]);
  >> A=freqresp(H,2*pi*592.7488)

  A =

     0.4328 - 0.5614i

  >> abs(A)

  ans =

     0.7088

  >> 20*log10(ans)

  ans =

     -2.9890

  >> angle(A)*180/pi

  ans =

     -52.3726

fx >> |
```

2.49   Example 42: Step Response of Electric Circuits I

The step response of circuits can be calculated easily with Multisim. Assume that we want to obtain the step response of the circuit shown in Fig. 2.490. It is quite easy to show that the transfer function this circuit is $\frac{V_{out}(s)}{V_{in}(s)} = \frac{1}{(RC)^2 s^2 + 3RCs + 1}$.

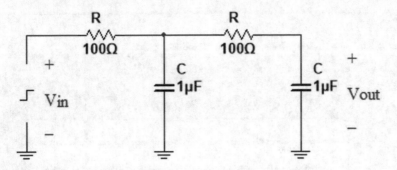

Fig. 2.490 Schematic of example 42

Let's use MATLAB to see the step response. The code shown in Fig. 2.491 calculates the step response of this function. Output of this code is shown in Fig. 2.492.

Fig. 2.491 Drawing the bode plot of H

```
Command Window                                    ⊙
    >> R=100;C=1e-6;
    >> H=tf(1,[(R*C)^2 3*R*C 1]);
    >> step(H)
    >> grid on
fx  >>
```

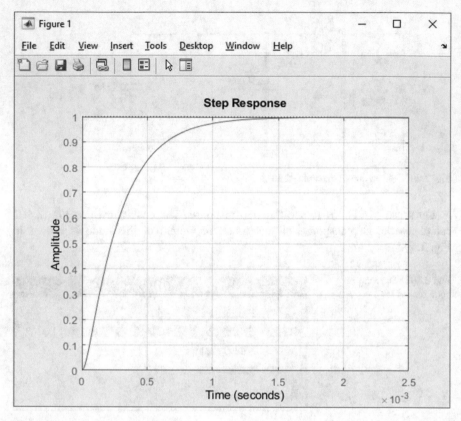

Fig. 2.492 Output of code in Fig. 2.491

The required Multisim schematic is shown in Fig. 2.493. This schematic uses a step voltage source block (Fig. 2.494).

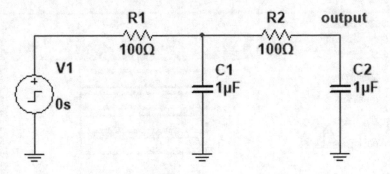

Fig. 2.493 Multisim schematic of example 42

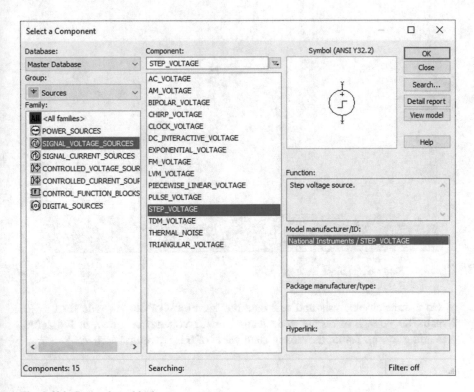

Fig. 2.494 Step voltage block

Double click on the step voltage source block and do the settings as shown in Fig. 2.495. These settings produce a step which goes from 0 to 1 at $t = 0$.

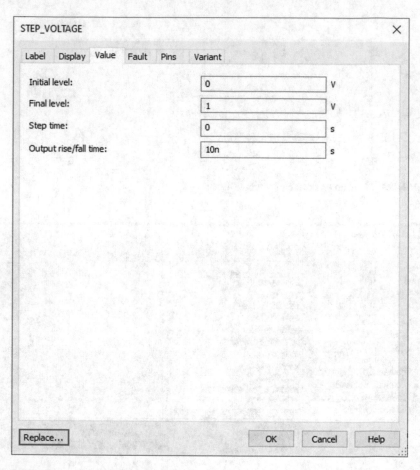

Fig. 2.495 Step voltage block settings

Do a transient analysis and calculate the waveform of output node for t = [0, 3 ms]. The voltage of output node (capacitor C2 voltage) is shown in Fig.2.496. According to Fig. 2.496, the steady state value of this waveform is 1 V.

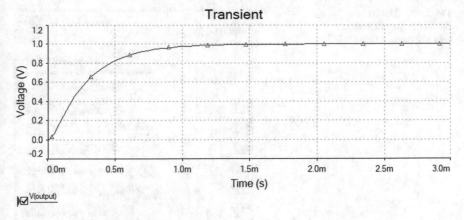

Fig. 2.496   Simulation result

Let's measure the rise time of the waveform. The rise time is the required time to go from 10% of output waveform to 90% of output waveform. Put cursor 1 at $0.1 \times 1 = 0.1$ V and cursor 2 at $0.9 \times 1 = 0.9$ V (Fig. 2.497). The dx row (Fig. 2.498) shows the time difference between the two cursors.

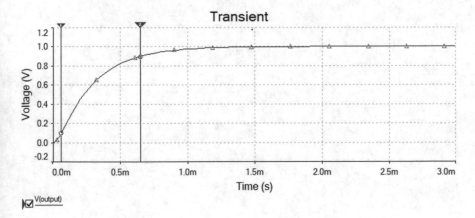

Fig. 2.497   Simulation result

Fig. 2.498 The dx row
shows the time difference
between the cursors

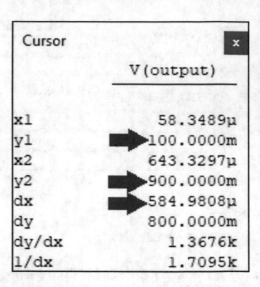

Let's check the obtained result. Figure 2.499 shows the 0.1 and 0.9 points.
Figure 2.500 calculates the time difference between these points in micro seconds.
The obtained result is quite close to Multisim result.

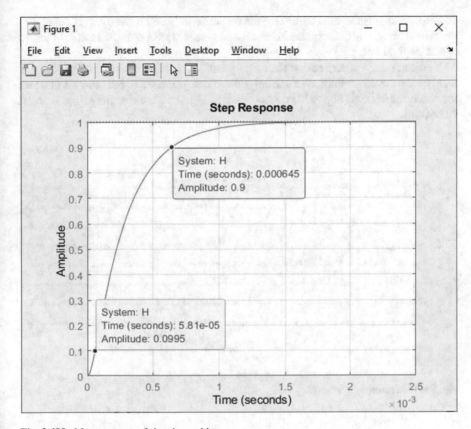

Fig. 2.499 Measurement of rise time with cursors

Fig. 2.500 Calculation of
time difference between the
cursors of Fig. 2.499

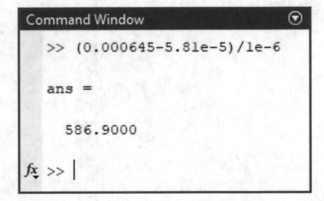

Settling time is the time required for a response to become steady. It is defined as the time required by the response to reach and steady within specified range of 2% to 5% of its final value.

You can measure the settling time easily with Multisim cursors. Assume that we want to use the 2% definition of settling time. Then you need to put cursor 1 at 0 and the other one at 0.98 (Fig. 2.501). According to Fig. 2.502, the settling time is about 1.0690 ms.

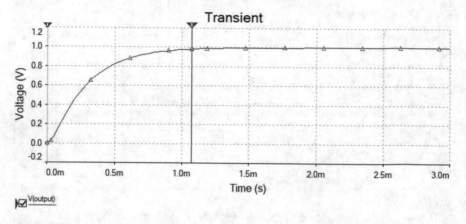

Fig. 2.501 Measurement of settling time

Fig. 2.502 The dx row shows the settling time of the system

Cursor	V(output)
x1	0.0000
y1	0.0000
x2	1.0690m
y2	980.4656m
dx	1.0690m
dy	980.4656m
dy/dx	917.1622
1/dx	935.4354

2.50   Example 43: Step Response of Electric Circuits II

The previous example used transient analysis to calculate the step response of the circuit. If apply a square wave to the circuit, then you can see the step response on the oscilloscope. Assume the schematic shown in Fig. 2.503. This schematic uses a pulse voltage block (Fig. 2.504) to stimulate the circuit.

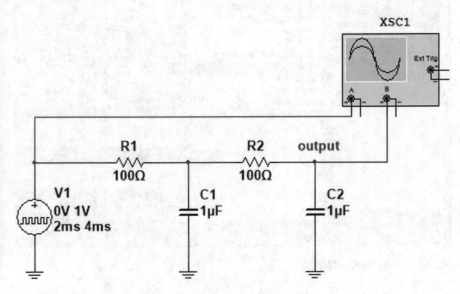

Fig. 2.503   Schematic of example 43

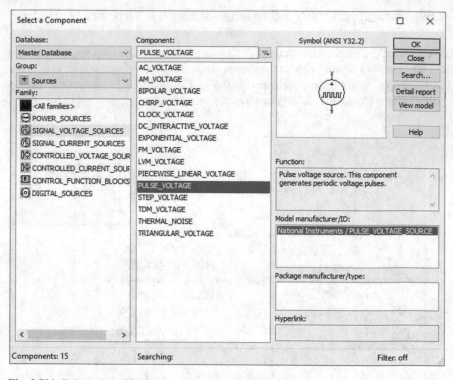

Fig. 2.504 Pulse voltage block

Double click on the pulse voltage block and do the settings as shown in Fig. 2.505. Period of square wave must be big enough to permit the output to reach its steady state. The settings shown in Fig. 2.505 produce the waveform which is shown in Fig. 2.506.

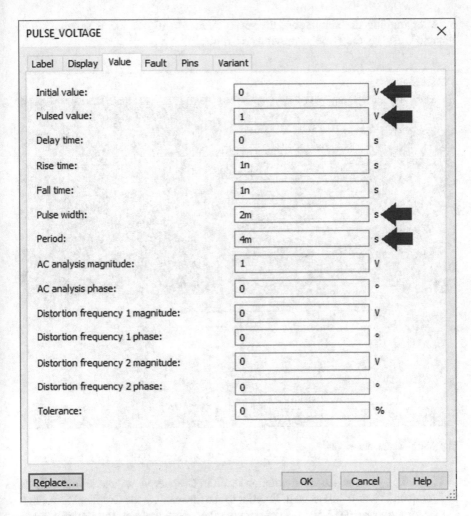

Fig. 2.505 Pulse voltage block settings

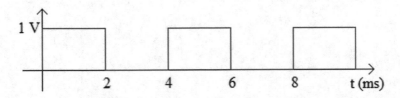

Fig. 2.506 Waveform generated with Fig. 2.505 settings

After running the simulation, the result shown in Fig. 2.507 is obtained. The obtained result is the same as previous example.

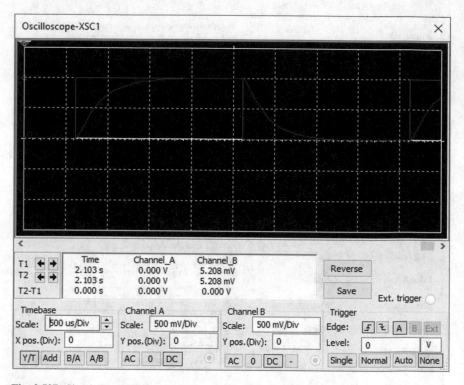

Fig. 2.507 Simulation result

If you like, you can see the simulation result in Grapher view (waveform analysis environment of Multisim) as well. In order to see the result is the Grapher view, click the view> grapher (Fig. 2.508). After clicking the view> grapher, the waveform will be shown in the Grapher environment. You can pause the simulation and use the cursors to measure the rise time, overshoot, settling time, etc.

Fig. 2.508 View> grapher

2.51   Example 44: Transfer Function Block

You can use the transfer function block and arbitrary Laplace function block (Fig. 2.509) to enter a transfer function to Multisim. The transfer function block can take a transfer function of form

$$T(s) = \frac{Y(s)}{X(s)} = K \frac{A_5 s^5 + A_4 s^4 + A_3 s^3 + A_2 s^2 + A_1 s + A_0}{B_5 s^5 + B_4 s^4 + B_3 s^3 + B_2 s^2 + B_1 s + B_0}$$

So, the transfer function block can be used to enter a transfer function which is ratio of two polynomials in s and maximum degree of polynomials is less than or equal to 5. The user enters the value of coefficients to the block.

The arbitrary Laplace function block can take any transfer function. There is no limitation on the degree of transfer function and the user enters the equation of transfer function to the block.

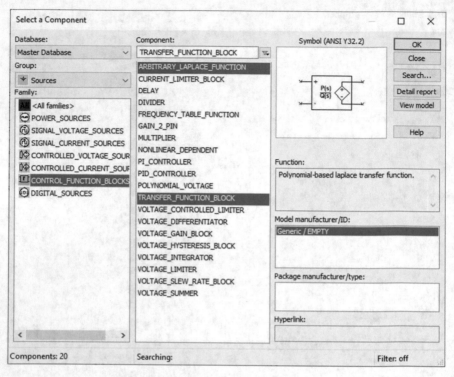

Fig. 2.509 Arbitrary Laplace function and transfer function blocks

Let's see an example. Assume the circuit shown in Fig. 2.510. The transfer function of this circuit is calculated in Fig. 2.511.

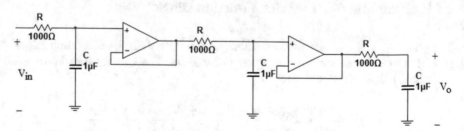

Fig. 2.510 Schematic of example 44

Fig. 2.511 MATLAB
calculations

```
Command Window                                              ⊙

    >> R=1000;C=1e-6;
    >> s=tf('s');
    >> H=(1/(R*C*s+1))^3


    H =


                          1

    -------------------------------------
    1e-09 s^3 + 3e-06 s^2 + 0.003 s + 1

    Continuous-time transfer function.

fx >>
```

We want to simulate the output of this circuit for the input signal which is shown
in Fig. 2.512. We use the transfer function block instead of drawing the circuit
schematic. Assume the schematic shown in Fig. 2.513.

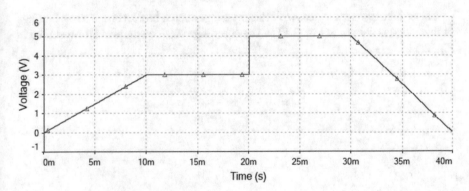

Fig. 2.512 Input signal waveform

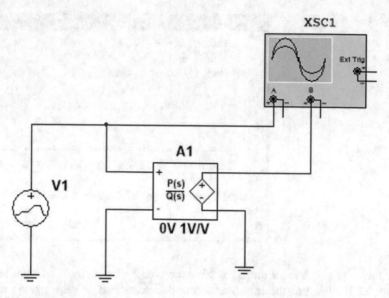

Fig. 2.513 Simulation of circuit in Fig. 2.510 with transfer function block

This schematic uses a piecewise linear voltage source block (Fig. 2.514). Settings of this block is shown in Fig. 2.515. These settings produce the waveform shown in Fig. 2.516.

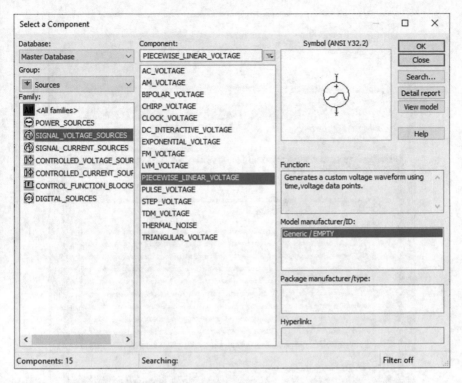

Fig. 2.514 Piecewise Linear Voltage (PLW) block

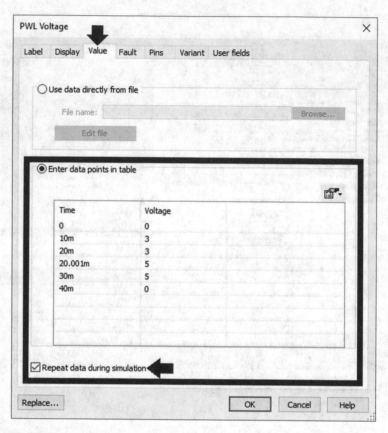

Fig. 2.515 PWL block settings

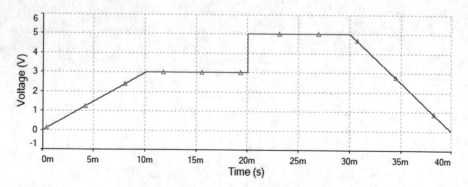

Fig. 2.516 The signal which is generated with Fig. 2.515 settings

Double click on the A1 transfer function block in Fig. 2.513 and do the settings as shown in Fig. 2.514. These settings (Fig. 2.517) model the transfer function shown in Fig. 2.511.

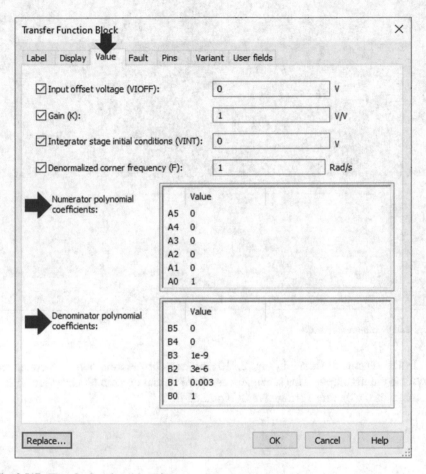

Fig. 2.517 Transfer function A1 settings

Run the simulation. The result is shown in Fig. 2.518.

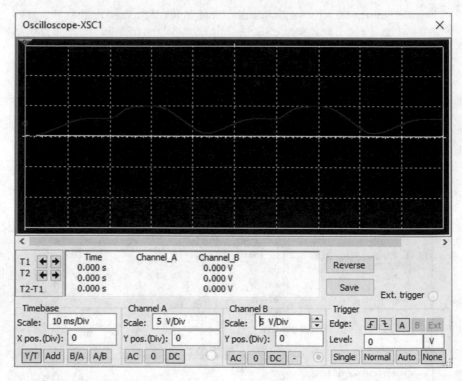

Fig. 2.518 Simulation result

In the schematic shown in Fig. 2.519, the transfer function output is compared with the circuit output. This schematic uses two virtual op amp blocks (Fig. 2.520). Settings of the op amps are shown in Fig. 2.521.

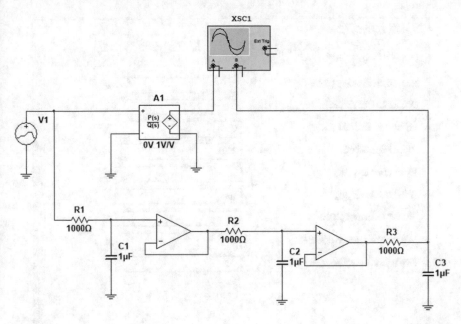

Fig. 2.519 Comparison of outputs

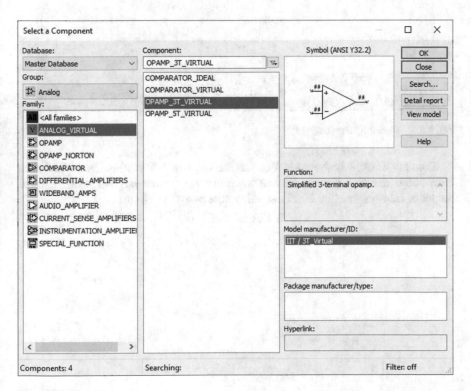

Fig. 2.520 Virtual op amp

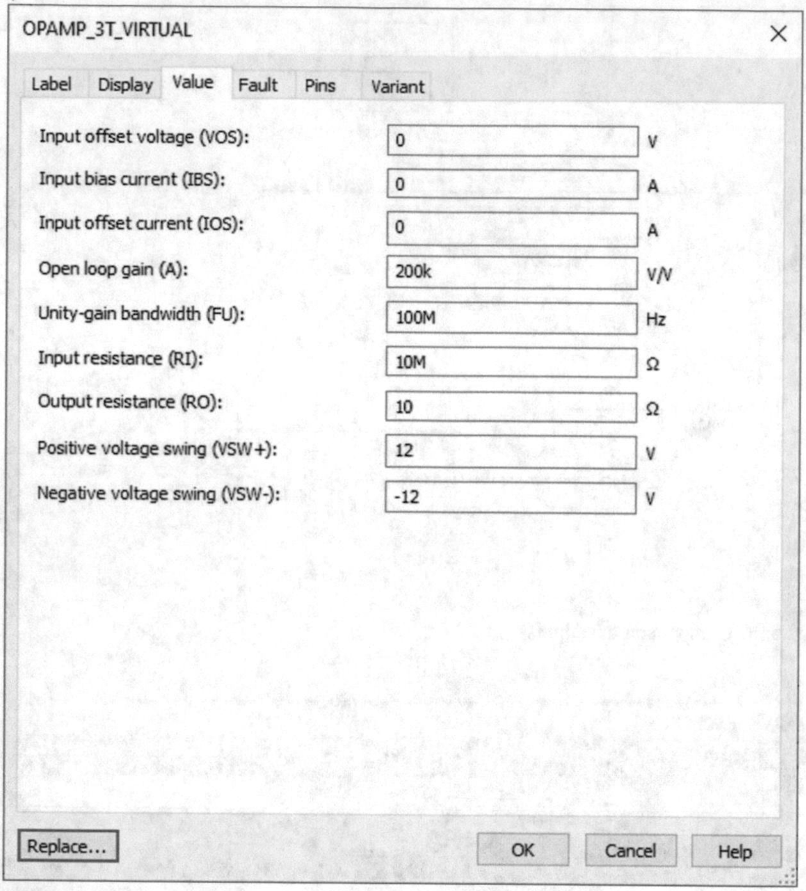

OPAMP_3T_VIRTUAL   ✕

Label   Display   Value   Fault   Pins   Variant

Input offset voltage (VOS): | 0 | V

Input bias current (IBS): | 0 | A

Input offset current (IOS): | 0 | A

Open loop gain (A): | 200k | V/V

Unity-gain bandwidth (FU): | 100M | Hz

Input resistance (RI): | 10M | Ω

Output resistance (RO): | 10 | Ω

Positive voltage swing (VSW+): | 12 | V

Negative voltage swing (VSW-): | -12 | V

Replace...   OK   Cancel   Help

Fig. 2.521 Virtual op amp settings

Simulation result is shown in Fig. 2.522. The two waveforms overlapped each other and because of that we see one waveform on the screen. This shows that the output of transfer function blocks equals to the output of circuit.

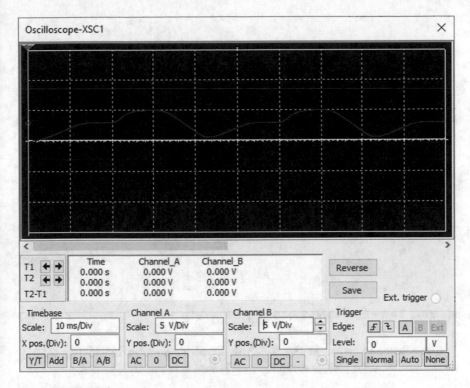

Fig. 2.522 Simulation result

There is another way to ensure that output of transfer function block equals to the output of circuit. We can subtract the output of circuit from output of transfer function block. The result of this subtraction must be around zero if the signals are equal. Schematic shown in Fig. 2.523 uses this method.

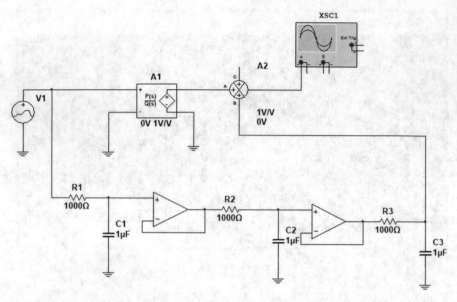

Fig. 2.523 Measurement of difference between the two outputs

The schematic shown in Fig. 2.523 uses voltage summer block (Fig. 2.524). Settings of the used voltage summer block is shown in Fig. 2.525.

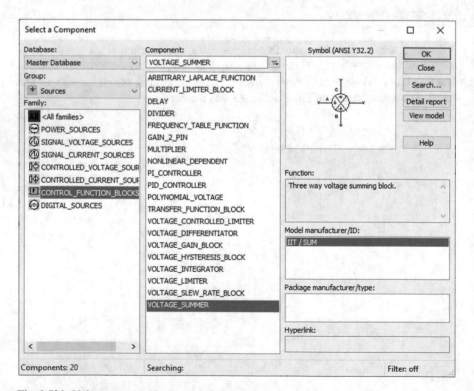

Fig. 2.524 Voltage summer block

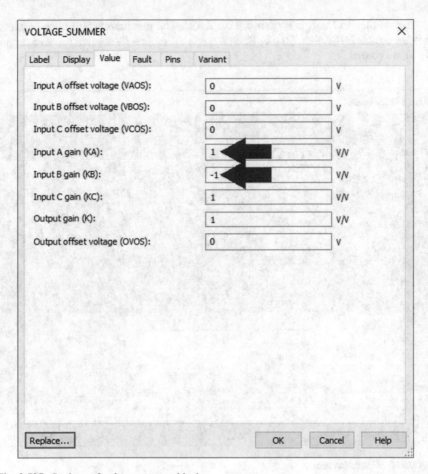

Fig. 2.525 Settings of voltage summer block

The simulation result is shown in Fig. 2.526. The maximum difference between the two signals is less than 50 μV. Such a small difference shows that the two output are almost equal.

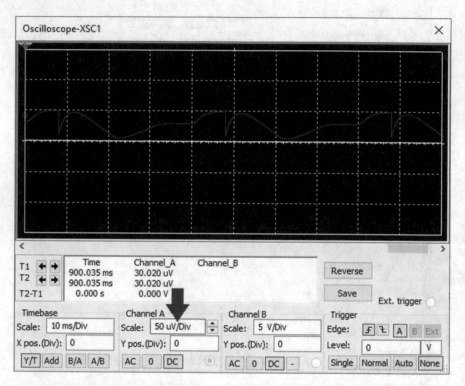

Fig. 2.526 Simulation result

2.52   Exercises

1. Set up a Multisim simulation to measure the RMS of a triangular wave. Use MATLAB or hand calculation to verify the result.
2. Use a transient analysis to see the voltage of the capacitor in Fig. 2.527. Initial values are shown on the figure.

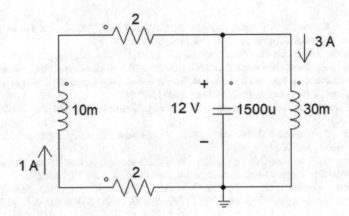

Fig. 2.527 Schematic of exercise 2

3. **a)** Show that Vth = 12 V and Rth = 13.6 for the circuit shown in Fig. 2.528. Vth and Rth show the Thevenin voltage and resistance seen from a and b, respectively.
   b) Use Multisim to measure the Thevenin equivalent of this circuit.

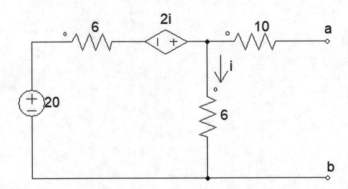

Fig. 2.528 Schematic of exercise 3

4. Use two wattmeter blocks to measure the power of a three phase circuit shown in Fig. 2.381. **Hint:** use the two-wattmeter method.
5. Replace the inductor of Fig. 2.396 with a 100 μF capacitor and resolve the example 36.

Further Readings

1. W. Hayt, J. Kemmerly, S. Durbin, *Engineering Circuit Analysis*, 9th edn. (McGraw-Hill, 2021)
2. J. Nilsson, S. Riedel, *Electric circuits*, 11th edn. (Pearson, 2018)
3. C. Alexander, M.N.O. Sadiku, *Fundamentals of Electric Circuits*, 6th edn. (McGraw-Hill, 2016)
4. Download page for student version of Multisim: https://www.ni.com/en-tr/shop/electronic-test-instrumentation/application-software-for-electronic-test-and-instrumentation-category/what-is-multisim/multisim-education.html (Visiting date: 15.07.201)
5. NI Multisim User Manual https://www.ni.com/pdf/manuals/374483d.pdf (Visiting date: 15.07.2021)
6. Multisim Component Reference guide: https://www.ni.com/pdf/manuals/374485a.pdf (Visiting date: 15.07.2021)
7. Multisim default keyboard shortcuts: https://www.ni.com/en-tr/support/documentation/supplemental/18/multisim-default-keyboard-shortcuts.html (Visiting date: 15.07.2021)
8. Simulation Errors in Multisim and Solutions: https://knowledge.ni.com/KnowledgeArticleDetails?id=kA00Z0000019RqVSAU&l=en-TR (Visiting date: 15.07.2021)

Chapter 3
Simulation of Electronic Circuits with Multisim™

3.1 Introduction

Previous chapter studied the simulation of electric circuits with Multisim. This chapter focus on the simulation of electronic circuits (i.e., circuits which contain diode, transistor, IC's, etc.) with Multisim.

3.2 Example 1: Diode I–V Characteristics

We want to extract the I-V characteristics of a diode in this example. Remember that for a diode $I_D = I_s \times \left(e^{\frac{V_D}{\eta V_T}} - 1 \right)$ where I_D, V_D, I_s, η and V_T show the diode current, diode voltage, saturation current of diode, emission coefficient and thermal voltage (25.8 mV at 300 K), respectively. Assume the schematic shown in Fig. 3.1.

© The Author(s), under exclusive license to Springer Nature Switzerland AG 2022   429
F. Asadi, *Essential Circuit Analysis using NI Multisim™ and MATLAB®*,
https://doi.org/10.1007/978-3-030-89850-2_3

Fig. 3.1 Schematic of example 1

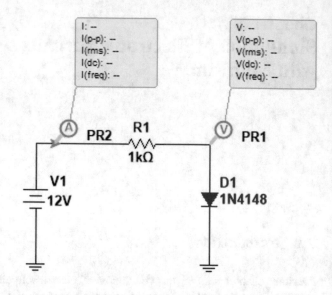

Double click on the voltage source and change its value to 0.5 V (Fig. 3.2).

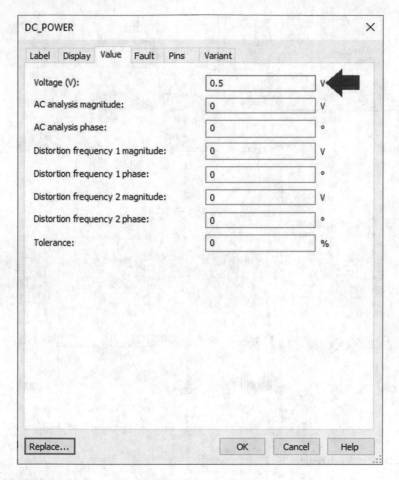

Fig. 3.2 Settings of voltage source V1

Run the simulation. According to Fig. 3.3, for V1 = 0.5 V, the diode voltage and current are 486 mV and 14.3 μA, respectively. Repeat, these steps for V1 = 1 V, 1.5 V, 2 V, ..., 7 V. Results are shown in Figs. 3.3, 3.4, 3.5, 3.6, 3.7, 3.8, and 3.9.

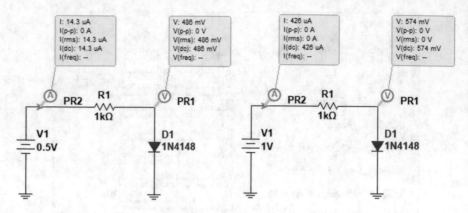

Fig. 3.3 Simulation result

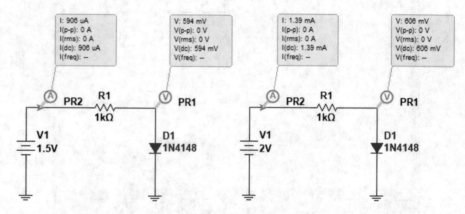

Fig. 3.4 Simulation result

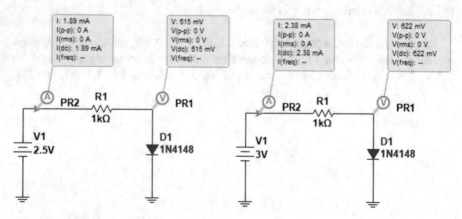

Fig. 3.5 Simulation result

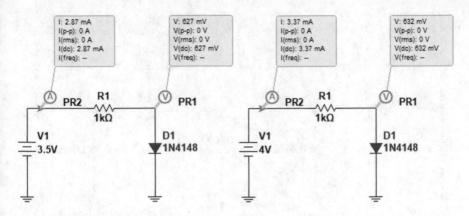

Fig. 3.6 Simulation result

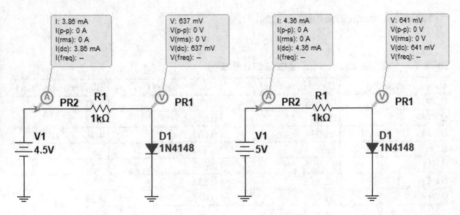

Fig. 3.7 Simulation result

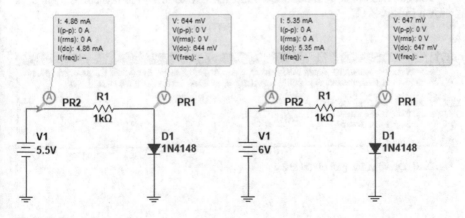

Fig. 3.8 Simulation result

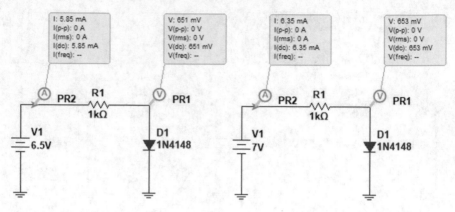

Fig. 3.9 Simulation result

Table 3.1 summarizes the results.

Table 3.1 Voltages and currents of the 1 N4148 diode

Diode voltage (V)	Diode current (mA)	Diode voltage (V)	Diode current (mA)
0.486	0.0143	0.632	3.37
0.574	0.426	0.637	3.87
0.594	0.906	0.641	4.36
0.606	1.39	0.644	4.86
0.615	1.89	0.647	5.35
0.622	2.38	0.651	5.85
0.627	2.87	0.653	6.35

You can use the commands shown in Fig. 3.10 to draw the I-V curve. Output is shown in Fig. 3.11.

```
Command Window
>> v=[0.486 0.574 0.594 0.606 0.615 0.622 0.627 0.632 0.637 0.641 0.644 0.647 0.651 0.653];
>> i=[0.0143 0.426 0.906 1.39 1.89 2.38 2.87 3.37 3.86 4.36 4.86 5.35 5.85 6.35];
>> plot(v,i,v,i,'r*')
>> xlabel('Voltage (V)')
>> ylabel('Diode current (mA)')
>> title('Diode I-V Curve')
fx >> |
```

Fig. 3.10 Drawing the plot of Table 3.1

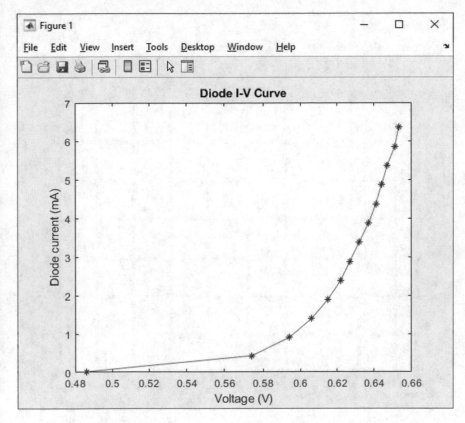

Fig. 3.11 Result of code in Fig. 3.10

You can add grids to the graph by using the grid on command (Fig. 3.12). Figure 3.13 shows the output of this code.

```
Command Window
 >> v=[0.486 0.574 0.594 0.606 0.615 0.622 0.627 0.632 0.637 0.641 0.644 0.647 0.651 0.653];
 >> i=[0.0143 0.426 0.906 1.39 1.89 2.38 2.87 3.37 3.86 4.36 4.86 5.35 5.85 6.35];
 >> plot(v,i,v,i,'r*')
 >> xlabel('Voltage (V)')
 >> ylabel('Diode current (mA)')
 >> title('Diode I-V Curve')
 >> grid on
fx >>
```

Fig. 3.12 Grid on command adds grid to the plot

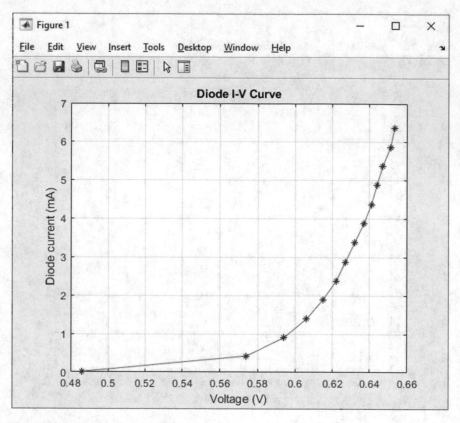

Fig. 3.13 Result of code in Fig. 3.12

You can use the edit> copy the figure into clipboard (Fig. 3.14). After coping the figure into the clipboard, you can paste (using Ctrl+V) into other programs. This is very useful when you want to write a report and you need to add the plot of data to your report.

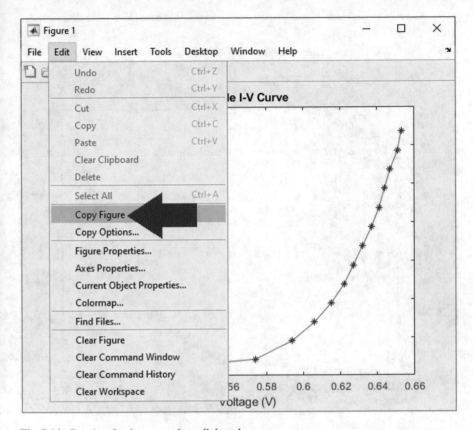

Fig. 3.14   Copying the drawn graph to clipboard

You can save the obtained graph as a graphical file as well. In order to do this, click the file>save as (Fig. 3.15).

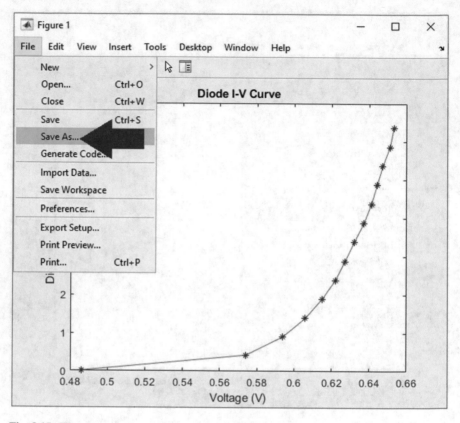

Fig. 3.15 File> save as

Now select the desired output file format from the save as type drop down list (Fig. 3.16).

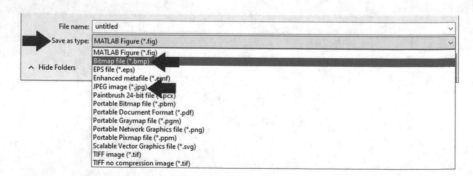

Fig. 3.16 Selection of output format

You can use the MATLAB's Curve Fitting Toolbox™ to fit the obtained data on an exponential curve. Run the MATLAB's Curve Fitting Toolbox by using the cftool command (Fig. 3.17).

```
Command Window
   >> v=[0.486 0.574 0.594 0.606 0.615 0.622 0.627 0.632 0.637 0.641 0.644 0.647 0.651 0.653];
   >> i=[0.0143 0.426 0.906 1.39 1.89 2.38 2.87 3.37 3.86 4.36 4.86 5.35 5.85 6.35];
   >> plot(v,i,v,i,'r*')
   >> xlabel('Voltage (V)')
   >> ylabel('Diode current (mA)')
   >> title('Diode I-V Curve')
   >> cftool
fx >> |
```

Fig. 3.17 cftool command

After running the cftool command, the window shown in Fig. 3.18 appears.

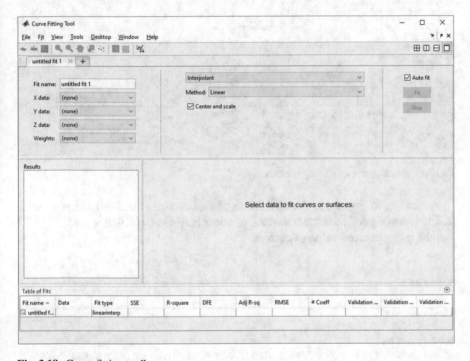

Fig. 3.18 Curve fitting toolbox

Do the settings similar to Fig. 3.19. According to Fig. 3.19, the $I = 6.928 \times 10^{-9}e^{31.61 \times v}$ is the best $I = ae^{b \times v}$ function which describes our data. Note that $I = 6.928 \times 10^{-9}e^{31.61 \times v}$ gives the current in mA because data entered into cftool has unit of V and mA. If you want to have the current in Amps, then you need to use $I = 6.928 \times 10^{-12}e^{31.61 \times v}$. Note that the exact equation of diode is $I_D = I_s \times \left(e^{\frac{V_D}{\eta V_T}} - 1\right)$, so what we obtained using cftool is an approximation since we ignored the -1 term.

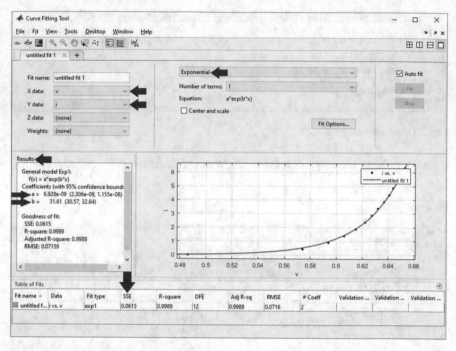

Fig. 3.19 Settings of curve fitting toolbox

According to Fig. 3.19, the SSE (Sum of Square Error) is 0.0615. Let's verify it. The following code calculate the sum of errors between $I = 6.928 \times 10^{-9}e^{31.61 \times v}$ and the real measured value of current.

```
clc
clear all

Vt=25.8e-3;

%measured data
vD=[0.486 0.574 0.594 0.606 0.615 0.622 0.627 0.632 0.637 0.641 0.644
0.647 0.651 0.653];
iD=[0.0143 0.426 0.906 1.39 1.89 2.38 2.87 3.37 3.87 4.36 4.86 5.35 5.85
6.35];

%approximated Is and n
Is_app=6.928e-9;
n_app=1/(31.61*Vt);

iD_app=Is_app*exp(1/n_app/Vt*vD);
Error_app=(iD-iD_app)*(iD-iD_app)';
disp('Approximate model error:')
disp(Error_app)
```

Figure 3.20 shows the output of this code. The obtained result is quite close to MATLAB result.

Fig. 3.20 Output of code

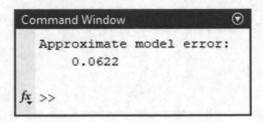

3.3 Example 2: IV Analyzer

Multisim has a device called IV analyzer (Fig. 3.21). You can obtain the IV curve of semiconductor devices easily with this element.

Fig. 3.21 IV analyzer

Connect a 1 N4148 diode to the IV analyzer (Fig. 3.22).

Fig. 3.22 1 N4148 is
connected to IV analyzer

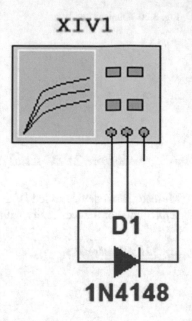

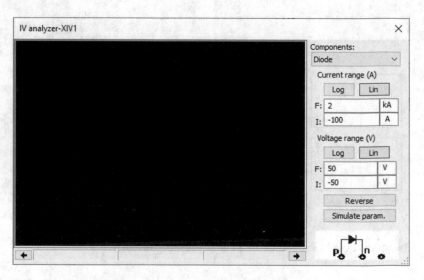

Double click the IV analyzer. The window shown in Fig. 3.23 appears.

Fig. 3.23 IV analyzer window

You can use the components drop down list (Fig. 3.24) to select the type of component which is connected to the IV analyzer. The reveres button changes the background color to white. The simulate param. Button determine the range of voltages that are used to test the device.

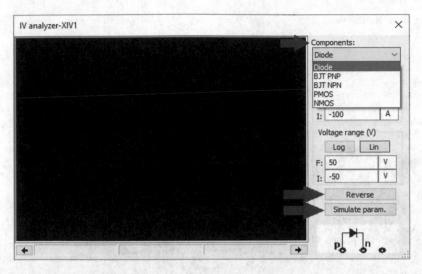

Fig. 3.24 IV analyzer window

Click the simulate param. button. The window shown in Fig. 3.25 appears and you can enter the desired range of voltages there. For instance, assume that we want to draw the IV curve for $-1 < V < 0.8$. Then the start box must be filled with -1 and stop box must be filled with 0.8. The increment box determines the step size. For instance, increment is 10 mV in Fig. 3.25. This means that -1 V, -1 V + 10 mV, -1 V + 20 mV, -1 V + 30 mV, ..., 0.8 V is applied to the diode.

Fig. 3.25 Simulation parameters window

Run the simulation by clicking the run button (Fig. 3.26). Result of this simulation is shown in Fig. 3.27. The F and I in Fig. 3.27 determines the maximum and minimum of voltage/current axis. The log and lin button determines the type of axis (i.e. logarithmic or linear). You can use the cursor to read the graph. When you move the cursor, the value of voltage and currents will be shown in the bottom of the window (Fig. 3.28).

Fig. 3.26 Run button

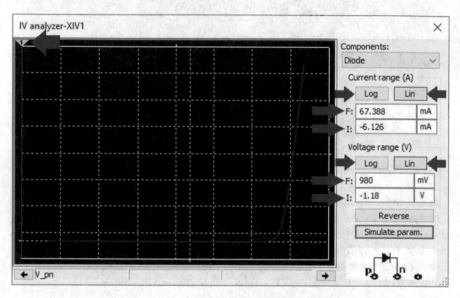

Fig. 3.27 Simulation result

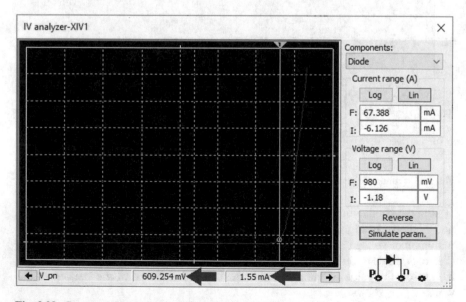

Fig. 3.28 Cursor can be used to read the graph

The IV analyzer can be used to draw the IV curves for transistors as well. For instance, assume that we want to draw the collector current vs. collector emitter voltage for base currents of 10 µA, 20 µA, 30 µA, 40 µA and 50 µA. Assume that the transistor under test is 2 N2222 (Fig. 3.29).

Fig. 3.29 2 N2222 is connected to IV analyzer

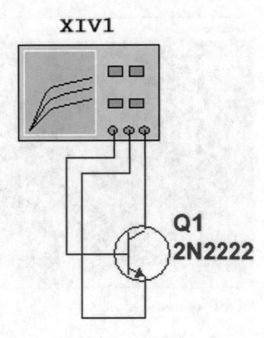

Double click the IV analyzer and click the simulate param. Button. Do the settings similar to what shown in Fig. 3.30.

Simulate Parameters   ✕

Source name V_ce   Source name I_b

Start:	0	V
Stop:	10	V
Increment:	50	mV

Start:	10	µA
Stop:	50	µA
Num steps:	5	

OK   Cancel

Fig. 3.30 Simulate parameters window

Run the simulation. The result shown in Fig. 3.31 is obtained. The collector current is not constant and increases as the collector emitter voltage increases. This is called Early effect. Remember that for forward active region, the collector current is $I_C = I_S e^{\frac{V_{BE}}{V_T}} \left(1 + \frac{V_{CE}}{V_A}\right)$ where V_A shows the Early voltage.

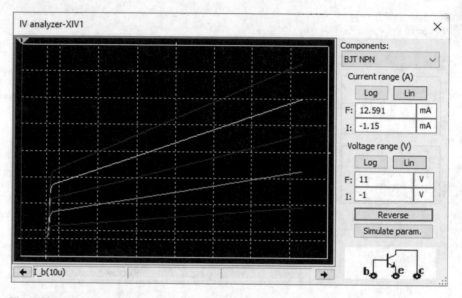

Fig. 3.31 Simulation result

In order to read a specific curve, click on it, then move the cursor to the desired point. This cause the cursor read the intersection with the selected curve. Note that the name of selected curve is shown in the left side of the window (Fig. 3.32).

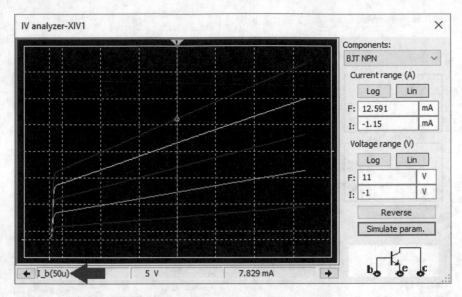

Fig. 3.32 Click on the graph to read it with cursor

3.4 Example 3: Effect of Temperature on the Voltage Drop of Diode

In this section, we want to study the effect of temperature on the voltage drop of diode. Assume the schematic shown in Fig. 3.33. The diode current is about 10 mA.

Fig. 3.33 Schematic of example 3

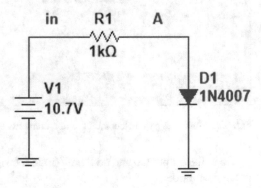

Click the interactive button (Fig. 3.34).

Fig. 3.34 Interactive button

Select the temperature sweep. Do the settings similar to Fig. 3.35.

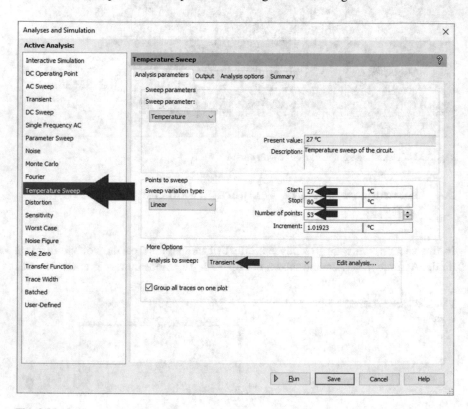

Fig. 3.35 Analyses parameters tab of temperature sweep analysis

Go to the output tab and add the V(a) to the right list (Fig. 3.36). V(a) is the anode voltage of diode.

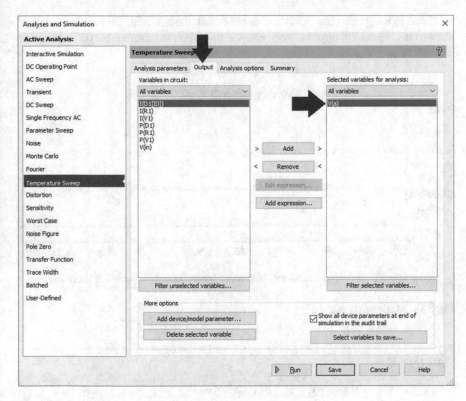

Fig. 3.36 Output tab of temperature sweep analysis

Run the simulation. Result is shown in Fig. 3.37.

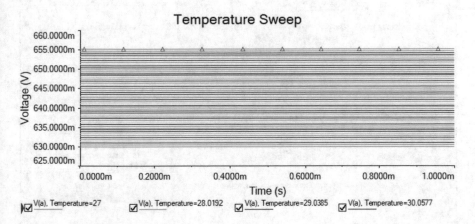

Fig. 3.37 Simulation result

Zoom in the graph (Fig. 3.38) and double click the maximum curve. The window shown in Fig. 3.39 appears and shows that this maximum curve belongs to T = 27 °C.

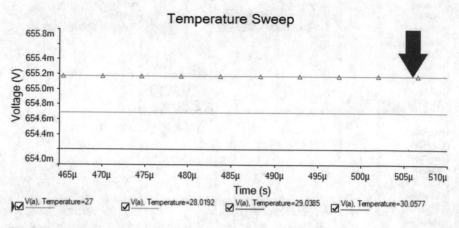

Fig. 3.38 Maximum is selected

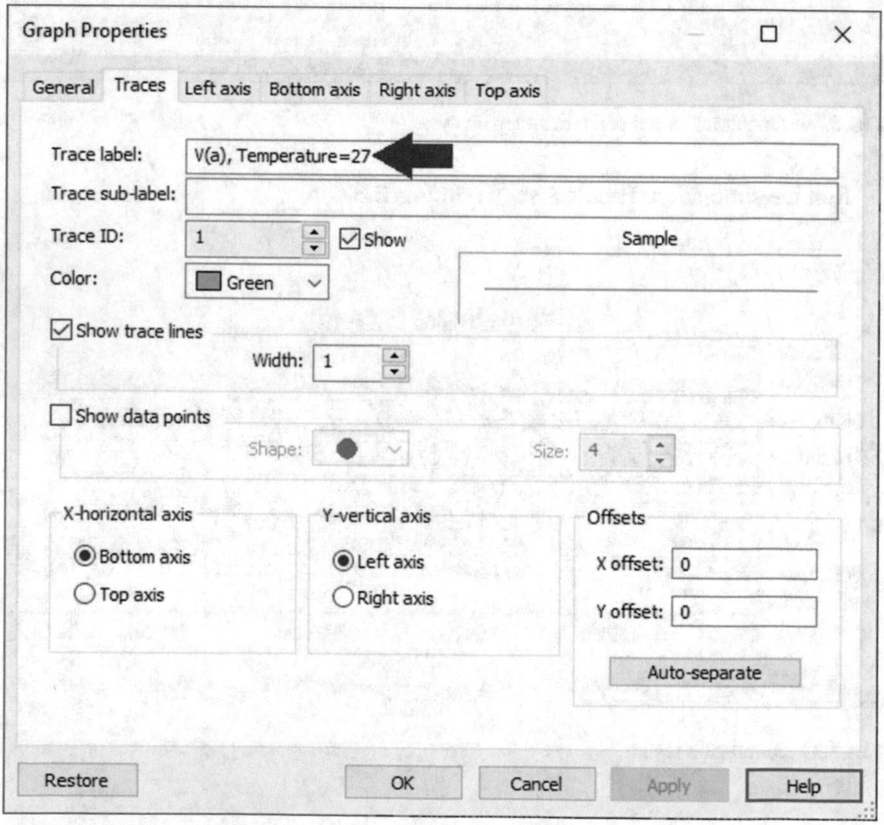

Fig. 3.39 The maximum graph belongs to temperature of 27 °C

Now zoom in again (Fig. 3.40). Double click on the minimum curve. The window shown in Fig. 3.41 appears and shows that this minimum curve belongs to $T = 80\,°C$.

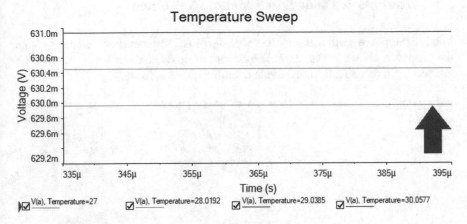

Fig. 3.40   Minimum is selected

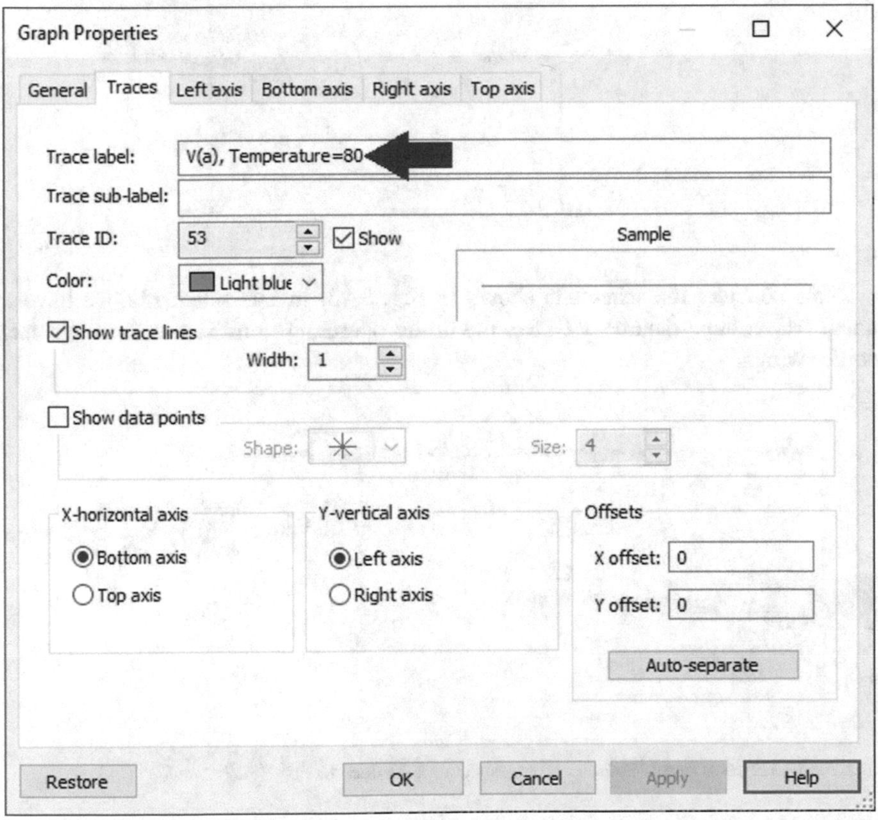

Fig. 3.41   The minimum graph belongs to temperature of 80 °C

So, we deduce that the as the temperature increases, the forward voltage drop of the diode decreases.

3.5   Example 4: Diode Small Signal Resistance

In this example, we will measure the small signal AC resistance of a diode. Assume the schematic shown in Fig. 3.42. The current of diode is about 1 mA. There is no AC source in this circuit, so the diode operating point is fixed.

Fig. 3.42 Schematic of example 4

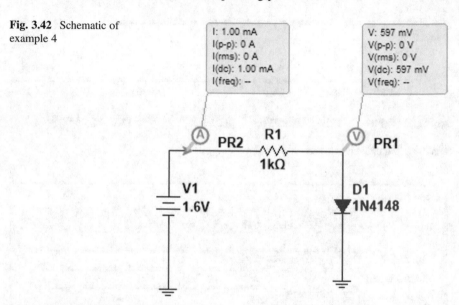

Now consider the schematic shown in Fig. 3.43. In this schematic, we have a small AC voltage (source V2). So, the diode operating point is not fixed and has some swings.

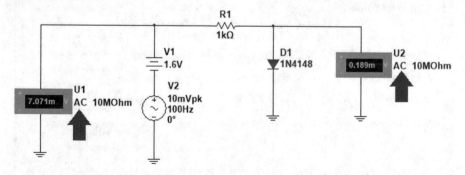

Fig. 3.43 Addition of AC source and measurement devices to Fig. 3.42

The DC source V1, makes the large signal part of the diode current and the source V2, makes the small signal part of the current. The source V2 sees the circuit like what is shown in Fig. 3.44.

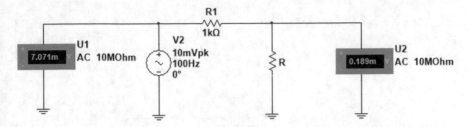

Fig. 3.44 Equivalent circuit from V2 viewpoint

Calculations shown in Fig. 3.45 shows that the value of resistor R is 27.4629 Ω.

Fig. 3.45 MATLAB calculations

```
Command Window

>> syms R
>> solve(R/(R+1e3)*7.071e-3==0.189e-3)

ans =

6972869259862211000/253900985430538269

>> eval(ans)

ans =

    27.4629

fx >>
```

3.6   Example 5: Zener Diode Characteristics

In this example we want to obtain the IV characteristics of a zener diode. Assume the schematic shown in Fig. 3.46. This schematic uses the dc interactive voltage block. Place of this block is shown in Fig. 3.47. Note that the zener diode is reverse biased.

Fig. 3.46 Schematic of
example 5

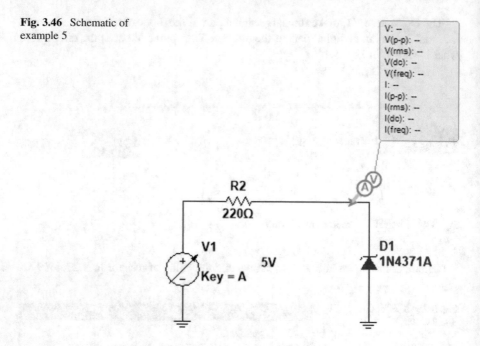

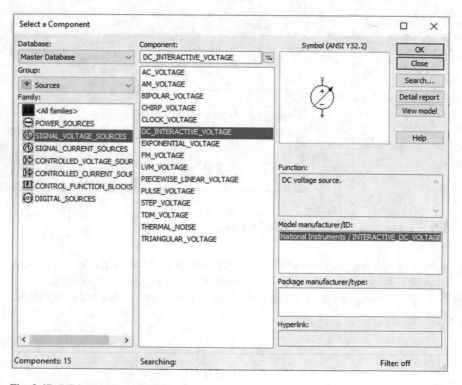

Fig. 3.47 DC interactive voltage block

Double click the dc interactive voltage block and do the settings similar to
Fig. 3.48. When you press the key A of keyboard, the value of this variable voltage
source increases. You can press the Shift+A to decrease the value of voltage source.

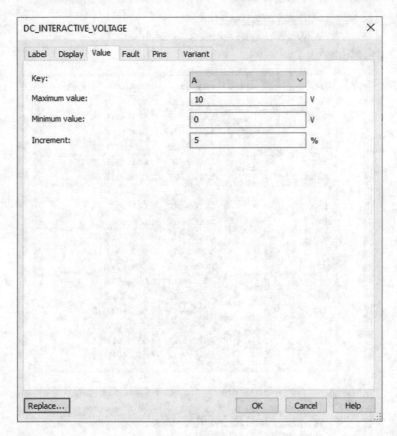

Fig. 3.48 Settings of DC interactive voltage block

We need to measure the DC value of diode voltage and current only. So, double
click the voltage and current probe in Fig. 3.46 and do its settings similar to
Fig. 3.49.

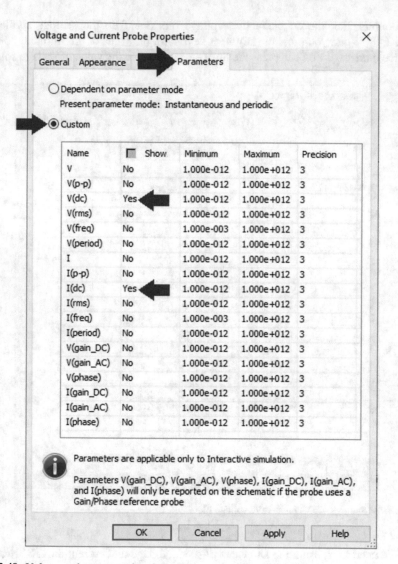

Fig. 3.49 Voltage and current probe, shows the average value of voltage and current only

Measurements are shown in Figs. 3.50, 3.51, and 3.52.

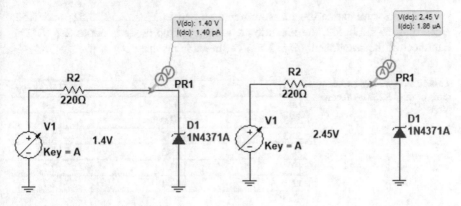

Fig. 3.50 Simulation result

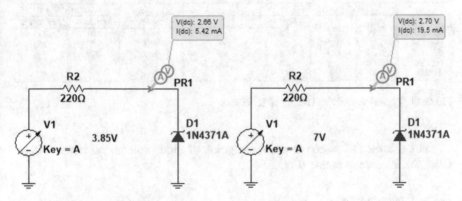

Fig. 3.51 Simulation result

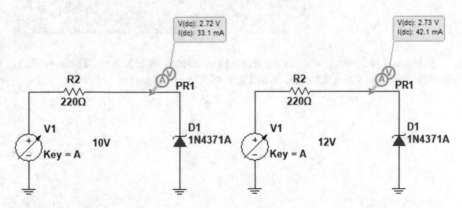

Fig. 3.52 Simulation result

Table 3.2 summarize the measurements shown in Figs. 3.50, 3.51, and 3.52. According to Table 3.2, the breakdown voltage of the diode is about 2.7 V. The datasheet of the component (Fig. 3.53) verifies this result.

Table 3.2 Voltages and currents of the 1N4371A zener diode

V1 (V)	I (mA)	VD1 (V)
1.4	0	1.4
2.45	0.00186	2.45
3.85	5.42	2.66
7	19.5	2.70
10	33.1	2.72
12	42.1	2.73

Electrical Characteristics T_A=25°C unless otherwise noted

Device	V_Z (V) @ I_Z = 20mA (Note 1)			Z_Z (Ω) @ I_Z = 20mA	I_{ZM} (mA) (Note 2)	I_R (μA) @ V_R = 1V	
	Min.	Typ.	Max.			Ta = 25°C	Ta = 125°C
1N4370A	2.28	2.4	2.52	30	150	100	200
1N4371A	2.57	2.7	2.84	30	135	75	150
1N4372A	2.85	3.0	3.15	29	120	50	100
1N746A	3.14	3.3	3.47	28	110	10	30
1N747A	3.42	3.6	3.78	24	100	10	30

Fig. 3.53 Typical breakdown voltage of 1N4371A

Let's use the DC sweep to draw the graph of diode voltage vs. input voltage. Click the interactive button (Fig. 3.54).

Fig. 3.54 Interactive button

Select the DC sweep and do the settings as shown in Fig. 3.55. These settings change the input voltage from 0 V to 10 V with 0.5 V steps.

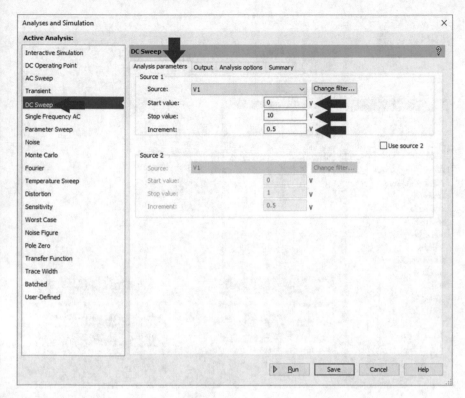

Fig. 3.55   Analysis parameters tab of DC sweep simulation

Go to the output tab and use the remove button (Fig. 3.56) to remove the I(PR1) from the right list. I(PR1) is the current of voltage and current probe and we don't need it in this measurement. After removing the I(PR1), V(PR1) remains in the right list (Fig. 3.57).

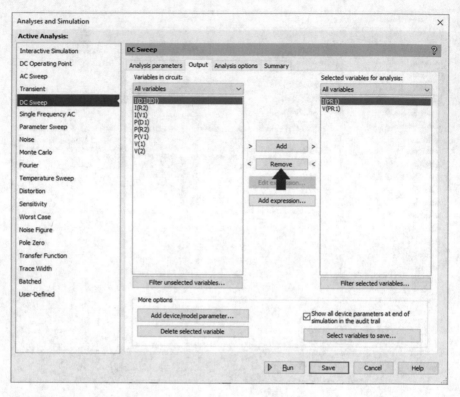

Fig. 3.56 Output tab of DC sweep simulation

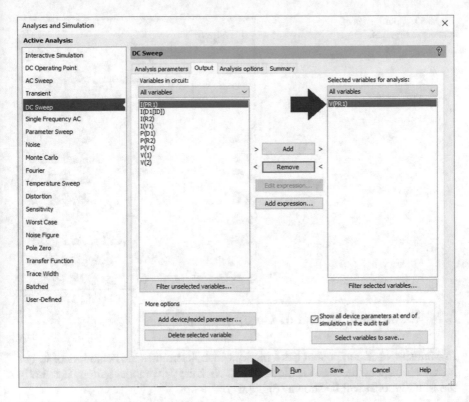

Fig. 3.57 V(PR1) is added to the right list

Click the run button in Fig. 3.57. The simulation result is shown in Fig. 3.58. You can use the cursors to read different points of the curve. The knee of Fig. 3.58 is about 2.5 V and the break down voltage of the diode is 2.72 V (Fig. 3.59).

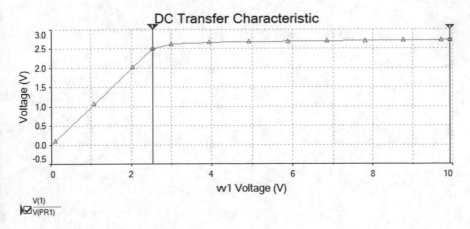

Fig. 3.58 Simulation result

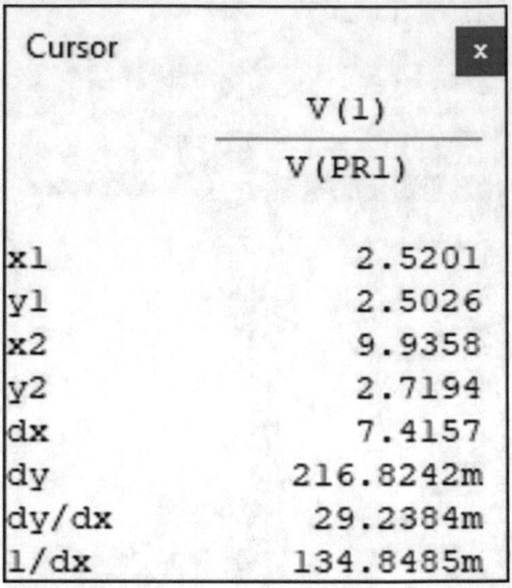

Fig. 3.59 Coordinates of
cursors in Fig. 3.58

Cursor	x
	V(1)
	V(PR1)
x1	2.5201
y1	2.5026
x2	9.9358
y2	2.7194
dx	7.4157
dy	216.8242m
dy/dx	29.2384m
1/dx	134.8485m

3.7   Example 6: Model of Components

In Multisim, each component has a model which determines the behaviors of it. For instance, consider the 1 N4007 diode. If you click the view model button (Fig. 3.60), the SPICE model will be shown (Fig. 3.61).

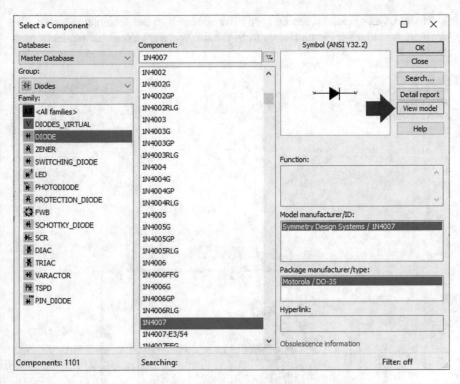

Fig. 3.60 View model button shows the SPICE model of the component

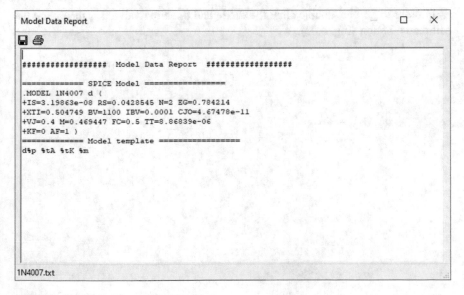

Fig. 3.61 SPICE model of 1 N4007

Model of a component can be seen after adding it to the schematic as well. For instance, consider the circuit shown in Fig. 3.62. We want to see the model of used diode.

Fig. 3.62 A simple circuit to measure the IV curve of the zener diode

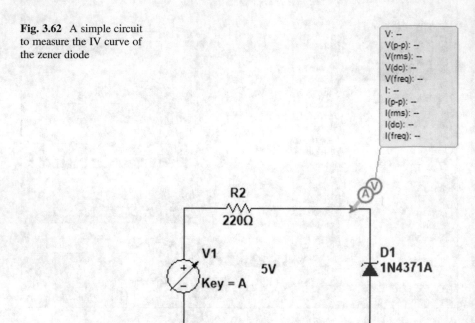

In order to do this, double click the diode and open the value tab, then click the edit model button (Fig. 3.63).

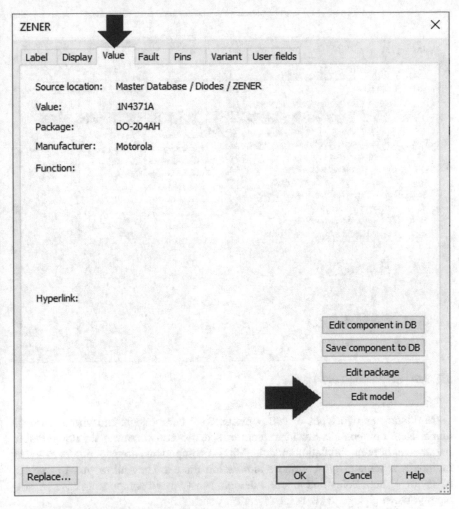

Fig. 3.63 Edit model button shows the SPICE model of the component

Now the model of component will be shown. According to Fig. 3.64, the break down voltage is 2.7 V.

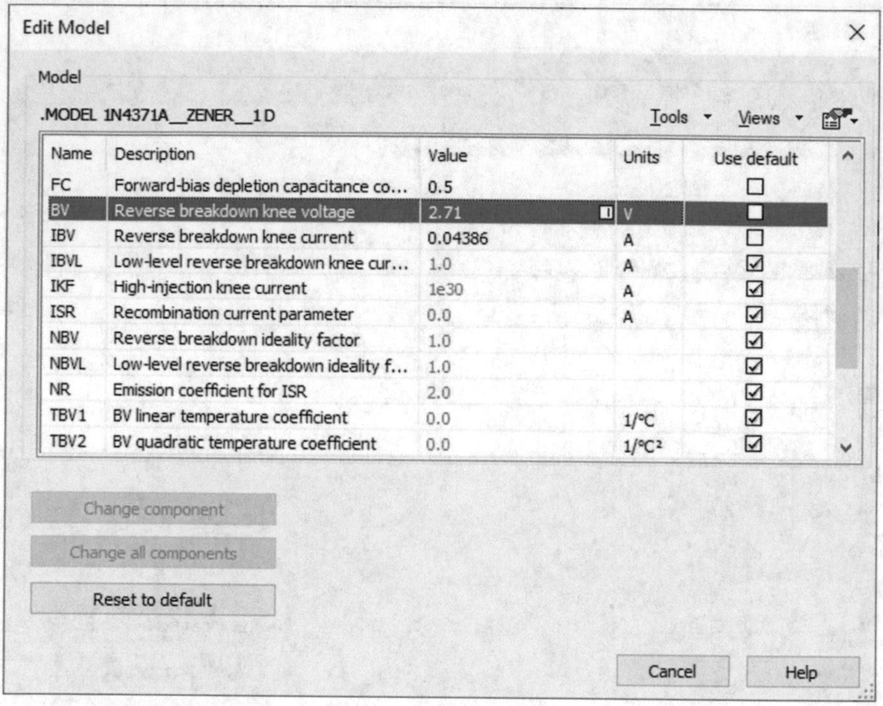

Edit Model   ×

Model

.MODEL 1N4371A__ZENER__1 D   Tools ▾   Views ▾   🔧▾

Name	Description	Value	Units	Use default	^
FC	Forward-bias depletion capacitance co...	0.5		☐	
BV	Reverse breakdown knee voltage	2.71	▯ V	☐	
IBV	Reverse breakdown knee current	0.04386	A	☐	
IBVL	Low-level reverse breakdown knee cur...	1.0	A	☑	
IKF	High-injection knee current	1e30	A	☑	
ISR	Recombination current parameter	0.0	A	☑	
NBV	Reverse breakdown ideality factor	1.0		☑	
NBVL	Low-level reverse breakdown ideality f...	1.0		☑	
NR	Emission coefficient for ISR	2.0		☑	
TBV1	BV linear temperature coefficient	0.0	1/°C	☑	
TBV2	BV quadratic temperature coefficient	0.0	1/°C²	☑	∨

Change component

Change all components

Reset to default

Cancel   Help

Fig. 3.64 SPICE model of 1N4371A

Multisim has two types of components: Real components and virtual components. Real components are manufacturer-specific components with non-editable parameters in their simulation models. Virtual components are generic devices that include a schematic symbol and a simulation model. They allow you to explore behavior without being tied to a real device. Most virtual components have parameters you can change to study the effect on your design.

Place of virtual diode and transistor are shown in Figs. 3.65 and 3.66, respectively.

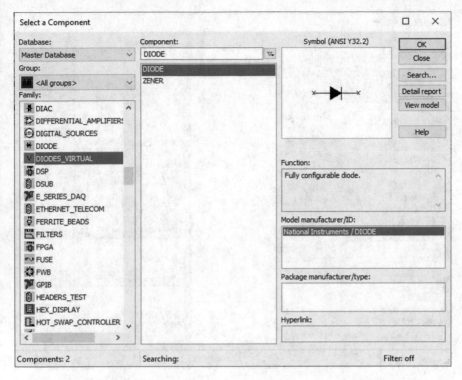

Fig. 3.65 Virtual diode

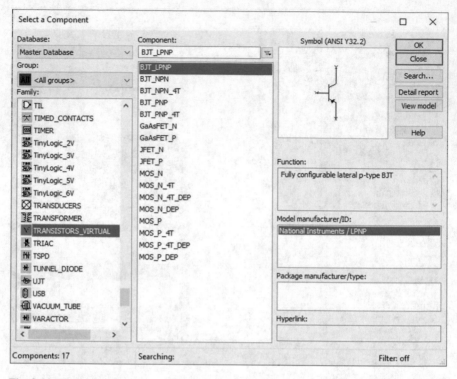

Fig. 3.66 Virtual transistor

Virtual components are very useful to simulate the textbooks circuits. Because you can make the required custom component for the question in hand. For instance, assume that the question told you that the current gain of the transistor is 150. In order to model such a transistor, add a virtual transistor to the schematic, then double click on it. Open the value tab and click the edit model button (Fig. 3.67).

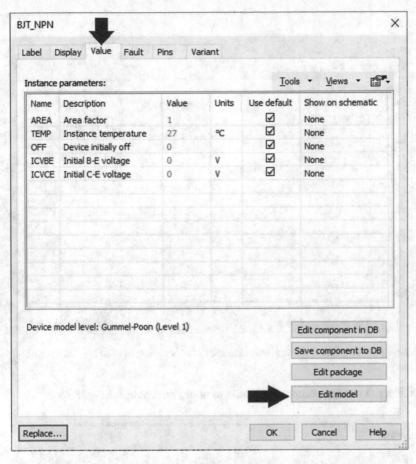

Fig. 3.67 Edit model button permits you to change the SPICE model of the component

Enter the given value of current gain to the ideal maximum forward beta box and press the enter key (Fig. 3.68).

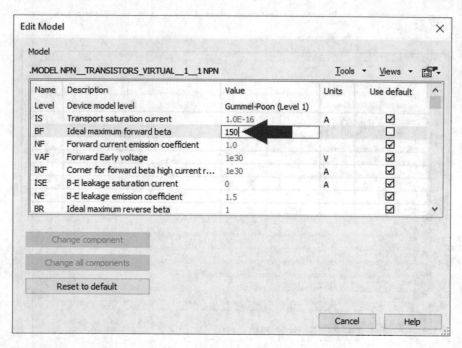

Fig. 3.68 Ideal maximum forward beta determines the current gain (β or h_{FE}) of transistor

Click the change component button to apply the changes (Fig. 3.69).

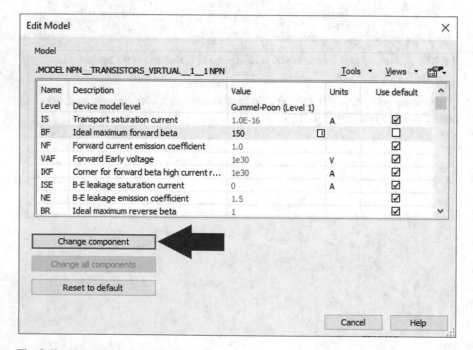

Fig. 3.69 Change component button

Click OK button and the component is ready (Fig. 3.70).

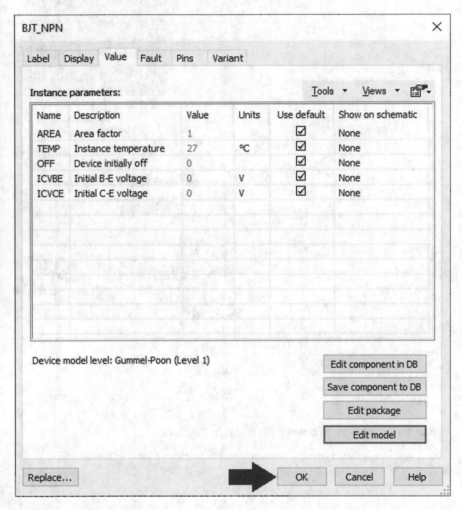

Fig. 3.70 BJT_NPN window

3.8 Example 7: Half Wave Rectifier

We want to simulate a half wave rectifier in this example. Assume the schematic shown in Fig. 3.71.

Fig. 3.71 Schematic of
example 7

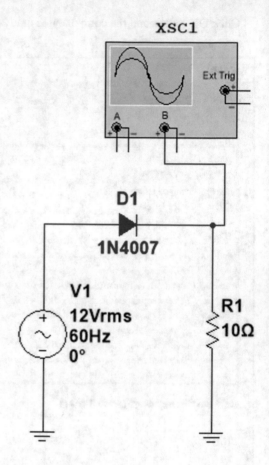

Run the simulation. The result is shown in Fig. 3.72.

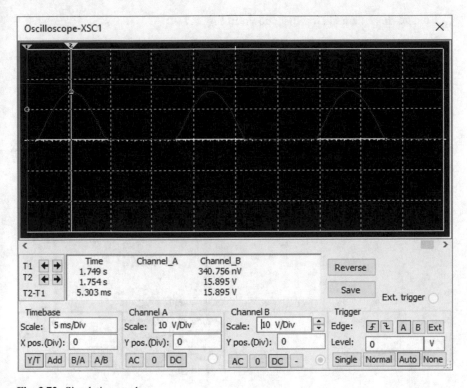

Fig. 3.72 Simulation result

According to Fig. 3.73, the peak voltage is 15.895 V. So, the forward voltage drop of diode is 1.0756 V.

Fig. 3.73 Calculation of voltage drop of diode

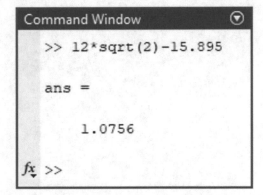

Change the schematic to what shown in Fig. 3.74.

Fig. 3.74 Measurement of
voltage drop of diode and
the load voltage

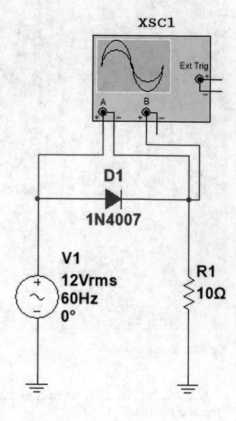

Left click on the wire that connects the oscilloscope to the voltage source V1
(Fig. 3.75).

Fig. 3.75 Select the wire
from channel 1 to circuit

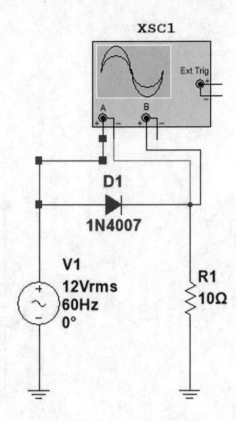

Right click on the selected wire. Then click the segment color (Fig. 3.76).

Fig. 3.76 Segment color determines the color of waveform of channel 1

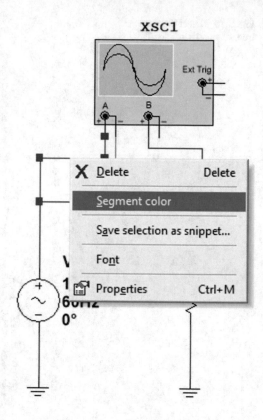

The color window shown in Fig. 3.77 appears and permits you to select a color for the signal that will be shown on the channel 1 of oscilloscope.

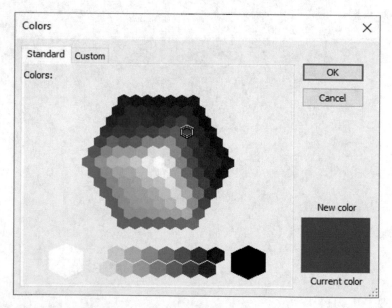

Fig. 3.77 Colors window

Now run the simulation. The result is shown in Fig. 3.78.

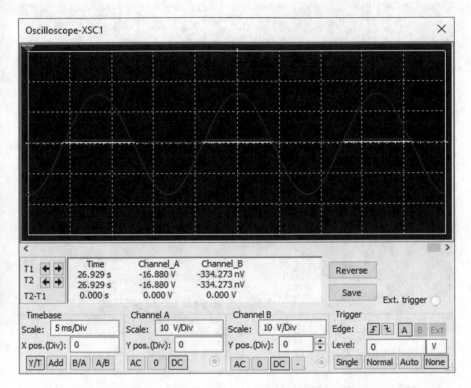

Fig. 3.78 Simulation result

You can move the ground line of channel 2 to see the voltage waveform of channel 1 better. To do this, change the Y pos. (Div) of channel 2 to 1 (Fig. 3.79).

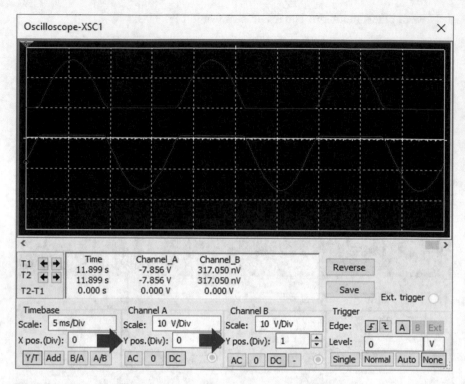

Fig. 3.79 Moving the Y position of waveform to have a better view

You can use the cursors to read the value of forward voltage drop of the diode and the reverse voltage applied to it (Fig. 3.80). The applied reverse voltage must be less than the allowed maximum revers voltage that is given in the datasheet.

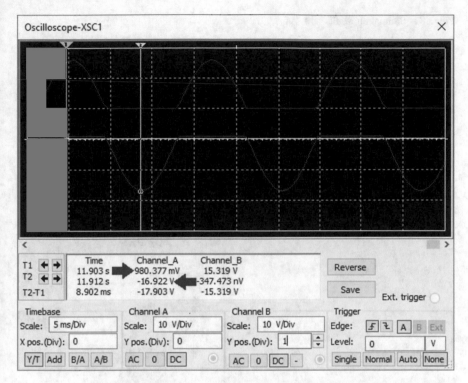

Fig. 3.80 Cursors are used to measure the voltage drop of diode and peak of applied reverse voltage

Add probes to measure the RMS of current and dissipated power in diode. According to Fig. 3.81, the RMS of current is 788 mA, the average value of current is 495 mA and average of dissipated power in the diode is 473 mW.

Fig. 3.81 Simulation result

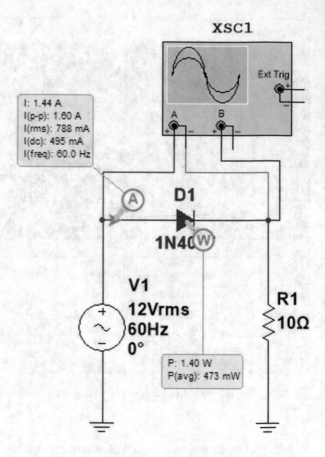

Let's verify the obtained result. If we ignore the diode voltage drop, then the load voltage DC and RMS values are $\frac{V_m}{\pi}$ and $\frac{V_m}{2}$, respectively. The average and RMS current are $\frac{V_m}{\pi R}$ and $\frac{V_m}{2R}$, respectively. Calculations in Fig. 3.82 ignores the voltage drop of diode and calculates the average and RMS current. The obtained results are bigger than Multisim results, since the voltage drop of diode is ignored.

Fig. 3.82 MATLAB calculations

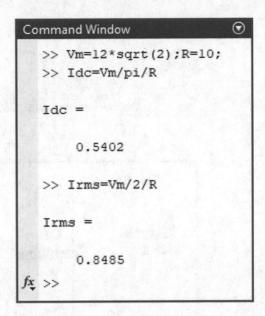

Average of dissipated power in the diode can be approximated by multiplying the diode voltage drop to average of current through the diode (Fig. 3.83).

Fig. 3.83 Calculation of average power dissipated in the diode

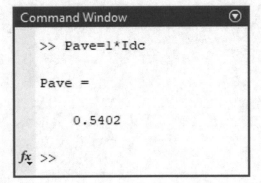

3.9   Example 8: Half Wave Rectifier with Filer Capacitor

In this example we add a filter capacitor to the circuit of previous example. Assume the schematic shown in Fig. 3.84.

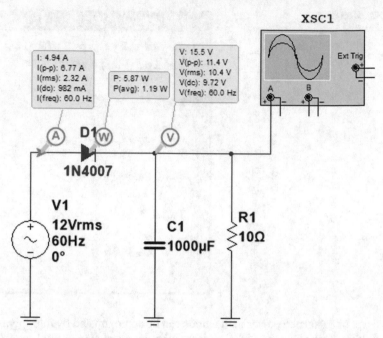

Fig. 3.84 Simulation result

The load voltage is shown in Fig. 3.85. Maximum of the waveform is 15.883 V and minimum is 4.466 V. So, the output voltage ripple is 15.883–4.466 = 11.417 V.

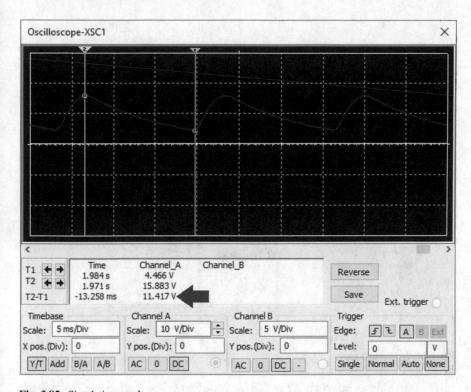

Fig. 3.85 Simulation result

Increase the capacitor to 5000 μF. RMS and average value of current and diode dissipation is increased (Fig. 3.86).

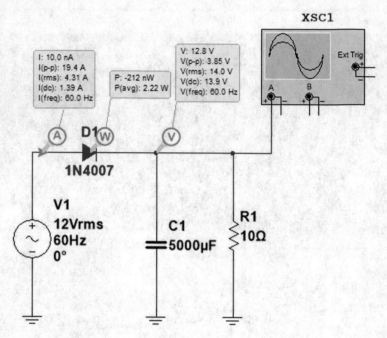

Fig. 3.86 Simulation result

The load voltage is shown in Fig. 3.87. According to Fig. 3.87, the output voltage ripple is 3.826 V. So, increasing the capacitor value decrease the ripple voltage.

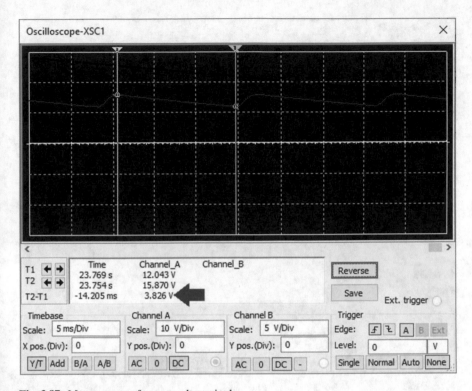

Fig. 3.87 Measurement of output voltage ripple

Let's see the diode current waveform. In order to do this, click the interactive button (Fig. 3.88).

Fig. 3.88 Interactive button

Select the transient and do the settings as shown in Fig. 3.89.

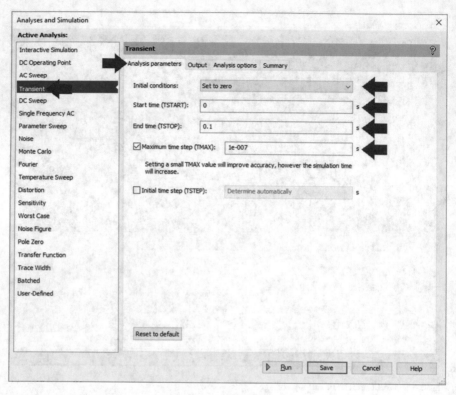

Fig. 3.89 Analyses parameters tab of transient analysis

Open the output tab and add the I(D1[ID]) to the right list and click the run button (Fig. 3.90). I(D1[ID]) is the current of diode D1.

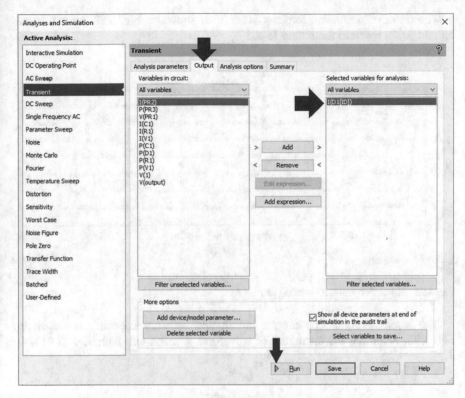

Fig. 3.90 Output tab of transient analysis

The result of simulation for C = 1000 μF is shown in Fig. 3.91. Note that the current peaks reach to more than 6 A.

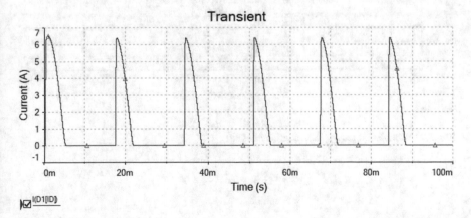

Fig. 3.91 Simulation result (C = 1000 μF)

The result of simulation for C = 5000 µF is shown in Fig. 3.92. Note that the current peaks reach to more than 15 A.

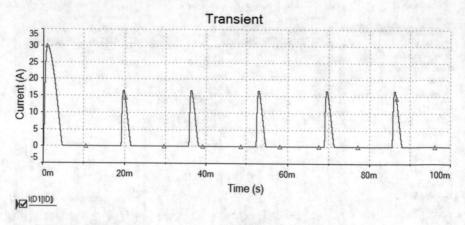

Fig. 3.92 Simulation result (C = 5000 µF)

The result of simulation for the circuit without filter capacitor is shown in Fig. 3.93. Note that the maximum of waveform is much less than Figs. 3.91 and 3.92.

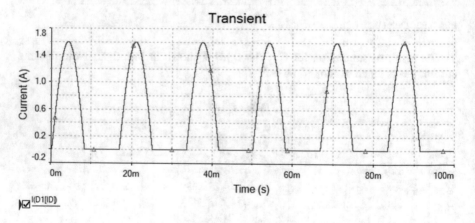

Fig. 3.93 Simulation result (Capacitor is removed)

3.10   Example 9: Full Wave Rectifier

In this example we want to simulate a full wave rectifier circuit. Assume the schematic shown in Fig. 3.94. This schematic uses a transformer with 1 primary winding and 1 secondary winding (Fig. 3.95).

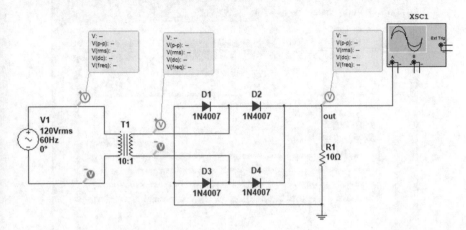

Fig. 3.94 Schematic of example 9

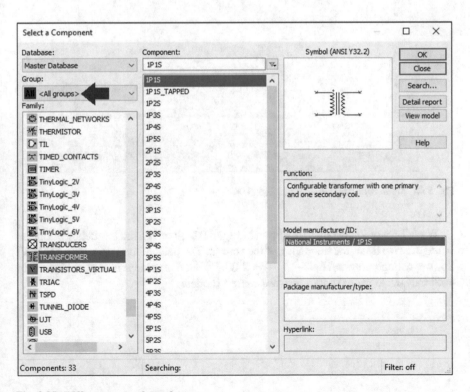

Fig. 3.95 Different types of transformer

Settings of the transformer is shown in Fig. 3.96. This transformer decreases the input AC voltage to 12 Vrms. The peak of secondary voltage is $12\sqrt{2} = 16.97$ V.

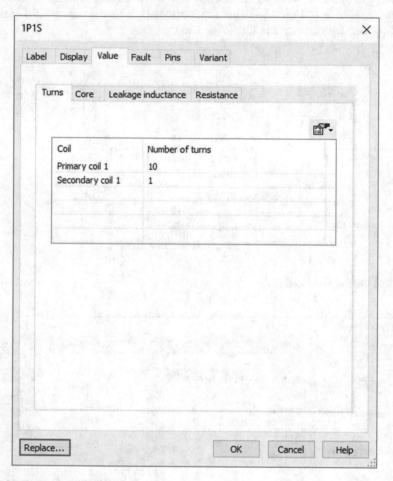

Fig. 3.96 Settings of transformer T1 in Fig. 3.94

Waveform of load voltage is shown in Fig. 3.97. Note that the frequency of output voltage is two times the frequency of the source. The peak of load voltage is 14.94 V. So, we voltage drop is 16.97–14.94 = 2.03 V. Such a voltage is expected since in each half cycle we have two forward biased diodes.

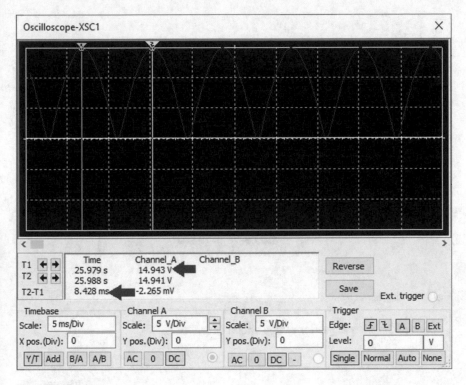

Fig. 3.97 Simulation result

You can use the differential probe to measure the voltages according to desired nodes. According to Fig. 3.98, the DC component of load voltage is 9.01 V. If we ignore the voltage drop of diodes, the DC value of output voltage for full wave rectifier can be obtained using the $\frac{2}{\pi} V_m$ equation, where V_m shows the peak of input voltage to the rectifier. For this example, $\frac{2}{\pi} \times 16.97 = 10.80$ V.

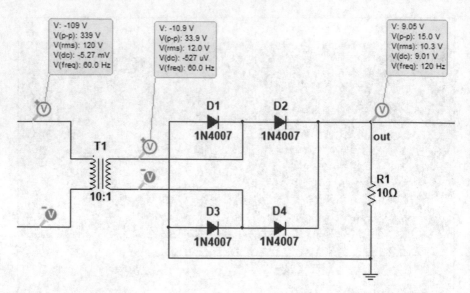

Fig. 3.98 Simulation result

Let's add a filter capacitor to this circuit (Fig. 3.99).

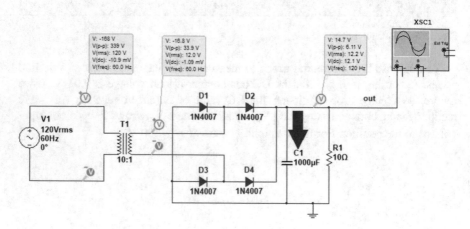

Fig. 3.99 A filter capacitor is added to the circuit

The output voltage is shown in Fig. 3.100. Note that the frequency of output voltage is 120 Hz.

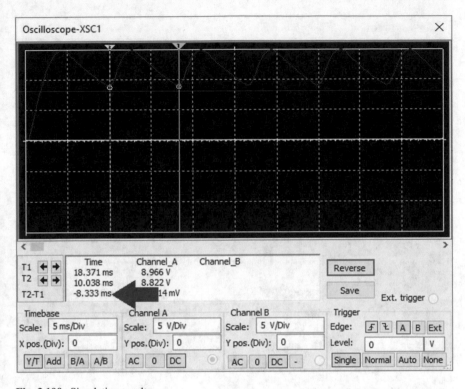

Fig. 3.100 Simulation result

The ripple of output voltage can be measured using the cursors. According to Fig. 3.101, the output voltage ripple is 6.073 V.

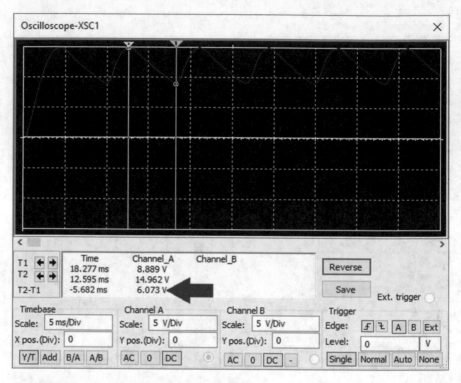

Fig. 3.101 Measurement of output voltage ripple

3.11 Example 10: Voltage Regulators

Voltage regulator helps us to obtain fixed output voltage despite of input voltage changes or load changes. Most important regulator IC's are available in Multisim. Let's add a voltage regulator IC to the circuit of previous example. Assume the schematic shown in Fig. 3.102. This schematic uses LM 7805 regulator IC which gives the output voltage of 5 V for 7 V < Vin <35 V. Maximum of load current for LM 7805 is about 1.5 A. Place of LM7805 is shown in Fig. 3.103.

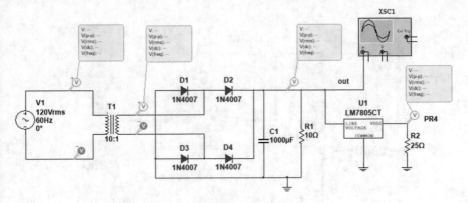

Fig. 3.102 Schematic of example 10

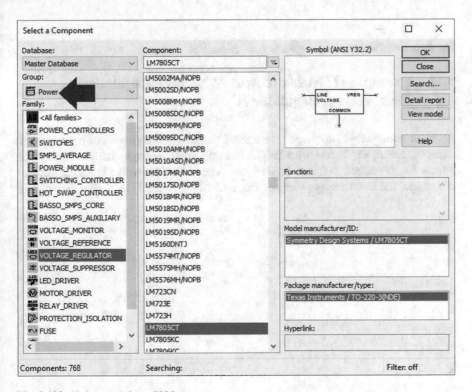

Fig. 3.103 Voltage regulator 7805

Run the simulation. The output of regulator is 5 V (Fig. 3.104).

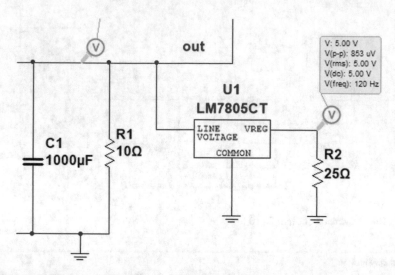

Fig. 3.104 Addition of voltage regulator to the circuit

3.12 Example 11: Measurement of Voltage Regulation for a Voltage Regulator IC

An ideal voltage regulator must give a fixed output voltage despite of input voltage changes. However, in real world regulators, the input voltage changes have some effects on the output voltage. Assume the schematic shown in Fig. 3.105. In this schematic, the input voltage changes from 7.9289 V to 22.0711 V (Fig. 3.106).

Fig. 3.105 Schematic of example 11

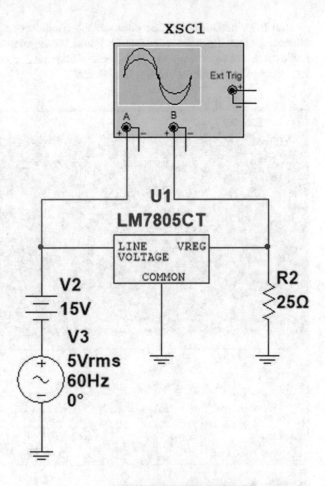

Fig. 3.106 Calculation of minimum and maximum of input voltage

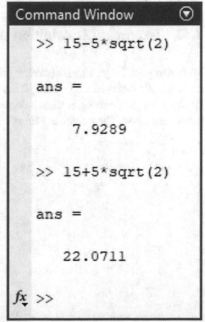

Run the simulation. Double click on the oscilloscope and select AC mode for channel 2 of oscilloscope (Fig. 3.107). Pause the simulation by pressing the F6 key and use cursors to measure the peak-peak of channel 2. According to Fig. 3.107, the AC component of output voltage is 1.41 mV.

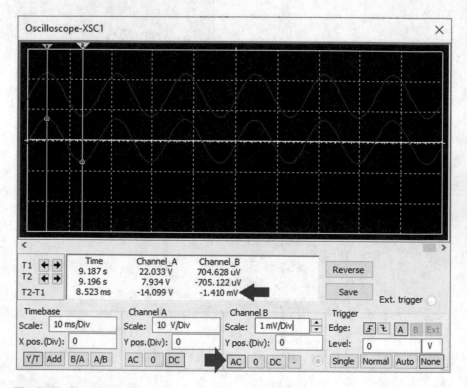

Fig. 3.107 Simulation result

3.13   Example 12: Addition of Fault to Components

Multisim permits you to add fault to the component and study the behavior of the circuit with faulty components. This example shows the fault tab of Multisim. Assume full wave rectifier circuit shown in Fig. 3.108. This schematic uses a center tap transformer. The center tap transformer place is shown in Fig. 3.109.

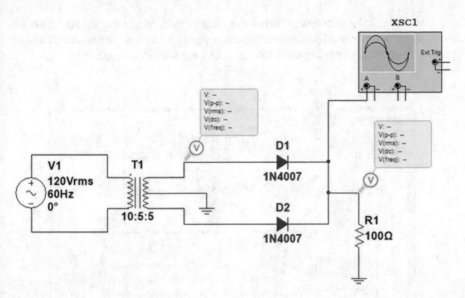

Fig. 3.108 Schematic of example 12

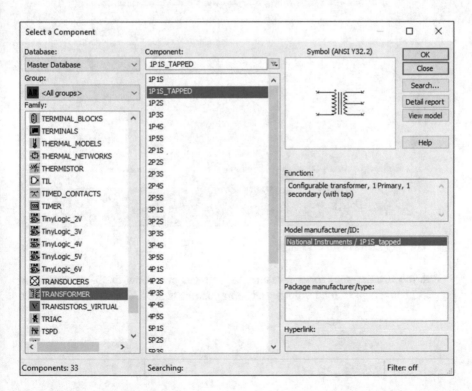

Fig. 3.109 Tapped transformer

Double click the transformer and do the settings as shown in Fig. 3.110. The core, leakage inductance and resistance tabs permit you to make a more realistic model of your transformer. Changing these tabs are not necessary in this example.

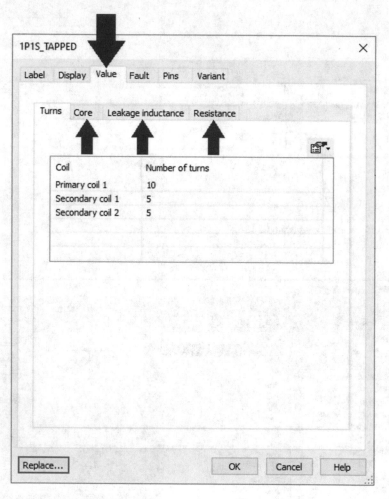

Fig. 3.110 Tapped transformer settings

Run the simulation. Probes measure the voltages and show them. According to Fig. 3.111, average of load voltage (DC component) is 53.1 V.

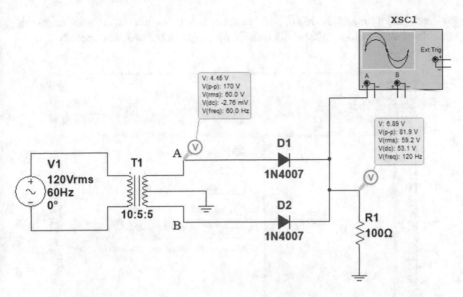

Fig. 3.111 Simulation result

Double click the oscilloscope. The waveform shown in Fig. 3.112 appears.

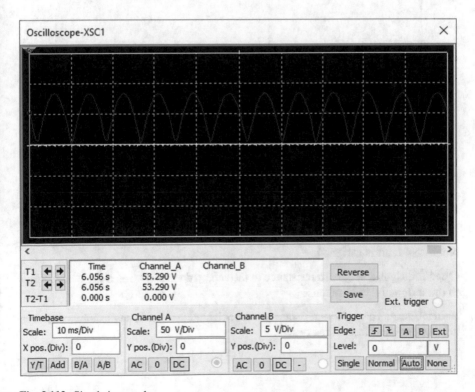

Fig. 3.112 Simulation result

Let's add a fault to diode D1 and see its effect on the circuit. In order to do this, double click the diode D1 and open the fault Table A list of faults appears.

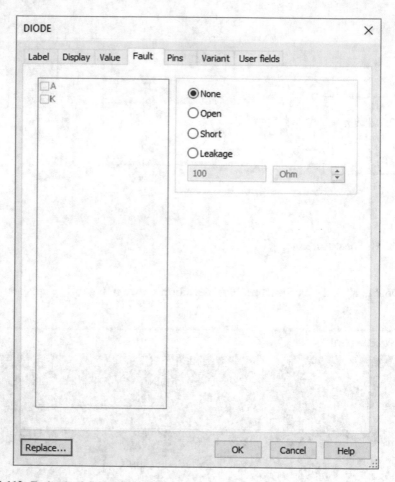

Fig. 3.113 Fault tab of diode window

Here is the description of faults shown in Fig. 3.113.

None No fault is assigned.

Open Assigns a very high resistance to the selected terminals, as if the wire leading to the terminals was broken.

Short Assigns a very low resistance to the selected terminals, so the component has no measurable effect on the design.

Leakage Assigns the resistance value specified in the fields below the option, in parallel with the selected terminals. This causes the current to leak past the terminals instead of going through them.

For instance, select open anode fault for diode D1 (Fig. 3.114). The addition of such a fault cause the anode of diode D1 to be isolated from the circuit (Fig. 3.115).

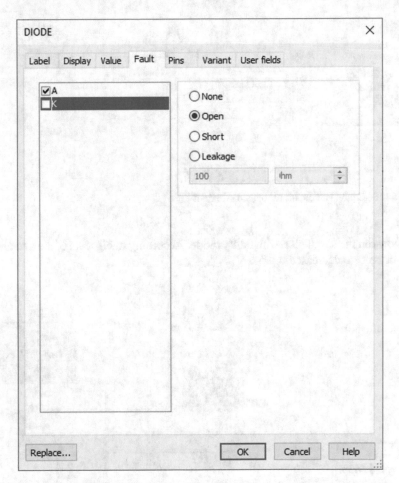

Fig. 3.114 Addition of open circuit fault to the circuit

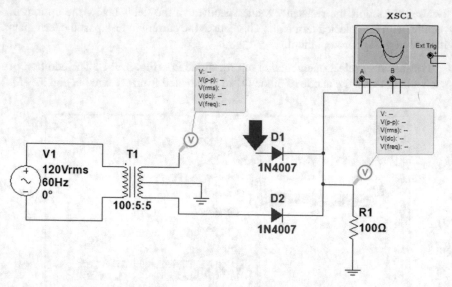

Fig. 3.115 Equivalent circuit for open diode fault

Now run the simulation with faulty diode. According to Fig. 3.116, the average of output voltage decreased to 26.6 V.

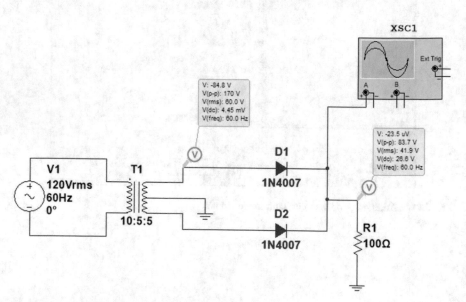

Fig. 3.116 Simulation result

Double click the oscilloscope block. The load waveform is shown Fig. 3.117. Since the diode D1 is removed from the circuit, we lose half of the waveform.

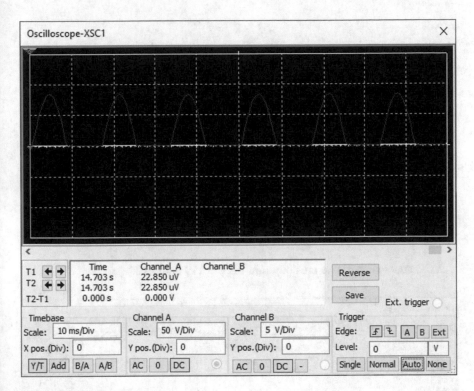

Fig. 3.117 Simulation result

3.14   Example 13: Common Emitter Amplifier

We want to simulate a common emitter amplifier and measure its voltage gain. Assume the amplifier shown in Fig. 3.118.

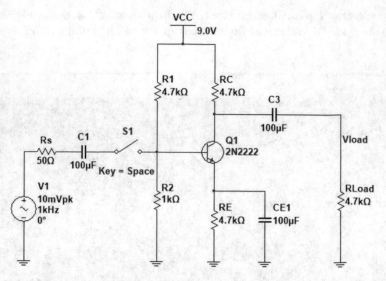

Fig. 3.118 Schematic of example 13

VCC and switch places are shown in Figs. 3.119 and 3.120.

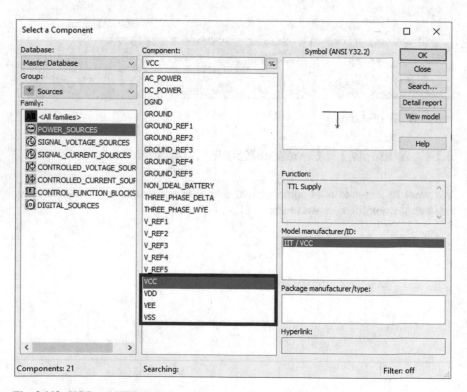

Fig. 3.119 VCC and VEE blocks

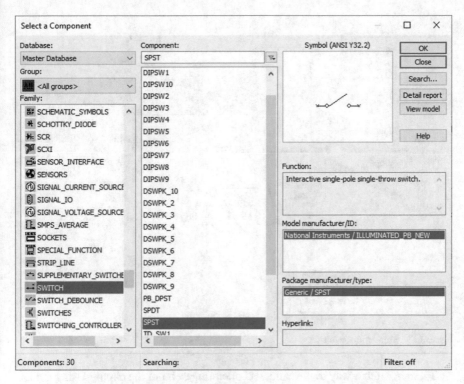

Fig. 3.120   SPST (Single Pole Single Through) switch

Run the simulation. The switch is open and therefore no AC signal enters the amplifier. So, we can measure the operating point. According to Fig. 3.121, the collector current is $\frac{9-8.02}{4.7} = 0.21 \, \text{mA}$ and collector emitter voltage is 8.02 V-0.994 V = 7.03 V.

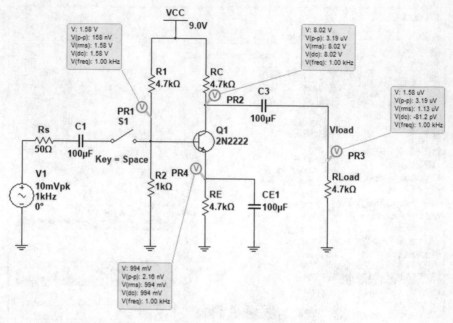

Fig. 3.121 Simulation result

There is another way to measure, DC operating point of the circuit: Using the DC operating point analysis. In DC operating point analysis, the capacitors are replaced with open circuit, inductors are replaced with short circuit and amplitude of AC sources are assumed to be zero. Assume the circuit shown in Fig. 3.122. B, C and E names are given to the base, collector and emitter nodes, respectively.

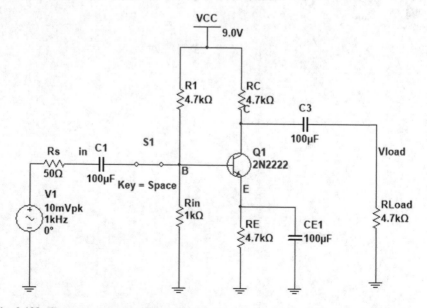

Fig. 3.122 Common emitter amplifier circuit

Click the interactive button (Fig. 3.123) and select the DC operating point. Then add I(Q1[IB]), I(Q1[IC]), I(Q1[IE]), V(b), V(c) and V(e) to the right list and click the run button (Fig. 3.124). I(Q1[IB]), I(Q1[IC]), I(Q1[IE]) shows the base current (current that enters the base), collector current (current that enters the collector) and emitter current (the current that enters the emitter), respectively. V(b), V(c) and V(e) shows the base, collector and emitter voltages (with respect to ground).

Fig. 3.123 Interactive button

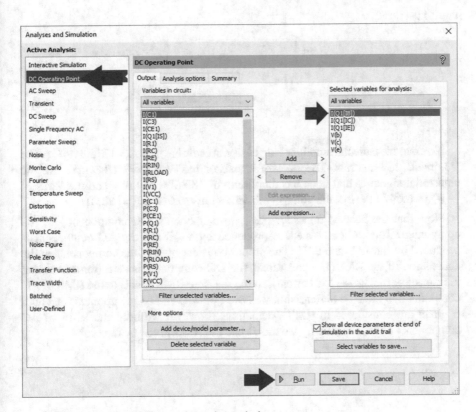

Fig. 3.124 Output tab of DC operating point analysis

The simulation result is shown in Fig. 3.125. The emitter current is negative since DC operating point analysis measures the current which enters the transistor terminals. +211.38882 μA exit from the emitter terminal or − 211.38882 μA enters the emitter terminal (Fig. 3.126).

Fig. 3.125 Simulation
result

	Variable	Operating point value
1	V(b)	1.57740
2	V(c)	8.01529
3	V(e)	993.52742 m
4	I(Q1[IB])	1.87632 u
5	I(Q1[IC])	209.51250 u
6	I(Q1[IE])	-211.38882 u

Fig. 3.126 Collector
current

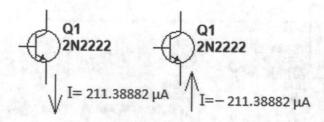

You can measure the amplifier gain easily. According to Fig. 3.127, the AC probe
V3, reads 133 mVrms. The base of transistor has V(p-p) = 18.8 mV. So, the
sinusoidal signal at that node has amplitude of 18.8/2 = 9.4 mV and the RMS of
$\frac{9.4\ \text{mV}}{\sqrt{2}}$ = 6.647 mV. So, the voltage gain is 133 mV/6.647 mV = 20.01.

Note that the base of transistor has two components: DC component and AC
component. The DC component is produced by VCC and the AC component is
produced by the AC source V1. The probes consider both of the components in the
calculation of the RMS. In node Vload, the DC component is zero (remember that
the capacitor blocks the DC current), so the measure RMS equals to the RMS of AC
component. In the base of transistor, we have both DC and AC components and both
of them are considered in RMS calculations, however we need the RMS of AC
components only. That is why we didn't use the RMS given by the probe.

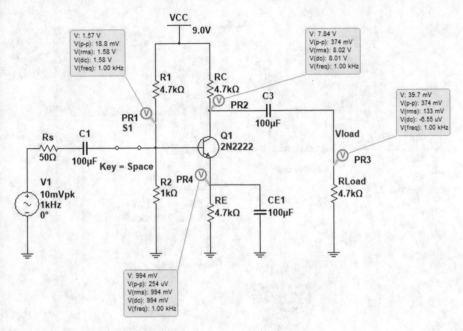

Fig. 3.127   Simulation result

   If you connect the probe to the left terminal of capacitor C1 (Fig. 3.128), then you can use the values given by the RMS section of probe PR1 (because the DC component at the left terminal is zero). According Fig. 3.128, the gain is 133 mV/ 6.65 mV = 20.

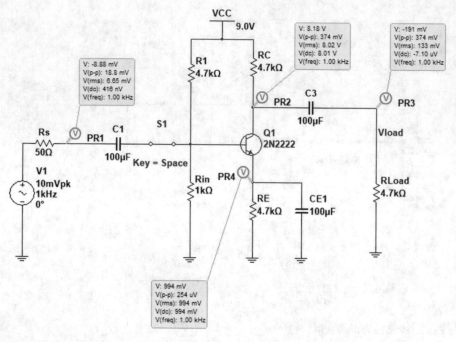

Fig. 3.128 Simulation result

3.15   Example 14: Measurement of Input/Output Impedance of Common Emitter Amplifier

In this example, we want to measure the input/output impedance of the common emitter amplifier of previous example. Let's start with input impedance. The input impedance is the impedance which is seen from the input of the amplifier. Impedance of capacitor C1 can be ignored. Figure 3.129 shows the equivalent circuit of the amplifier. Input impedance of the amplifier is shown with resistor Rin. Resistor Rin is unknown and must be calculated.

Fig. 3.129 Equivalent circuit of amplifier from source V1 viewpoint

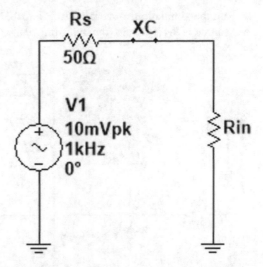

The resistor Rin can be calculated in two ways. We can measure the current I which is drawn from the source V1. Then V/I equals to Rs + Rin. So, V/I-Rs gives Rin. Another method is the measurement of Rin voltage. The voltage across Rin equals to Rin/(Rin + Rs) × V1. Rs and V1 values are known, so we can calculate the value of Rin. Let's use these methods to measure the input impedance oerf the amplifier. Assume the circuit shown in Fig. 3.130. The signal generator generates a sinusoidal signal with amplitude of 10 mV and frequency of 1 kHz. RMS of this input voltage equals to $\frac{10 \text{ mV}}{\sqrt{2}} = 7.07$ mV. You can use an AC voltage block if you prefer.

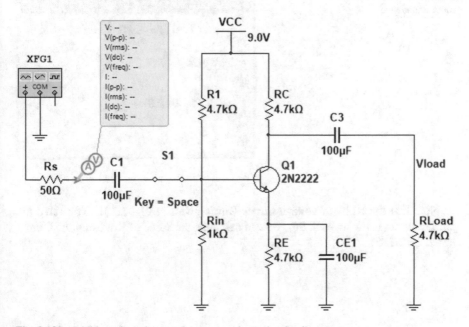

Fig. 3.130 Addition of a voltage and current probe to the circuit

Run the simulation. According to Fig. 3.131, the current that is drawn from the AC source is 8.41 µA. So, the input impedance is around 790 Ω (Fig. 3.132).

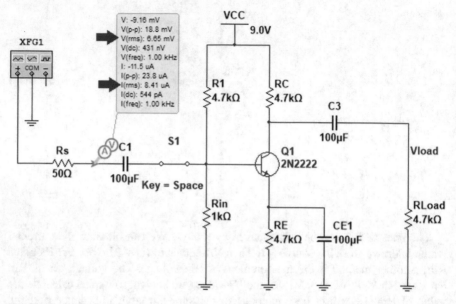

Fig. 3.131 Simulation result

Fig. 3.132 Calculation of input impedance

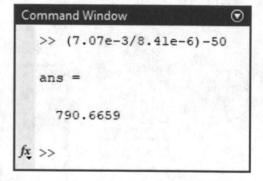

Note that the RMS of voltage across Rin is given in Fig. 3.131. You have the current through Rin as well. So, you can calculate the value of Rin using the Ohm's law (Fig. 3.133).

Fig. 3.133 Calculation of input impedance

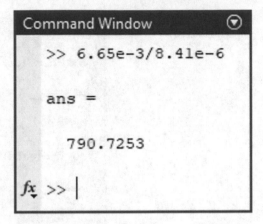

Let's use the second method to measure the input impedance as well. According to Fig. 3.134, the voltage across the Vin equals to 6.65 mVrms and input voltage V1 is 7.07 mVrms. So, we need to solve $\frac{R_x}{R_s+R_x} \times V_1 = V_{in}$. The value of Rx is found with the aid of commands shown in Fig. 3.135.

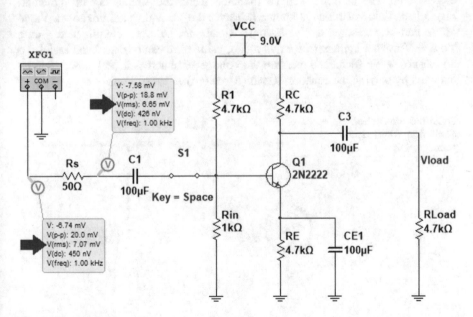

Fig. 3.134 Simulation result

Fig. 3.135 Calculation of input impedance

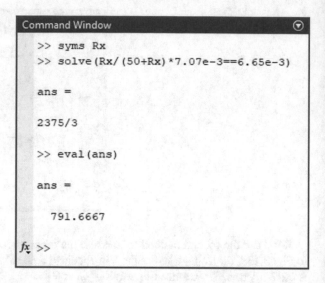

```
Command Window                                                    ⊙

>> syms Rx
>> solve(Rx/(50+Rx)*7.07e-3==6.65e-3)

ans =

2375/3

>> eval(ans)

ans =

   791.6667

fx >>
```

The output impedance of converter is the impedance which is seen from the output load. The amplifier can be modeled using the simple circuit shown in Fig. 3.136. Value of Rload is known however the Vo and ro are unknown. Value of Vo can be measured easily. Assume Rload=∞, i.e. open circuit. In this case Vout = Vo. After knowing the value of Vo, value of ro can be calculated easily. To do this connect a Rload and measure the voltage Vout across it. Value of ro can be obtained by solving the equation Rload/(Rload+ro) × Vo = Vout.

Fig. 3.136 Equivalent circuit from Rload view point

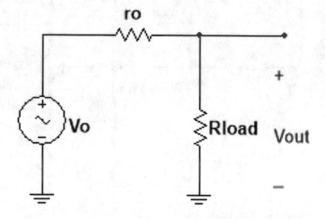

Let's apply this method to measure the output impedance of the amplifier. Assume the schematic shown in Fig. 3.137. The switch can be found in the switch family (Fig. 3.138). According to Fig. 3.137, the Vo = 258 mVrms.

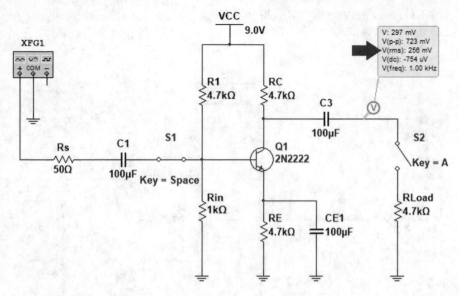

Fig. 3.137 Simulation result

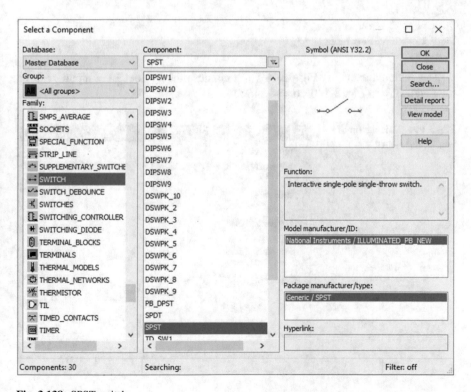

Fig. 3.138 SPST switch

Now close the switch, the voltage drop across the 4.7 kΩ resistor equals to 133 mVrms (Fig. 3.139).

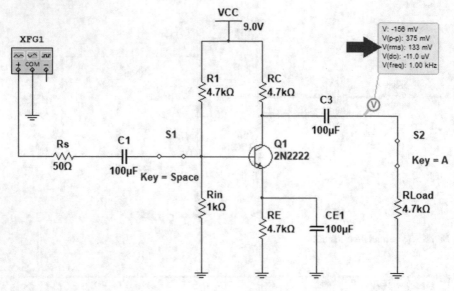

Fig. 3.139 Simulation result

The calculations in Fig. 3.140, calculates the value of output resistor ro. According to Fig. 3.140, ro = 4.3466 kΩ.

Fig. 3.140 Calculation of output impedance

```
Command Window

>> syms Rx
>> solve(4.7e3/(4.7e3+Rx)*256e-3==133e-3)

ans =

578100/133

>> eval(ans)

ans =

   4.3466e+03

fx >>
```

Let's check the obtained result. Assume that output load is 0.5 kΩ. According to calculation in Fig. 3.141, we expect the voltage drop across of it to be 26.4 mV. The Multisim verify this result (Fig. 3.142).

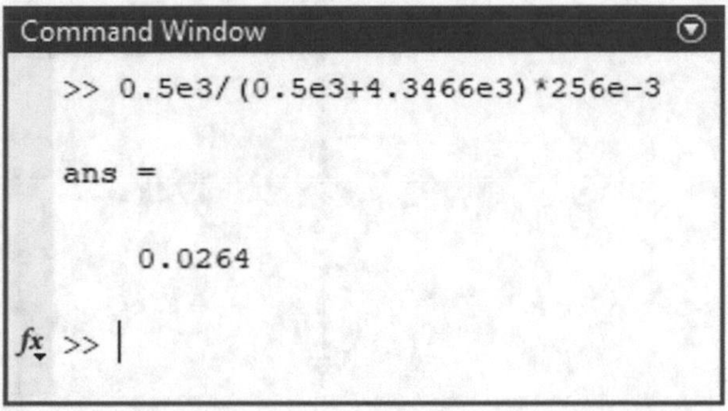

Fig. 3.141 Calculation of load voltage for 0.5 kΩ load

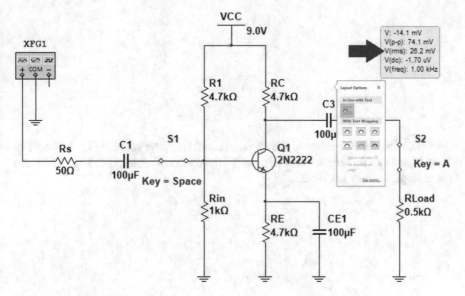

Fig. 3.142 Simulation result

3.16 Example 15: Measurement of Input/Output Impedance for Noninverting Op-Amp Amplifier

We want to measure the input and output impedance of the non-inverting op amp amplifier shown in Fig. 3.143.

Fig. 3.143 Schematic of example 15

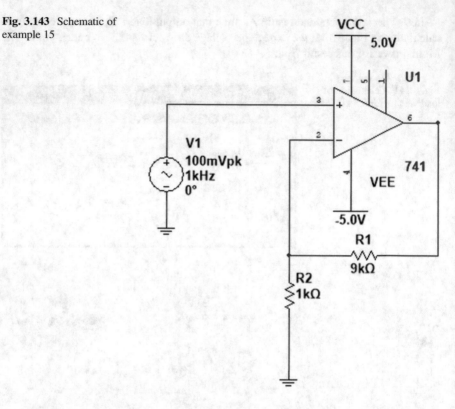

We use a probe to measure the current drawn from the source (Fig. 3.144). The ratio of voltage to current gives the input impedance. According to Fig. 3.145, the input impedance is 1.068 MΩ.

Fig. 3.144 Simulation result

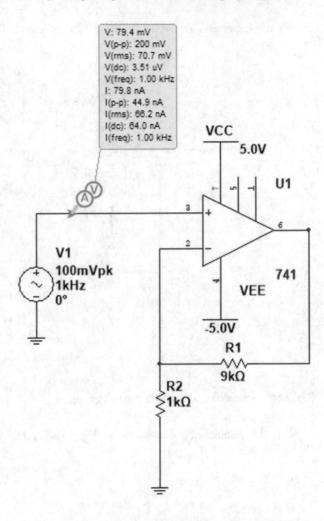

Fig. 3.145 Calculation of input impedance

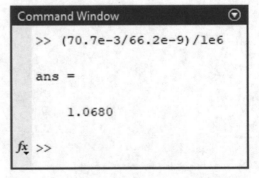

We use the schematic shown in Fig. 3.146 to measure the output impedance.

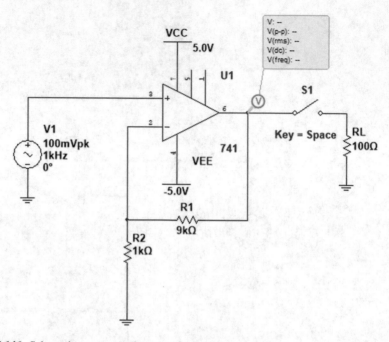

Fig. 3.146 Schematic to measure the output impedance

Run the simulation. According to Fig. 3.147, the open circuit voltage is 707 mVrms.

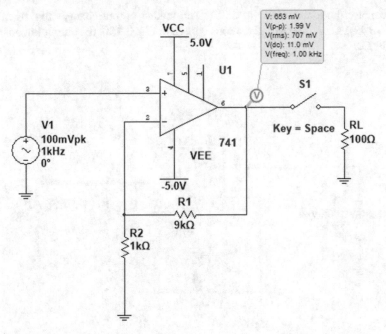

Fig. 3.147 Simulation result

Now load the output of amplifier. According to Fig. 3.148, the output doesn't show any significant change for 100 Ω load. This means that the load resistor is much bigger than the output impedance and we need to decrease it.

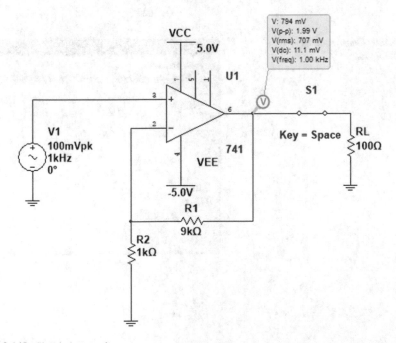

Fig. 3.148 Simulation result

Decrease the load resistor to 25 Ω. The loaded output changes to 516 mVrms (Fig. 3.149). According to calculations shown in Fig. 3.150, the output impedance is 9.2539 Ω.

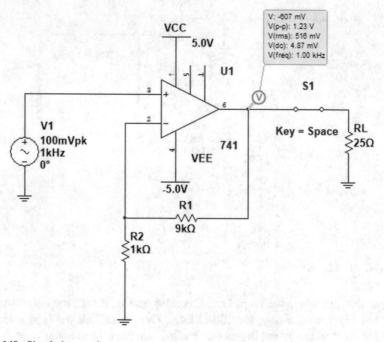

Fig. 3.149 Simulation result

Fig. 3.150 Calculation of output impedance

```
Command Window

>> syms Rx
>> solve(25/(25+Rx)*707e-3==516e-3)

ans =

4775/516

>> eval(ans)

ans =

    9.2539

fx >>
```

3.17   Example 16: Distortion Analyzer Block

The Total Harmonic Distortion (THD or THDi) is a measurement of the harmonic distortion present in a signal and is defined as the ratio of the sum of the powers of all harmonic components to the power of the fundamental frequency. For instance, for

$$v(t) = \sum_{n=1}^{\infty} V_n \sin(n\omega_0 t + \varphi_0), \quad THD = \frac{\sqrt{V_2^2 + V_3^2 + V_4^2 + \ldots}}{V_1}.$$

Multsim has a block called distortion analyzer (Fig. 3.151). You can measure the THD with this block easily. The distortion analyzer block is studied in this example.

Fig. 3.151 Distortion analyzer

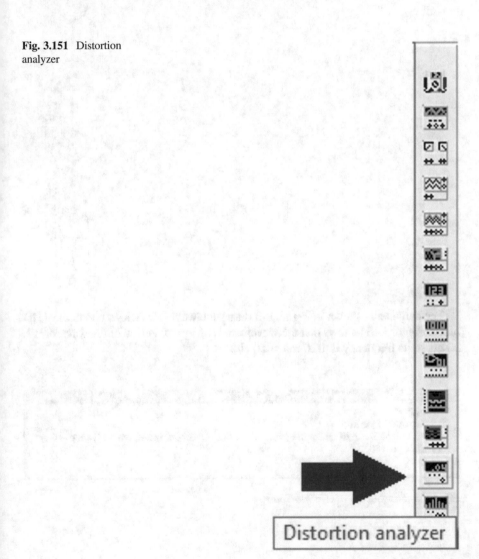

Distortion analyzer

Let's measure the THD for some simple examples. Assume the schematic shown in Fig. 3.152.

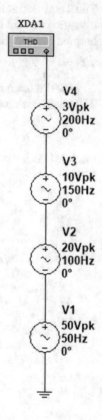

Fig. 3.152 Schematic of example 16

The commands shown in Fig. 3.153 draws the waveform which enters the THD block. Figure 3.154 shows that the waveform is distorted. Period of the waveform is 20 ms so its frequency is 1/20 ms = 50 Hz.

```
Command Window                                              ⊙
    >> syms t
    >> f=50;w0=2*pi*f;
    >> ezplot(50*sin(w0*t)+20*sin(2*w0*t)+10*sin(3*w0*t)+3*cos(4*w0*t),[0 40e-3])
    >> grid on
fx >> |
```

Fig. 3.153 Drawing the voltage waveform

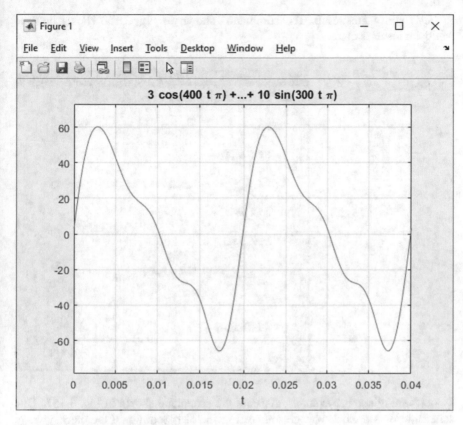

Fig. 3.154 Output of code in Fig. 3.153

Run the simulation. The result shown in Fig. 3.155 is obtained. According to Fig. 3.155, the THD is 45.122%.

Fig. 3.155 Distortion analyzer window

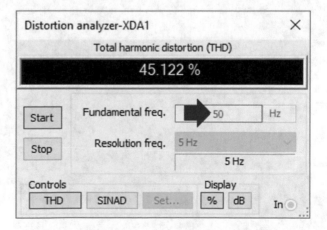

Let's verify this result. The commands shown in Fig. 3.156 shows that the Multisim result is correct.

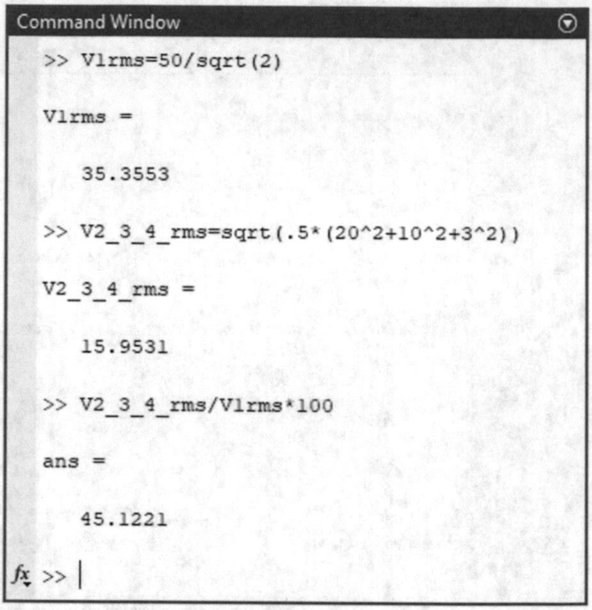

Fig. 3.156 MATLAB calculations

```
Command Window                                                        ⊙

  >> V1rms=50/sqrt(2)

  V1rms =

     35.3553

  >> V2_3_4_rms=sqrt(.5*(20^2+10^2+3^2))

  V2_3_4_rms =

     15.9531

  >> V2_3_4_rms/V1rms*100

  ans =

     45.1221

fx >> |
```

Let's study another example. Assume the schematic shown in Fig. 3.157. This schematic uses a clock voltage source (Fig. 3.158). Settings of the clock voltage block is shown in Fig. 3.159. These settings produce a square wave with frequency of 500 Hz, high level of 5 V and low level of 0 V (Fig. 3.160). The Fourier series of this signals is $\frac{V_p}{2} + \sum\limits_{n=1,3,5,\ldots}^{\infty} \frac{2 \times V_p}{n\pi} \sin(n\omega_0 t)$ where $V_p = 5$ V and $\omega_0 = 2\pi \times 500 = 3141.5 \frac{Rad}{s}$.

Fig. 3.157 Measurement of THD for a square wave

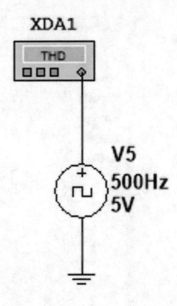

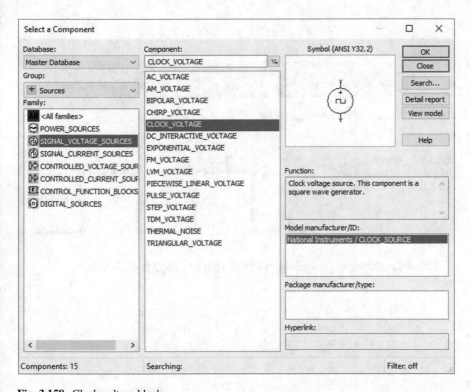

Fig. 3.158 Clock voltage block

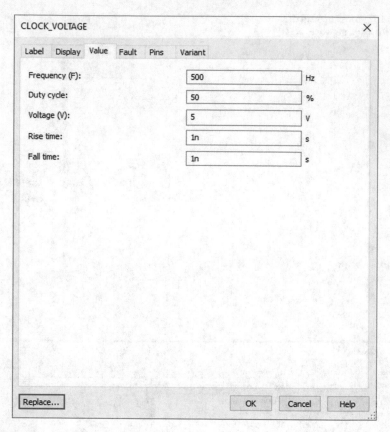

Fig. 3.159 Settings of clock voltage source

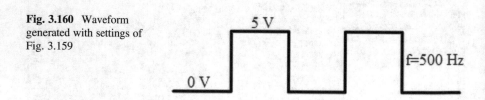

Fig. 3.160 Waveform
generated with settings of
Fig. 3.159

Run the simulation. According to Fig. 3.161, the THD is 42.902%.

Fig. 3.161 THD for square wave

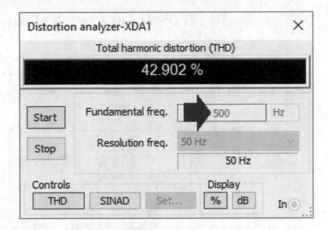

Let's verify the Multisim result. The code shown in Fig. 3.162, uses the first 17 harmonics to calculate the THD. The result is 45.3818%. The code shown in Fig. 3.163 uses the first 1001 harmonics to calculate the THD. The result is 48.2909%. As number of harmonics increases, the THD become closer to Multisim result.

```
Command Window
>> Vamplitude=5; n=[1 3 5 7 9 11 13 15 17]; V=2*Vamplitude/pi./n;
>> sqrt(0.5*sum(V(2:length(V)).^2))/(V(1)/sqrt(2))*100

ans =

    45.3818

fx >> |
```

Fig. 3.162 MATLAB calculations

```
Command Window
>> Vamplitude=5; n=[1:2:1001]; V=2*Vamplitude/pi./n;
>> sqrt(0.5*sum(V(2:length(V)).^2))/(V(1)/sqrt(2))*100

ans =

    48.2909

fx >> |
```

Fig. 3.163 MATLAB calculations

3.18   Example 17: Distortion Analysis of Amplifiers

We want to measure the THD of amplifiers in this example. Lower THD is better since it shows that the amplifier is linear and it does not produce harmonics during the amplification process. Assume the schematic shown in Fig. 3.164.

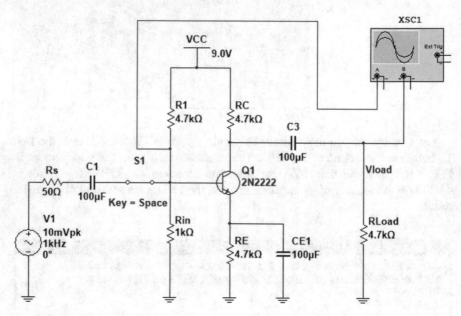

Fig. 3.164 Common emitter amplifier

Run the simulation. Put the cursor 1 at minimum of input and put the cursor 2 at maximum of input signal (Fig. 3.165).

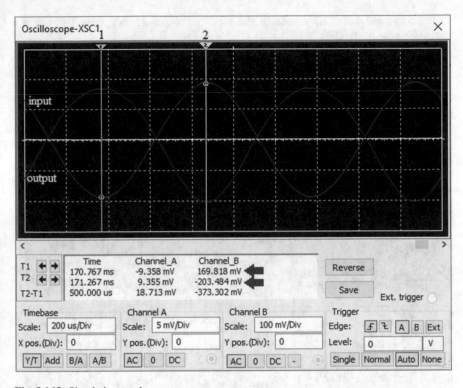

Fig. 3.165 Simulation result

Let's measure the voltage gain of the amplifier. The voltage gains are calculated in Fig. 3.166. The obtained gain values are not the same, this shows that the amplifier is not linear and distorts the input signal.

Fig. 3.166 MATLAB calculations

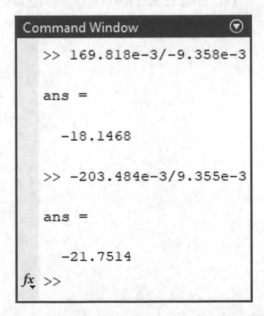

If you increase the amplitude of the input signal to 22 mV, you can see the distortion obviously (Fig. 3.167).

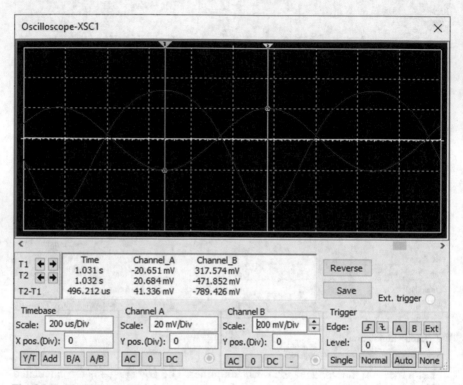

Fig. 3.167 Simulation result

Let's measure the THD of amplifier. Assume the schematic shown in Fig. 3.168.

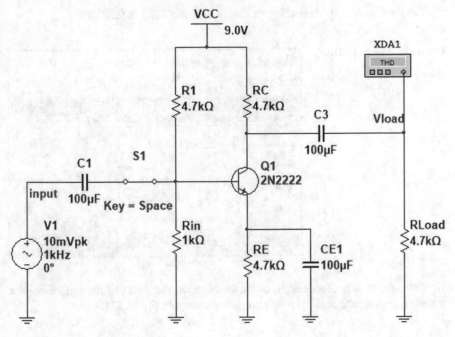

Fig. 3.168   Measurement of THD

Double click the distortion analyzer block and enter 1 kHz to the fundamental freq. Box. The frequency of input is 1 kHz, because of this fundamental freq. Box must be filled with 1 kHz (Fig. 3.169).

Fig. 3.169   Entering the fundamental frequency of the signal to the distortion analyzer

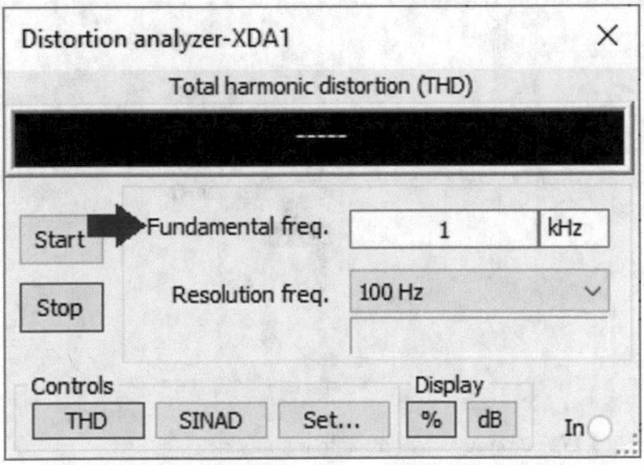

Run the simulation. According to Fig. 3.170, the THD is 9.802%.

Fig. 3.170 Measured THD

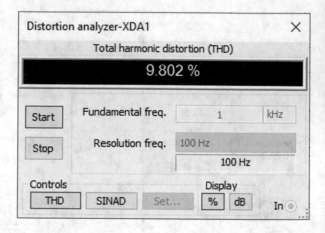

You need to use feedback in order to decrease the THD of the amplifier. For instance, assume the non-inverting amplifier shown in Fig. 3.171.

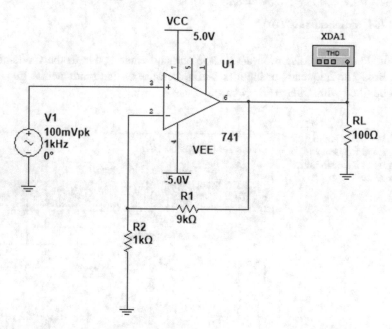

Fig. 3.171 Non inverting op amp amplifier

Run the simulation. According to Fig. 3.172, the THD is 0%.

Fig. 3.172 Measured THD
for amplifier in Fig. 3.171

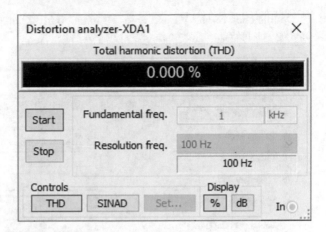

3.19   Example 18: Frequency Response of CE Amplifier

In example 13, we measured the voltage gain by simply dividing the load voltage to
the input voltage. We didn't consider the effect of frequency of input source there. In
this example, we want to obtain the frequency response (i.e. gain vs. frequency) of
the common emitter amplifier shown in Fig. 3.173. We use the AC sweep analysis to
obtain the frequency response of the amplifier.

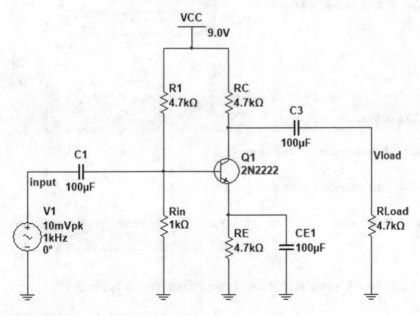

Fig. 3.173 Schematic of example 18

Double click on the V1 and enter 10 m and 0 to AC analysis magnitude and AC analysis phase boxes, respectively (Fig. 3.174).

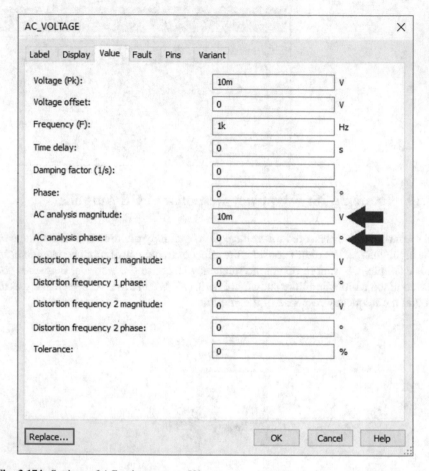

Fig. 3.174 Settings of AC voltage source V1

Click the interactive button (Fig. 3.175).

Fig. 3.175 Interactive button

Go to the AC sweep section and do the settings similar to Fig. 3.176.

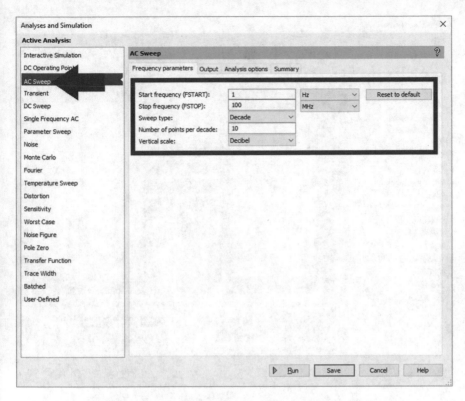

Fig. 3.176 Frequency parameter tab of AC sweep analysis

Go to the output tab, click the add expression button and enter V(vload)/V(input) to the expression box (Fig. 3.177) and click the OK button (Fig. 3.178). After clicking the OK button, the V(vload)/V(input) will be added to the right list.

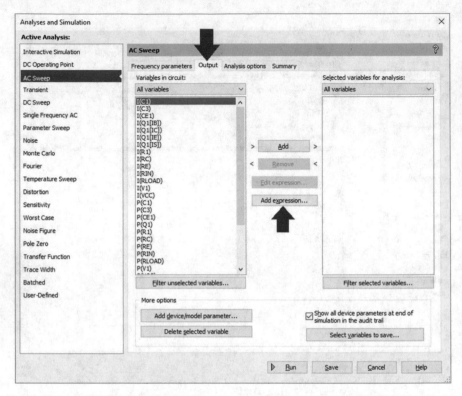

Fig. 3.177 Output tab of AC sweep analysis

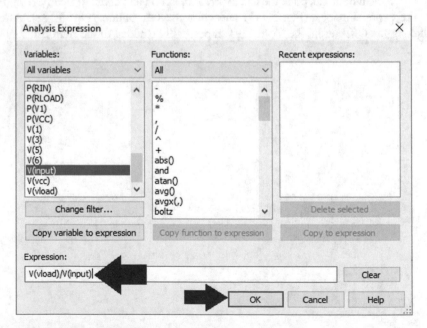

Fig. 3.178 Entering V(vload)/V(input) to the expression box

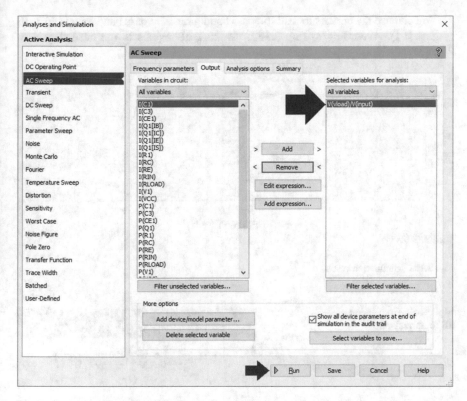

Fig. 3.179   V(vload)/V(input) is added to the right list

Run the simulation by clicking the run button in Fig. 3.179. The result of simulation is shown in Fig. 3.180. The behavior of the amplifier similar to a band pass filter.

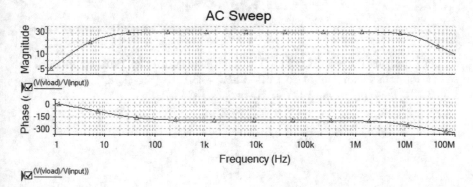

Fig. 3.180   Simulation result

You can use the cursors to read the pass band gain. The pass band gain is 26.1 dB (Fig. 3.181). According to Fig. 3.182, the 26.1 dB equals to the normal gain of 20.1837.

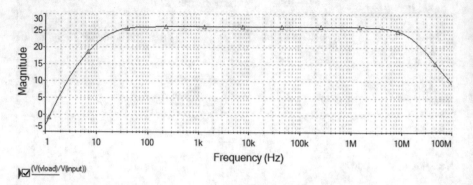

Fig. 3.181 Simulation result

Fig. 3.182 Conversion of dB to normal gain

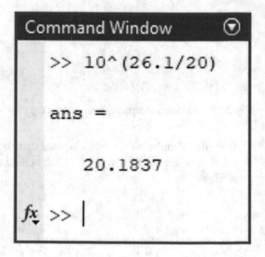

Let's measure the −3 dB points. The pass band gain is 26.1 dB, so we need to search for points with gain of 26.1 dB–3 dB = 23.1 dB (Fig. 3.183). According to Fig. 3.184, these points are 14.6691 Hz and 14.1434 MHz.

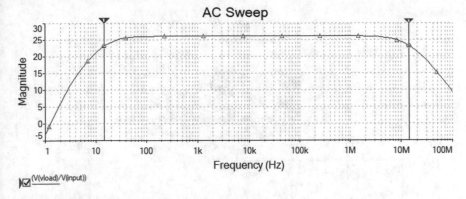

Fig. 3.183 Measurement of bandwidth of the amplifier

Fig. 3.184 Reading of
cursors in Fig. 3.183

Cursor	(V(vload)/V(input))
x1	14.6691
y1	23.1283
x2	14.1434M
y2	23.0975
dx	14.1434M
dy	−30.7507m
dy/dx	−2.1742n
1/dx	70.7043n

In real world, capacitance of Printed Circuit Board (PCB) traces and other parasitic capacitances makes it very difficult to reach upper cut off of 14.1434 MHz. For instance, assume that the parasitic capacitance between collector and emitter is 10 pF (Fig. 3.185). The frequency response of this amplifier is shown in Fig. 3.186. According to Fig. 3.187, the upper cut off frequency decreased to 4.6881 MHz. The pass band gain and lower cut off frequency is the same as the amplifier without capacitor C2.

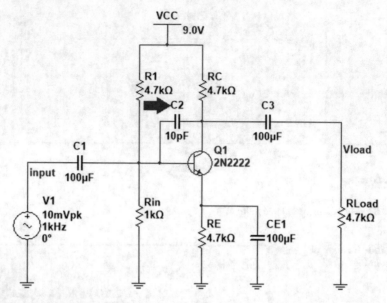

Fig. 3.185   Addition of a capacitor to the circuit

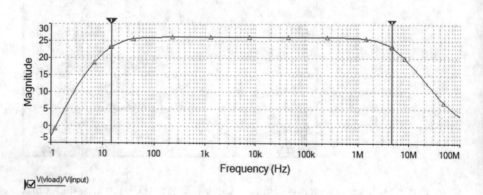

Fig. 3.186   Simulation result

Fig. 3.187 Reading of
cursors in Fig. 3.186

Cursor	x
	V(vload)/V(input)
x1	14.5506
y1	23.0915
x2	4.6881M
y2	23.0998
dx	4.6880M
dy	8.2629m
dy/dx	1.7626n
1/dx	213.3087n

3.20   Example 19: Input/Output Impedance of CE Amplifier

In example 14 we measured the input impedance by using the Ohm's law. We didn't
consider the effect of frequency of input source there. In this example, we want to
obtain the input impedance as a function of input source frequency. We use the AC
sweep analysis to obtain the input impedance of the amplifier. Assume the common
emitter amplifier shown in Fig. 3.188.

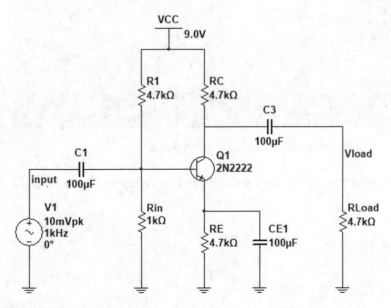

Fig. 3.188 Schematic of example 19

Click the interactive button (Fig. 3.189).

Fig. 3.189 Interactive
button

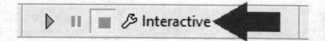

Go to AC sweep section and do the settings similar to Fig. 3.190. You can select logarithmic for the vertical scale box if you want to see the value of input impedance on the vertical axis.

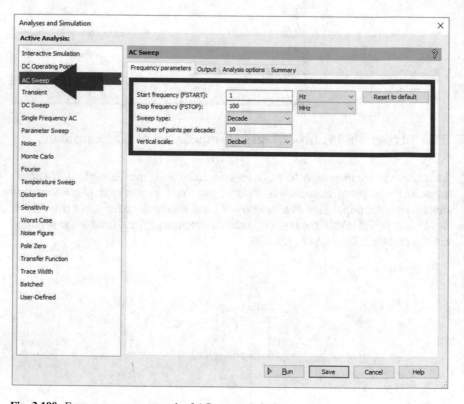

Fig. 3.190 Frequency parameters tab of AC sweep analysis

Go to the output tab and click the add expression button (Fig. 3.191).

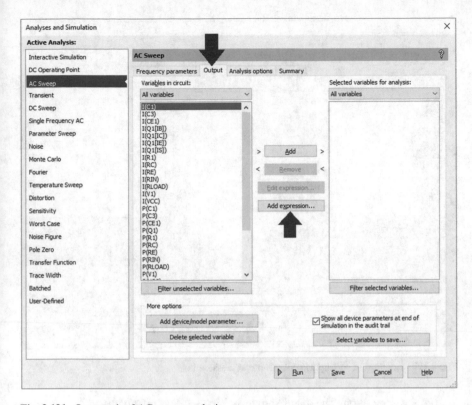

Fig. 3.191 Output tab of AC sweep analysis

Enter –V(input)/I(V1) to the expression box and click OK (Fig. 3.192). Note that I(V1) is the current that enters the positive terminal of V1. We need the current that exit from the positive terminal in order to calculate the input impedance. The current that exit from the positive terminal is –I(V1).

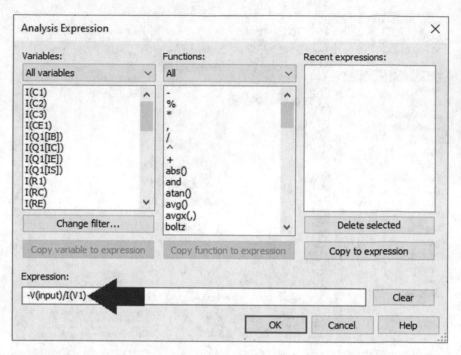

Fig. 3.192   Addition of −V(input)/I(V1) to the expression box

Run the simulation. The result shown in Fig. 3.193 is obtained.

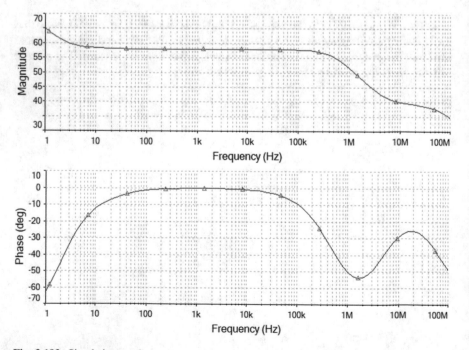

Fig. 3.193   Simulation result

You can use the cursors to read the graph. For instance, according to Fig. 3.194, the value of magnitude graph at 1 kHz is 57.9591 dB. If we convert this number from dB to Ohm (Fig. 3.195), we obtain 790.5967 Ω which is quite close to the result we obtained in example 14.

Fig. 3.194 Reading of cursor for 1 kHz

Cursor	x
	$-V(input)/I(V1)$
x1	1.0000k
y1	➤ 57.9591
x2	1.0000
y2	65.0770
dx	-999.0000
dy	7.1178
dy/dx	-7.1250m
1/dx	-1.0010m

Fig. 3.195 Conversion of dB to Ohm

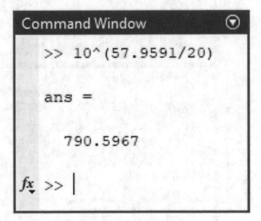

```
Command Window
    >> 10^(57.9591/20)

    ans =

       790.5967

fx >> |
```

The output impedance of the converter can be obtained in the same way. In order to measure the output impedance, connect the test source to the output node and rerun the simulation (Fig. 3.196).

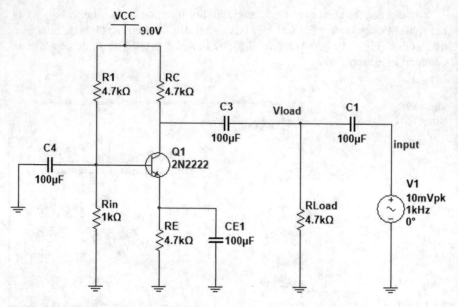

Fig. 3.196 Measurement of output impedance

The simulation result is shown in Fig. 3.197.

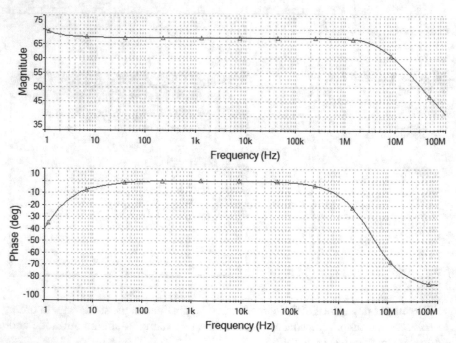

Fig. 3.197 Simulation result

You can use the cursors to read different points of the curve. For instance, according to Fig. 3.198, the impedance which is seen by test source V1 at 1 kHz is 67.109 dB. This value equals to 2.2670 kΩ (Fig. 3.199). The impedances which is seen by test source V1 is shown in Fig. 3.200. So, the parallel connection of Rload and Ro produced 2.2670 kΩ. According to calculations shown in Fig. 3.201, the value of Ro equals to 4.3793 kΩ. This value is quite close to the result we obtained in example 14.

Fig. 3.198 Reading of cursor for 1 kHz

Cursor		x
	−V(input)/I(V1)	
x1		1.0000k
y1		67.1090
x2		1.0000
y2		69.8985
dx		−299.0000
dy		2.7894
dy/dx		−9.3293m
1/dx		−3.3445m

Fig. 3.199 Conversion of dB to Ohm

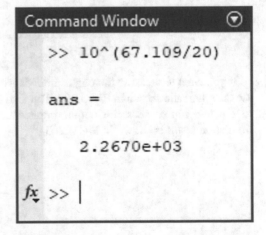

```
Command Window                  ⊙

    >> 10^(67.109/20)


    ans =


        2.2670e+03

fx >> |
```

Fig. 3.200 Equivalent circuit seen by source V1

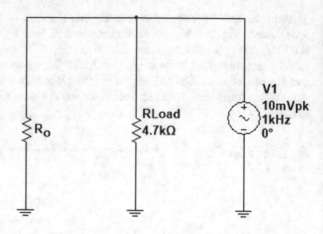

Fig. 3.201 MATLAB calculations

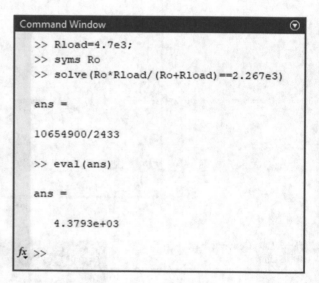

If you want to measure the output impedance of the amplifier only, then you can use the schematic shown in Fig. 3.202. The load resistor is removed from this circuit, so the test source sees the output impedance of the amplifier (Ro) only. The simulation result is shown in Fig. 3.203.

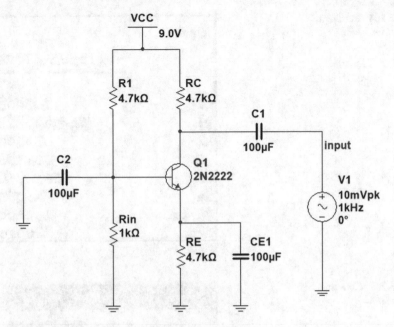

Fig. 3.202 Measurement of output impedance of amplifier

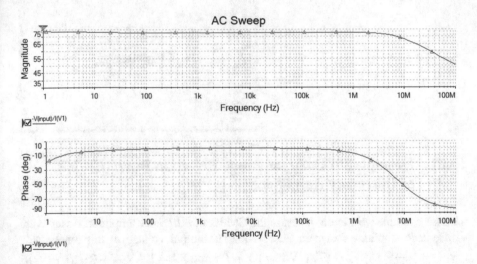

Fig. 3.203 Simulation result

According to Fig. 3.204, value of output impedance at 1 kHz is 72.8266 dB which equals to 4.3785 kΩ (Fig. 3.205). The obtained value is quite close to the number in Figs. 3.140 and 3.201.

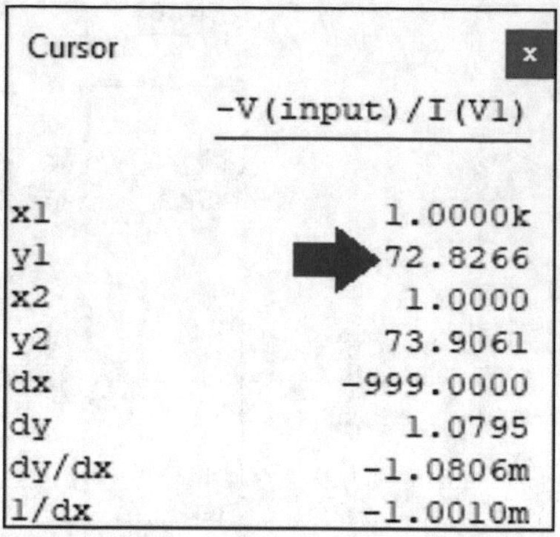

Fig. 3.204 Reading of
cursor for 1 kHz

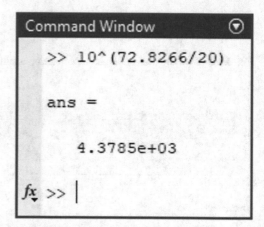

Fig. 3.205 Conversion of
dB to Ohm

3.21   Example 20: Common Mode Rejection Ratio (CMRR) of Difference Amplifier

In this example we want to measure the CMRR of difference amplifier. Assume the difference amplifier shown in Fig. 3.206. The output voltage of this amplifier is $V_{out} = \left(\frac{R_1+R_2}{R_3+R_4}\right)\frac{R_4}{R_1}V_2 - \frac{R_2}{R_1}V_1$. When R1 = R3 and R2 = R4, $V_{out} = \frac{R_2}{R_1}(V_2 - V_1)$.

Fig. 3.206 Difference amplifier

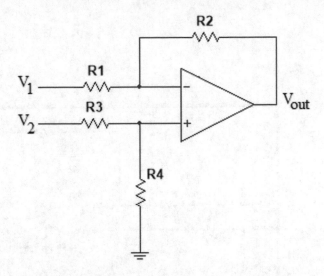

Assume that we want to measure the CMRR of the schematic shown in Fig. 3.207.

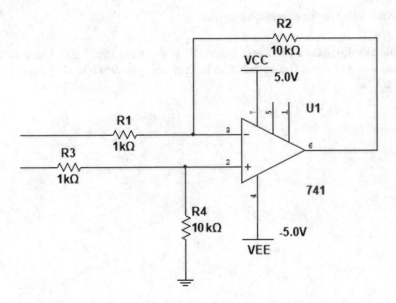

Fig. 3.207 Difference amplifier with $V_{out} = (V_2 - V_1)$

Let's measure the common mode gain with the aid of schematic shown in Fig. 3.208.

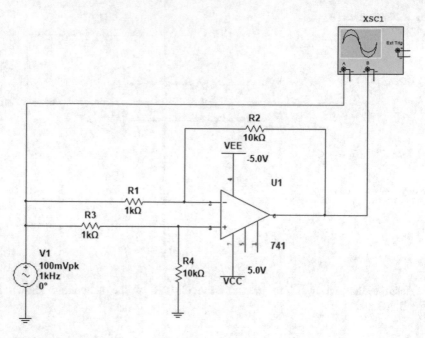

Fig. 3.208 Measurement of common mode gain

Run the simulation. The simulation result is shown in Fig. 3.209. Use the cursor to read value of input and output. The voltage gain is 3.0831×10^{-4} according to Fig. 3.210.

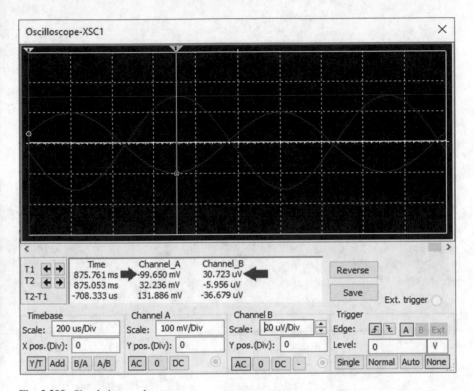

Fig. 3.209 Simulation result

Fig. 3.210 Calculation of common mode gain

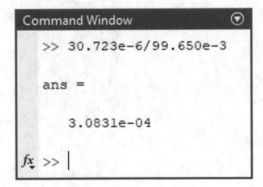

Use the schematic shown in Fig. 3.211 to measure the differential voltage gain. Note that there is 180° of phase difference between the AC sources V1 and V2. Settings of V1, V2 and voltage gain blocks are shown in Figs. 3.212, 3.213, and 3.214.

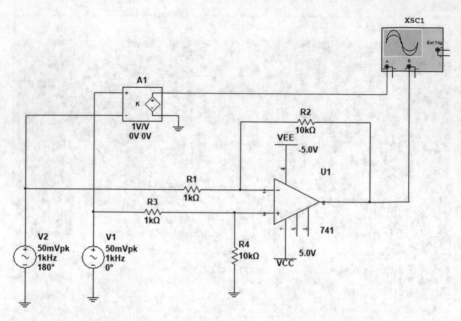

Fig. 3.211 Measurement of differential mode gain

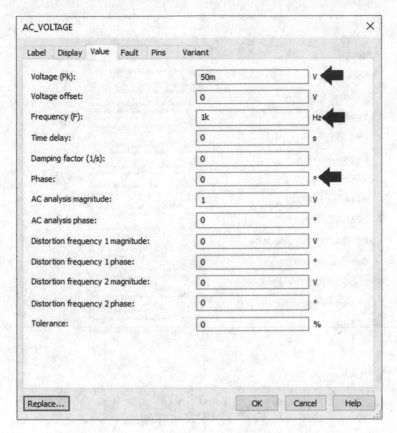

Fig. 3.212 Settings of AC voltage source V1

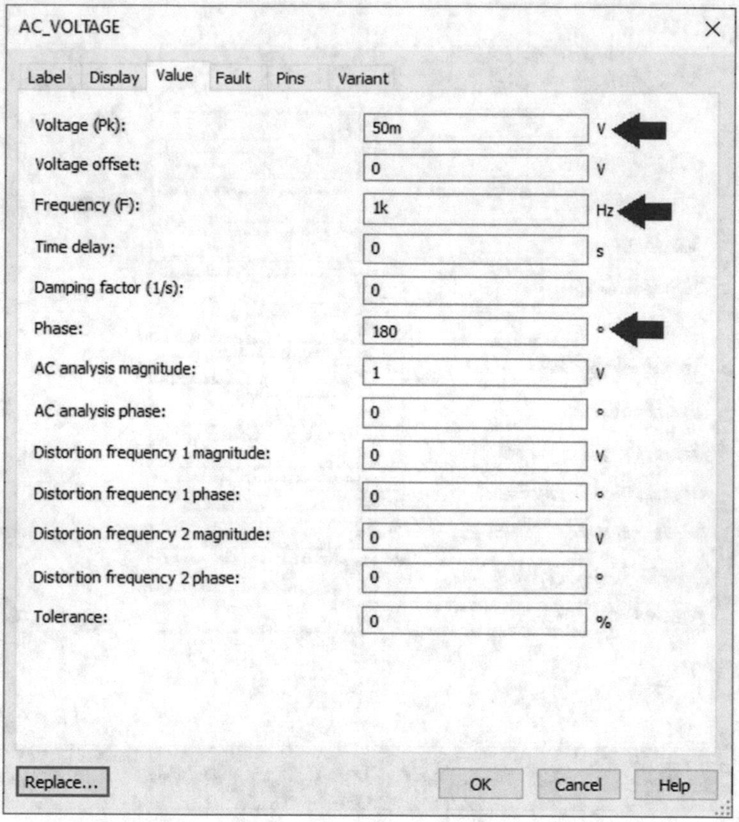

Fig. 3.213 Settings of AC voltage source V2

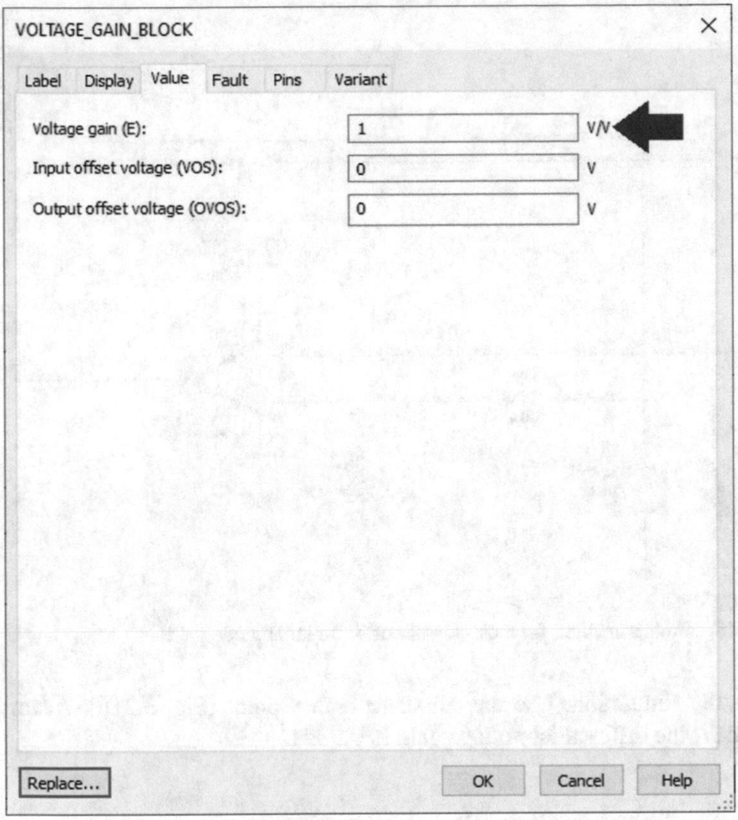

Fig. 3.214 Settings of voltage gain block

You can use the schematic shown in Fig. 3.215 to measure the differential voltage gain as well. This schematic is simpler than Fig. 3.211.

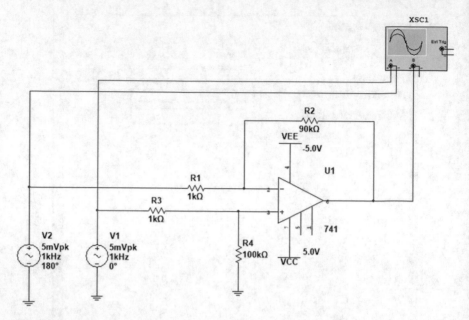

Fig. 3.215 Another method for measurement of differential mode gain

Run the simulation. Use the cursor to read a point (Fig. 3.216). According to Fig. 3.217, the differential voltage gain is 9.9894.

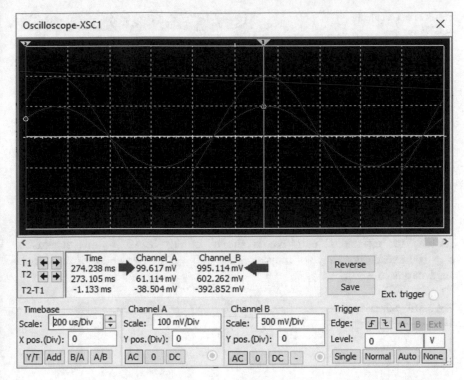

Fig. 3.216 Simulation result

Fig. 3.217 Calculation of
differential mode gain

According to the calculation in Fig. 3.218, the CMMR is 90.2110 dB.

Fig. 3.218 Calculation of CMRR

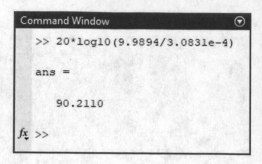

3.22   Example 21: CMRR of Differential Pair

We want to measure the CMRR of differential pair amplifier in this example. Assume the differential pair that is shown in Fig. 3.219. In this schematic, $\beta = 100$ and Early voltage of transistors are 100 V. The two collector resistances are accurate to within $\pm 1\%$.

Fig. 3.219 Schematic of example 21

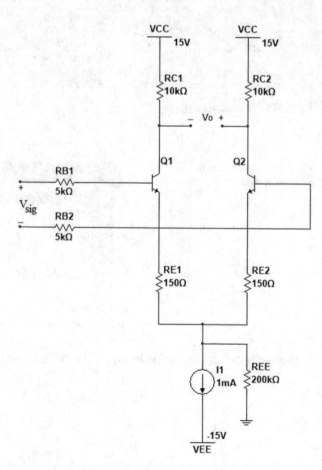

We need to use transistor virtual element (Fig. 3.220) since a specific transistor number is not given and instead β and Early voltage of transistors are given.

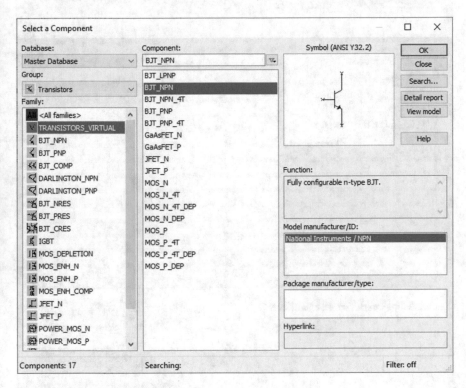

Fig. 3.220 Virtual NPN transistor

Double click the transistors in the schematic to set the β and Early voltage. Go to the value tab and click the edit model button (Fig. 3.221).

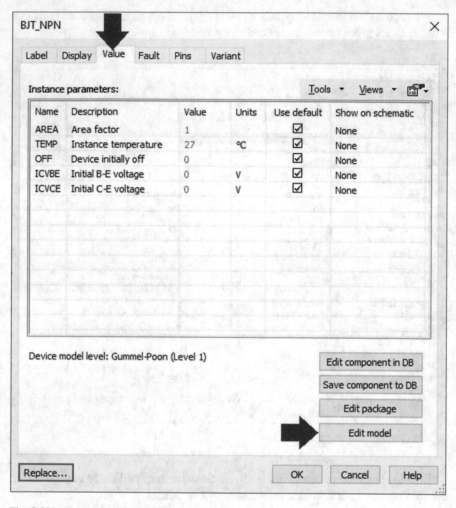

Fig. 3.221 Value tab of BJT_NPN window

Double click the BF field and enter the given value of β there. Double click the VAF field and enter the given value of Early voltage there. Then click the change component button (Fig. 3.222).

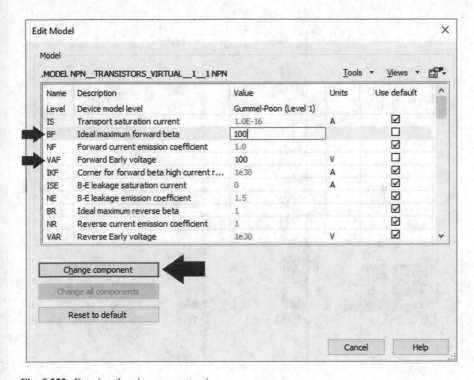

Fig. 3.222 Entering the given current gain

Figure 3.223 shows the DC analysis of the circuit.

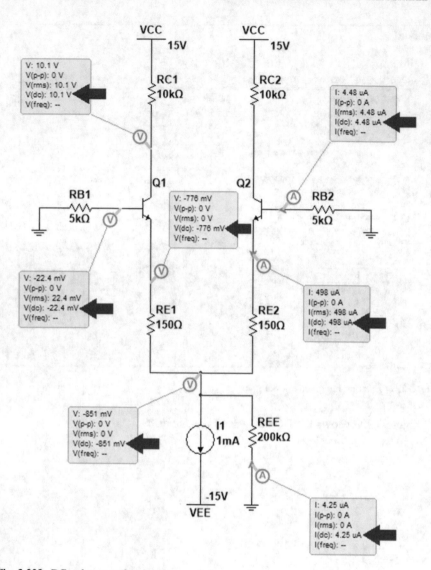

Fig. 3.223 DC voltages and currents

Let's measure the differential voltage gain first. Figure 3.224 shows the schematic which measures the differential voltage gain. Close up of input stage is shown in Fig. 3.225.

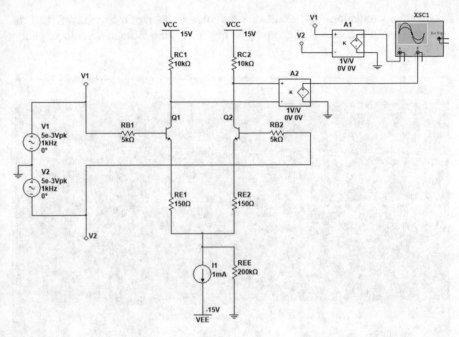

Fig. 3.224 Measurement of differential gain

Fig. 3.225 V1 and V2
sources used for differential
mode gain measurement

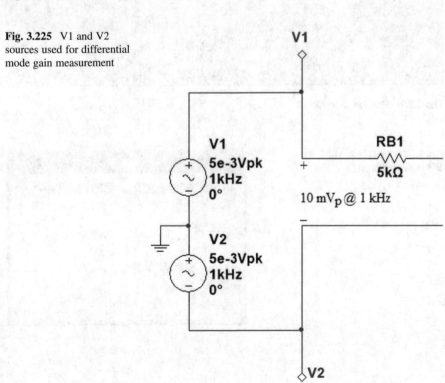

Run the simulation. The result of simulation is shown in Fig. 3.226. Use the cursor to read the channels. According to Fig. 3.227, the differential voltage gain is 39.3719.

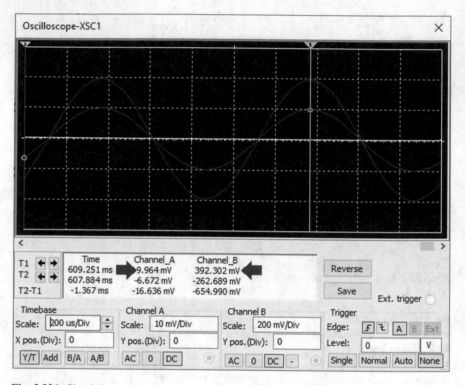

Fig. 3.226 Simulation result

Fig. 3.227 Calculation of differential mode gain

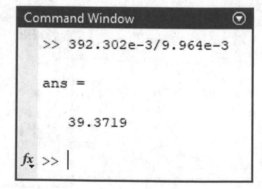

Let's measure the worst-case common-mode gain with the assumption that the two collector resistances are accurate to within $\pm 1\%$. Double click the resistor RC1 and change the resistance (R) box to 10 k*0.99 (Fig. 3.228). Double click the resistor RC2 and change the resistance (R) box to 10 k*1.01 (Fig. 3.229).

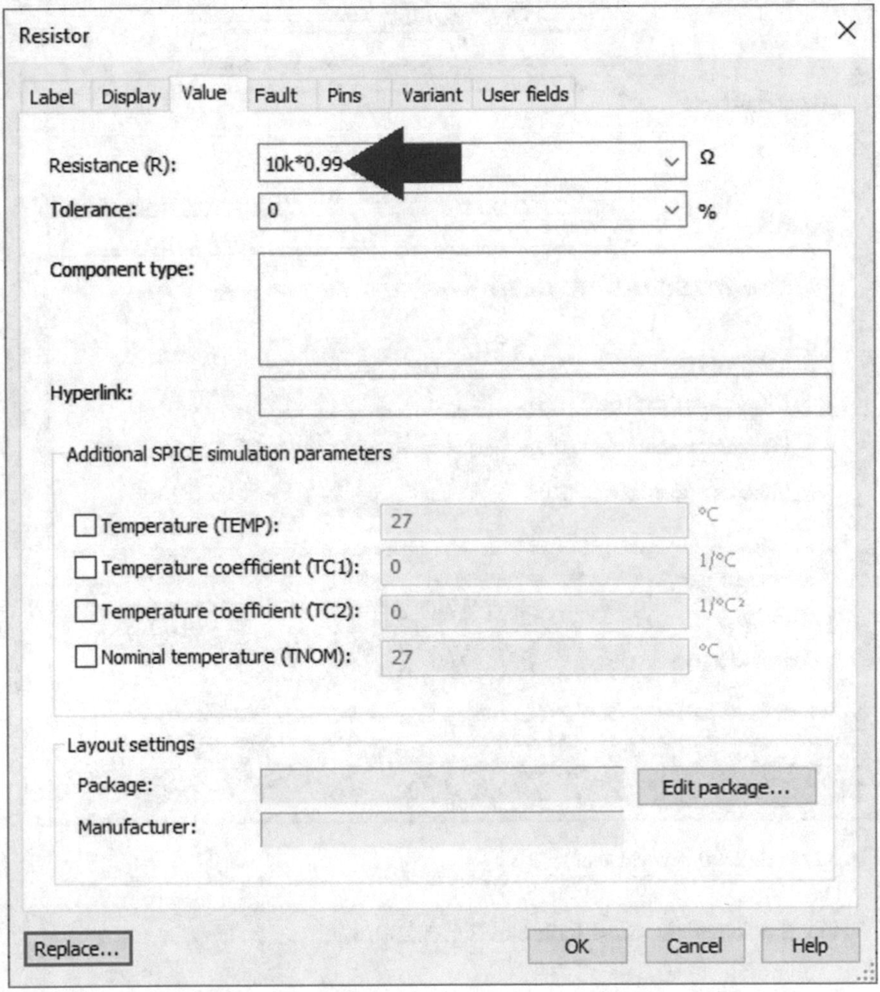

Fig. 3.228   Entering new value of RC1

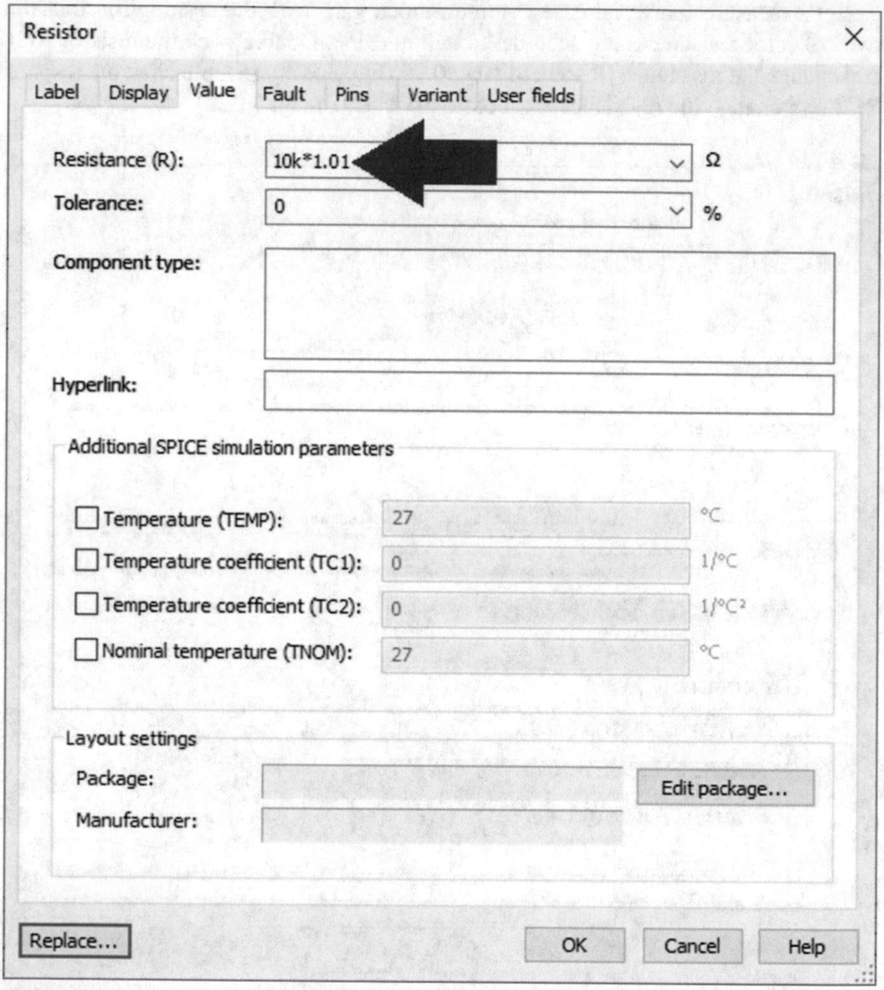

Fig. 3.229 Entering new value of RC2

Now the schematic must look like Fig. 3.230.

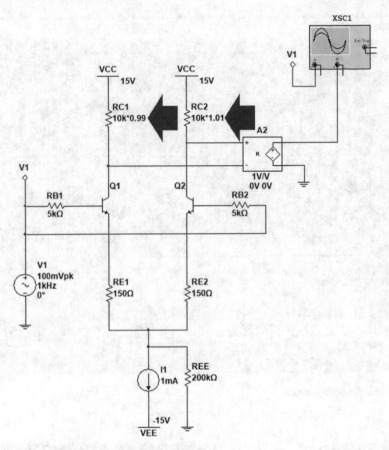

Fig. 3.230 New values of RC1 and RC2 are shown on the schematic

Run the simulation. Use the cursor to read channels (Fig. 3.231). According to Fig. 3.232, the worst common mode gain is 4.5211×10^{-4}. Note that this value shows the maximum value of common mode gain for 1% resistors. If the accuracy of resistors increases, then the worst common mode gain decrease.

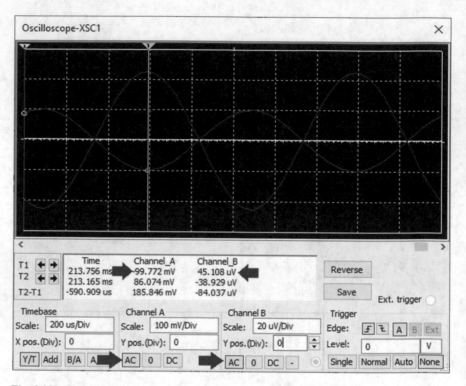

Fig. 3.231 Simulation result

Fig. 3.232 Calculation of
common mode gain

According to Fig. 3.233, the CMRR is 98.7988 dB.

Fig. 3.233 Calculation of CMRR

```
Command Window                        ⊙

  >> 20*log10(39.3719/4.5211e-4)

  ans =

      98.7988

fx >> |
```

3.23 Example 22: Measurement of Differential Mode Input Impedance for Differential Pair

In this example we want to measure the differential mode input impedance of previous example. The equivalent circuit for differential inputs is shown in Fig. 3.234.

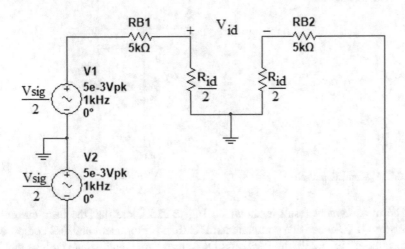

Fig. 3.234 Input stage of differential amplifier of Fig. 3.224

Let's use a current probe to measure the current drawn from the sources (Fig. 3.235).

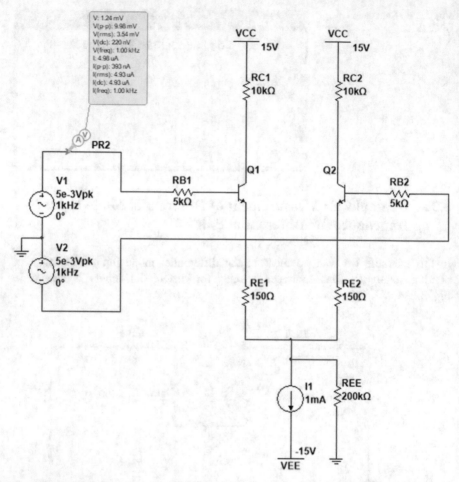

Fig. 3.235 Simulation result

The measurement result is shown in Fig. 3.236. Note that the base current of transistor Q1 (Q2) has two components: DC (bias) component and AC component. According to Fig. 3.236, the RMS and DC current are equal, so we deduce that the DC component is much bigger than the AC component. Therefore, we can't use probes to measure the AC component of the current.

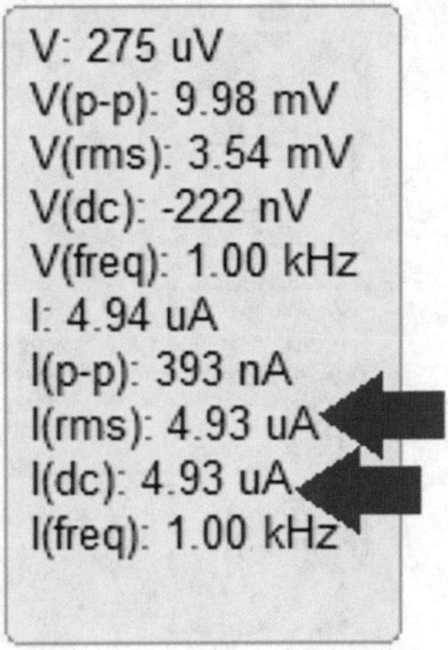

Fig. 3.236 Probe readings

V: 275 uV
V(p-p): 9.98 mV
V(rms): 3.54 mV
V(dc): -222 nV
V(freq): 1.00 kHz
I: 4.94 uA
I(p-p): 393 nA
I(rms): 4.93 uA
I(dc): 4.93 uA
I(freq): 1.00 kHz

Let's use the schematic shown in Fig. 3.237. We used a current to voltage converter to see the current on the oscilloscope. According to Fig. 3.238, the peak of AC component of current is 196 nA.

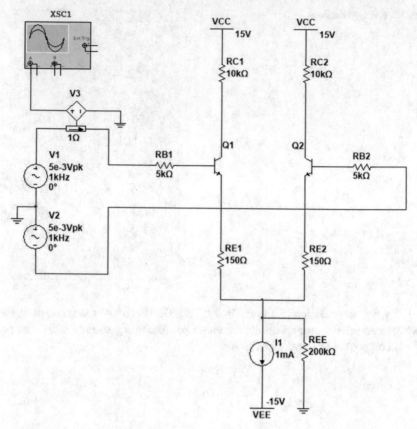

Fig. 3.237   Measurement of current for differential mode

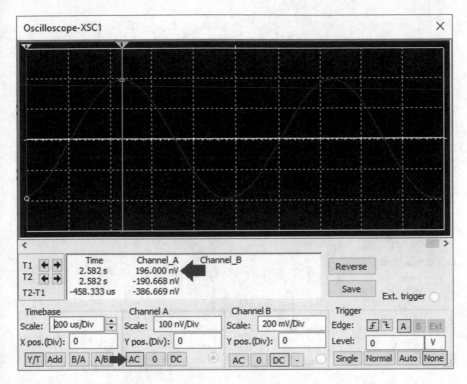

Fig. 3.238 Simulation result

The calculations shown in Fig. 3.239 shows that $\frac{R_{id}}{2} = 20.5102\ k\Omega$ or $R_{id} = 41.0204\ k\Omega$.

Fig. 3.239 Calculation of R_{id}

```
Command Window
>> (5e-3/196e-9)/1e3

ans =

    25.5102
>> (ans-5)*2

ans =

    41.0204

fx >>
```

Let's measure the common mode input impedance. We use the schematic shown in Fig. 3.240.

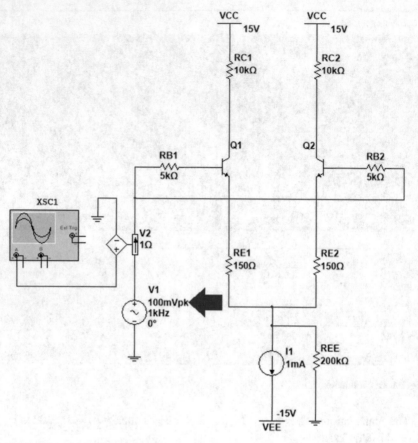

Fig. 3.240 Measurement of current for common mode

Simulation result is shown in Fig. 3.241. The peak of AC current is 12.739 nA. According to Fig. 3.242, the input impedance of common mode is 7.8499 MΩ.

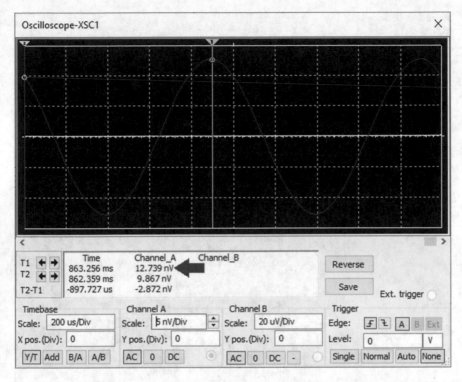

Fig. 3.241 Simulation result

Fig. 3.242 Calculation of input impedance

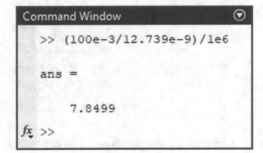

3.24 Example 23: Sub Circuit Block

Sub circuit permits you to hide details and permits you to produce more understandable schematics. The sub circuit is very useful for large designs.

Let's study this block with an example. Assume that you want to design an amplifier. Let's divide the schematic into two sub circuits: The power supply sub circuit and the amplifier sub circuit. Let's start with the power supply sub circuit. Click the place> new hierarchical block (Fig. 3.243).

Fig. 3.243 Place> new
hierarchical block

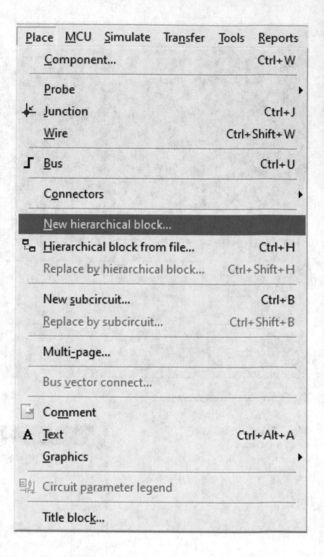

After clicking the place> new hierarchical block, the window shown in Fig. 3.244 appears. Enter the name of sub circuit and number of input and outputs to this window and click the OK button. The schematic changes to what shown in Fig. 3.245.

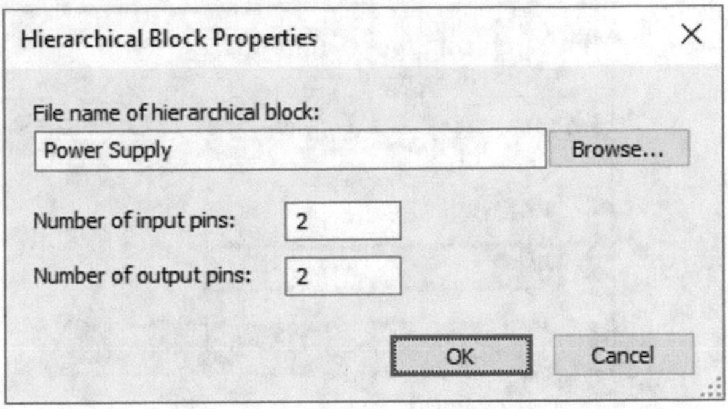

Fig. 3.244 Hierarchical
block properties window

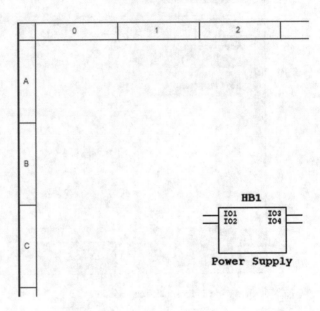

Fig. 3.245 Sub circuit
block is added to the
schematic

Double click the power supply sub circuit and click the open subsheet (Fig. 3.246).
Window shown in Fig. 3.247 appears.

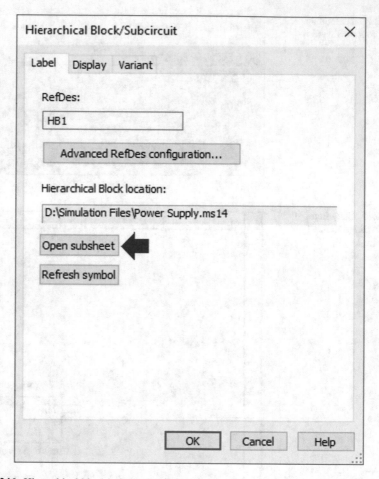

Fig. 3.246 Hierarchical block/subcircuit window

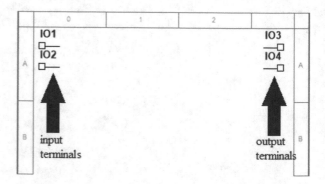

Fig. 3.247 Input and output terminals inside the subcircuit block

Double click on the terminals and rename them. For instance, double click the IO1 and enter the new name In1 to the name box (Fig. 3.248). Change the name IO2 to In2, IO3 to VCC and IO4 to VEE.

Fig. 3.248 Hierarchical connector window

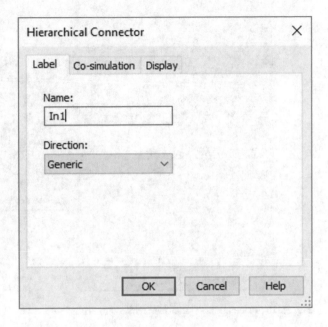

Draw the schematic of power supply (Fig. 3.249).

Fig. 3.249 Schematic in the power supply sub circuit

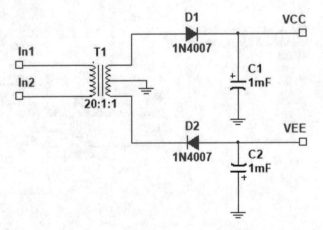

You can add input/output terminal to the schematic if you need. Use the place> connectors> hierarchical connector to add terminal (Fig. 3.250).

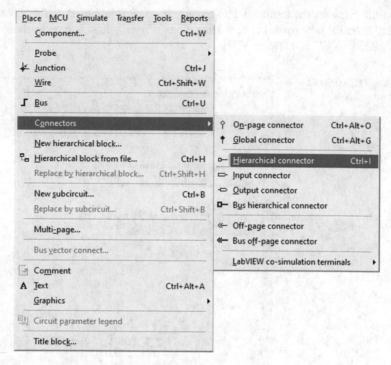

Fig. 3.250 Place> connectors> hierarchical connector

Connect the added new terminal to where you want (Fig. 3.251).

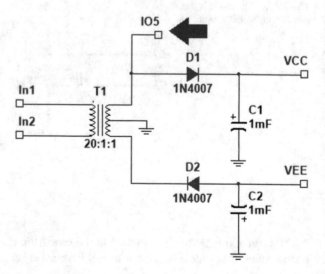

Fig. 3.251 Connecting a
hierarchical connector to
a node

Double click the added new terminal and set the name and type (input or output) of terminal and click the OK button (Fig. 3.252). The added terminal to Fig. 3.251 is of output type, so output must be selected. After clicking the OK button, the schematic changes to what shown in Fig. 3.253.

Fig. 3.252 Entering the name and type of terminal

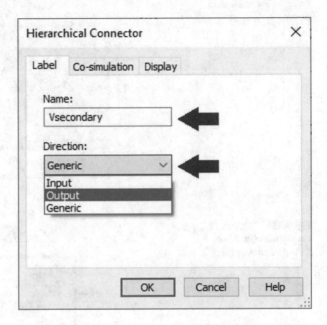

Fig. 3.253 The hierarchical connector name is shown on the schematic

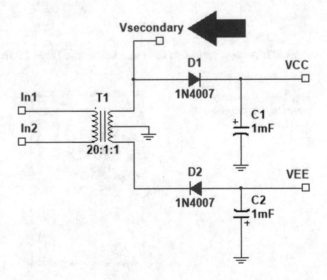

Click the close button (Fig. 3.254) to close the sub circuit and return to the main schematic. The main schematic looks like Fig. 3.255.

Fig. 3.254 Close button to return to the schematic editor

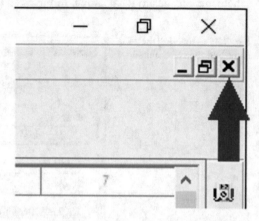

Fig. 3.255 The power supply sub circuit has 2 inputs and 3 outputs

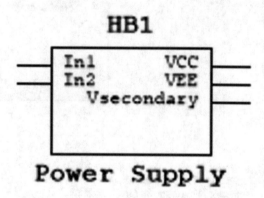

Add a voltage source to the main schematic (Fig. 3.256).

Fig. 3.256 Connecting the power supply sub circuit to an AC source

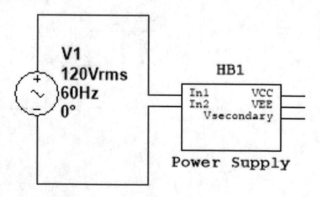

Add another sub circuit block to the main schematic (Fig. 3.257).

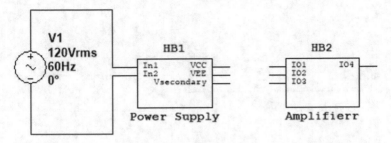

Fig. 3.257 Addition of a new sub circuit block to the schematic

Double click the amplifier sub circuit block and click the open sheet button (Fig. 3.258). Then draw the schematic shown in Fig. 3.259. After drawing the schematic, close it and return to the main schematic (Fig. 3.260).

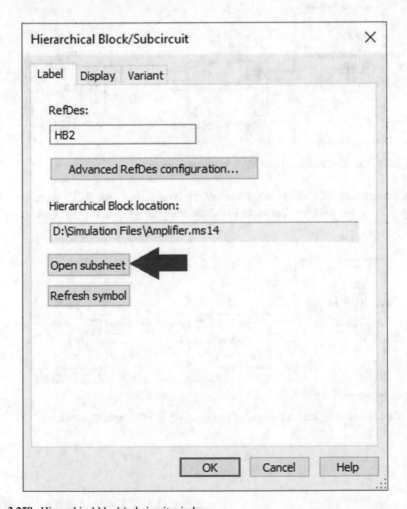

Fig. 3.258 Hierarchical block/subcircuit window

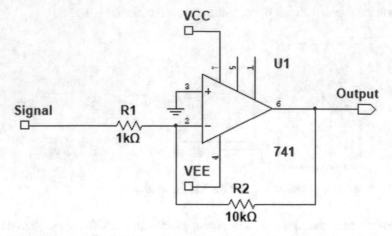

Fig. 3.259 Schematic in the amplifier sub circuit

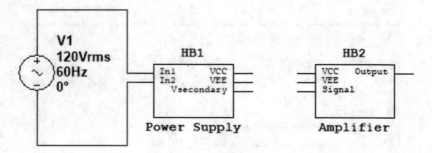

Fig. 3.260 The amplifier sub circuit has 3 inputs and 1 output

Connect the VCC lines together. After connecting the VCC lines together (Fig. 3.261), the window shown in Fig. 3.262 appears. Click the OK button.

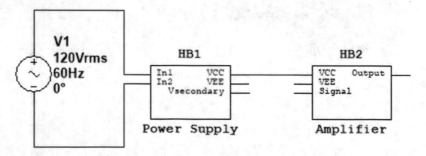

Fig. 3.261 Connecting the VCC output of power supply to VCC input of amplifier

Fig. 3.262 Resolve net
name conflict window

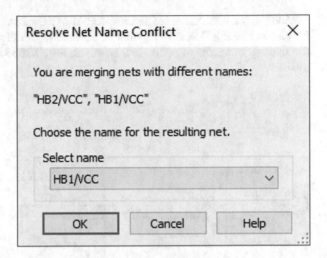

Connect the VEE lines together. After connecting the VEE lines together
(Fig. 3.263), the window shown in Fig. 3.264 appears. Click the OK button.

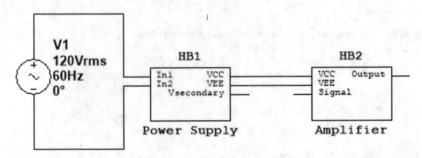

Fig. 3.263 Connecting the VEE output of power supply to VEE input of amplifier

Fig. 3.264 Resolve net
name conflict window

Add the V2 voltage source and resistor R1 to the main schematic. The schematic now looks like Fig. 3.265. Compare this schematic with Fig. 3.266. Both of them are the same from Multisim point view however, Fig. 3.265 is easier to understand for the user.

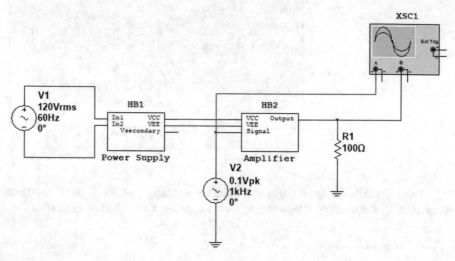

Fig. 3.265 The completed amplifier circuit with sub circuit blocks

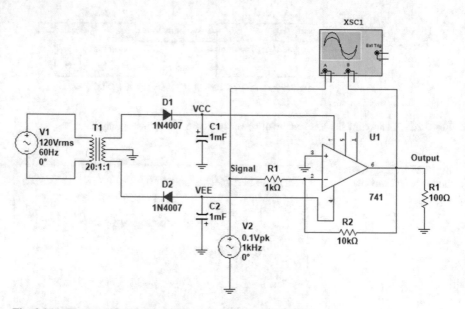

Fig. 3.266 The amplifier circuit without using sub circuit block

Run the simulation. The simulation result is shown in Fig. 3.267. According to Fig. 3.267, the voltage gain is about 10.

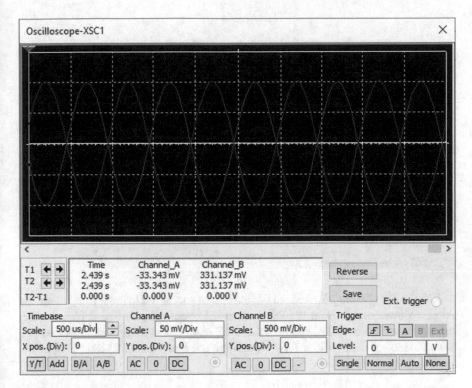

Fig. 3.267 Simulation result

3.25   Example 24: Efficiency of the Power Amplifier

In this example, we want to measure the efficiency of the power amplifier shown in Fig. 3.268.

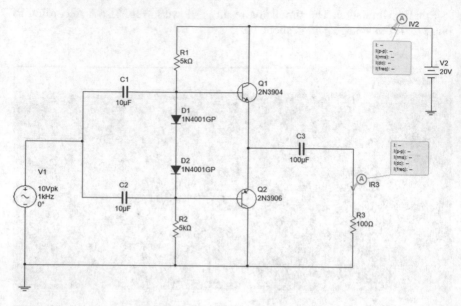

Fig. 3.268 Schematic of example 24

Run the simulation. We need the DC current of source V2 in order to calculate the average power drawn from the DC source V2. According to Fig. 3.269, the DC current drawn from the DC source V2 is 31.9 mA. The average load power can be calculated using the $P = R \times I_{RMS}^2$ equation. The load current RMS is 67.7 mA.

Fig. 3.269 Simulation
result

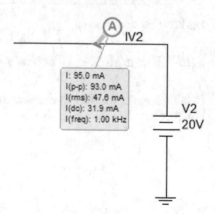

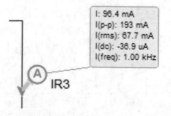

The efficiency is 71.84% according to Fig. 3.270.

Fig. 3.270 Calculation of efficiency

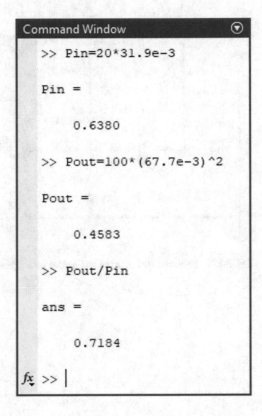

```
Command Window                                ⊙
>> Pin=20*31.9e-3

Pin =

     0.6380

>> Pout=100*(67.7e-3)^2

Pout =

     0.4583

>> Pout/Pin

ans =

     0.7184

fx >> |
```

You can measure the efficiency using the power probes as well (Fig. 3.271). According to Fig. 3.272, the DC source V2 average power is 638 mW and the average load power is 458 mW. These values are the same as shown in Fig. 3.270.

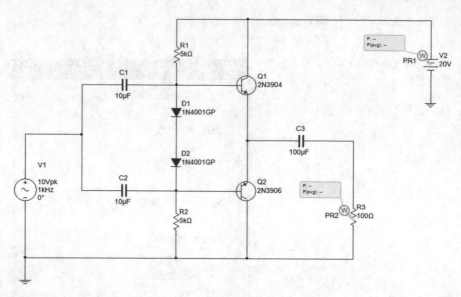

Fig. 3.271 Measurement of input and output power with power probes

Fig. 3.272 Simulation
result

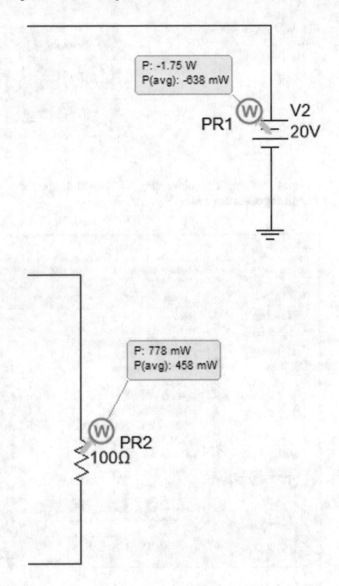

3.26   Example 25: Wien Bridge Oscillator and Colpitts Oscillator

Wien bridge oscillator and Colpitts oscillator circuits are available in the sample folder of Multisim. Click the open sample icon (Fig. 3.273).

Fig. 3.273 Open samples
button

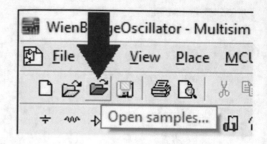

Open the Analog folder (Fig. 3.274) and open the ColpittsOscillator and
WienBridgeOscillator files (Fig. 3.275).

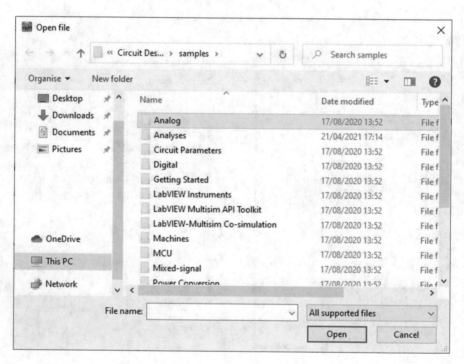

Fig. 3.274 Open file window

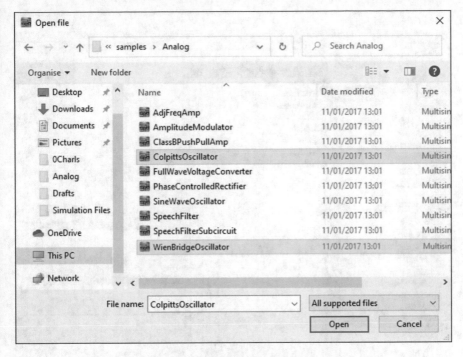

Fig. 3.275 Colpitts oscillator and Wien bridge oscillator samples

3.27 Example 26: Relaxation Oscillator

Schematic of a relaxation oscillator is shown in Fig. 3.276. This schematic uses a 2 N6027 PUT (Fig. 3.277).

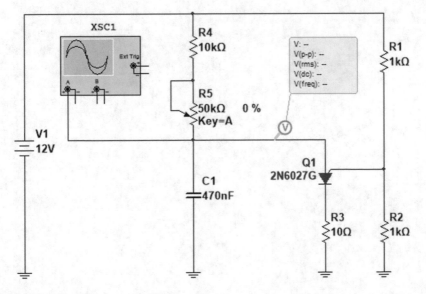

Fig. 3.276 Schematic of example 26

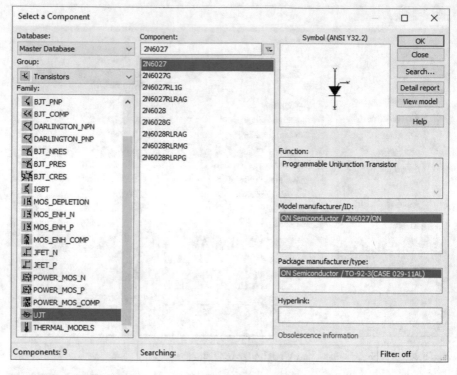

Fig. 3.277 2 N6027 Programmable Unijunction Transistor (PUT)

Run the simulation. Decrease the value of potentiometer R5 to 0%. The wave-form is shown in Fig. 3.278. The frequency of the load voltage is about 294 Hz.

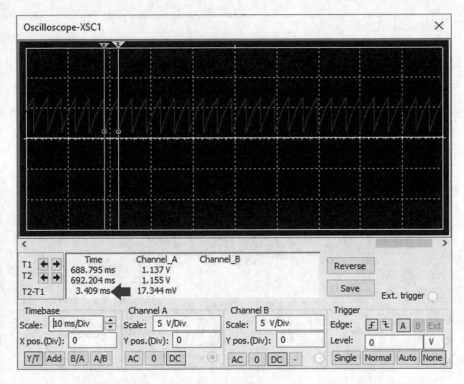

Fig. 3.278 Simulation result (Potentiometer R5 is 0%)

Increase the potentiometer to 100%. Frequency of the load voltage (Fig. 3.279) decreases. The frequency of the load voltage is about 50 Hz.

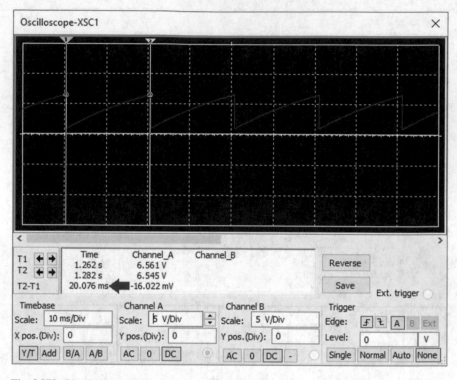

Fig. 3.279 Simulation result (Potentiometer R5 is 100%)

3.28   Example 27: Circuits Wizards

Multisim has a circuit wizard (Fig. 3.280) which permits you to simulate design specific type of circuits automatically.

Fig. 3.280 Tools > circuit
wizards

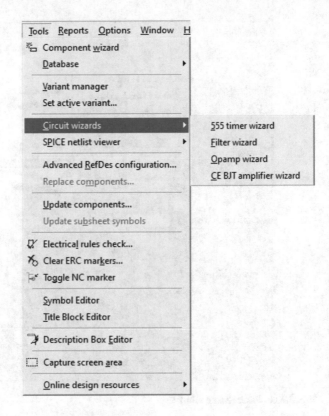

Let's study one of the wizards. For instance, assume that you want to design an active high pass filter. To do this, click the tools> circuit wizards> filter wizard. Do the settings similar to Fig. 3.281 and click the verify button. Wait until the "calculation was successfully completed" message appears (Fig. 3.282).

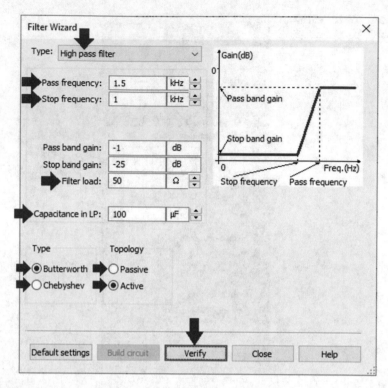

Fig. 3.281 Filter wizard window

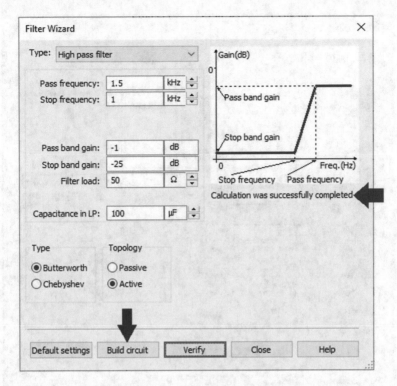

Fig. 3.282 Build circuit button make a circuit based on your inputs

Click the build circuit button in Fig. 3.282 and click on schematic editor to paste the designed circuit to schematic editor.

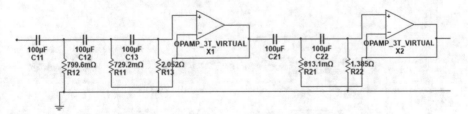

Fig. 3.283 Designed circuit (C = 100 μF)

Value of the resistor is very low in Fig. 3.283. If we decrease the value of capacitance box to 100 nF (Fig. 3.284), the resistor values increase (Fig. 3.285). We decreased the capacitor by factor of k = 1000 and this cause the resistor values to increase by factor of 1000.

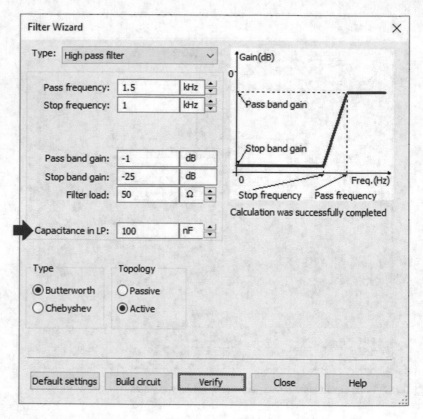

Fig. 3.284 Decreasing the capacitor value to 100 nF

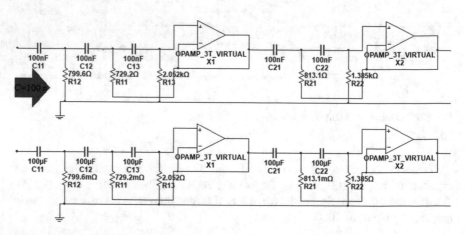

Fig. 3.285 Designed circuit (C = 100 nF)

If you multiply the values of resistor of a circuit to k, multiply the values of inductors of a circuit to k and divide the values of capacitors of a circuit to k, the transfer function of the circuit will not change. Let's see this subject with an example. Assume the simple RLC circuit shown in Fig. 3.286.

Fig. 3.286 Simple RLC circuit

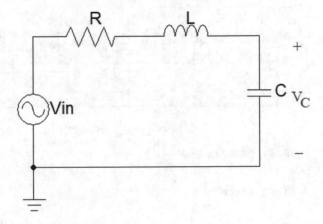

The transfer function of this circuit is $\frac{V_C(s)}{V_{in}(s)} = \frac{\frac{1}{Cs}}{R+Ls+\frac{1}{Cs}} = \frac{1}{LCs^2+RCs+1}$. Now, consider the circuit shown in Fig. 3.287.

Fig. 3.287 Scaled version of circuit in Fig. 3.286

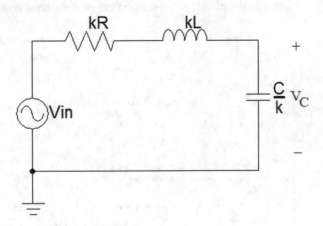

The transfer function of this circuit is $\frac{V_C(s)}{V_{in}(s)} = \frac{\frac{1}{\frac{C}{k}s}}{kR+kLs+\frac{1}{\frac{C}{k}s}} = \frac{1}{LCs^2+RCs+1}$. So, the transfer function of Figs. 3.286 and 3.287 are the same. You can use AC sweep to see the frequency response of designed filters (Fig. 3.288).

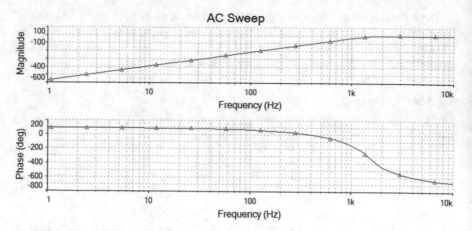

Fig. 3.288 Simulation result

3.29 Exercises

1. Simulate the half wave rectifier circuit (Fig. 3.71) with RL load (R = 10 Ω and L = 10 mH). Compare the result with purely resistive load.
2. Figure 3.289 shows an op-amp clamp circuit with a non-zero reference clamping voltage. The clamping level is at precisely the reference voltage. Use Multisim to simulate the circuit and see the effect of ReferenceVoltage source on output.

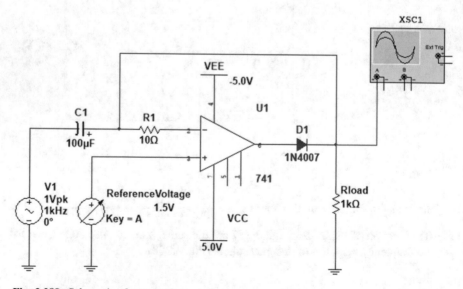

Fig. 3.289 Schematic of exercise 2

3. Measure the maximum output voltage swing of the circuit shown in Fig. 3.118.
4. Assume the amplifier shown in Fig. 3.290.

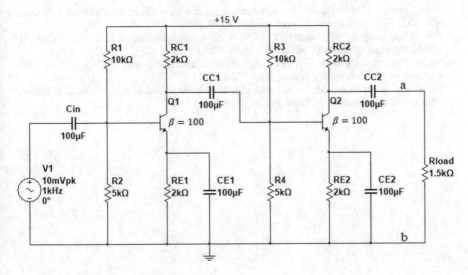

Fig. 3.290 Schematic of exercise 4

(a) Use hand analysis to calculate the DC voltages (operating point) of the circuit.
(b) Use Multisim to verify results of part (a).
(c) Use hand analysis to calculate the input impedance (impedance seen from source V1) and output impedance (impedance seen from points a and b) of the circuit.
(d) Use Multisim to verify part (c).
(e) Use hand analysis to calculate the overall gain ($\frac{V_{ab}}{V_1}$) of the circuit.
(f) Use Multisim to verify part (e).

5. Assume that both transistors in Fig. 3.290 are 2 N2222. Use Multisim to draw the:

   (a) Input impedance as a function of frequency.
   (b) Output impedance as a function of frequency.
   (c) Overall gain of the system a function of frequency.
   (d) Measure the THD of output voltage.

Further Readings

1. B. Razavi, *Fundamentals of Microelectronics*, 3rd edn. (Wiley, 2021)
2. M.H. Rashid, *Microelectronic Circuits: Analysis and Design* (Cengage Learning, 2016)
3. A. Sedra, K. Smith, T.C. Carusone, V. Gaudet, *Microelectronic Circuits*, 8th edn. (Oxford University Press, 2019)

4. F. Asadi, K. Eguchi, *Electronic Measurements: A Practical Approach* (Morgan & Claypool, 2021)
5. Crystal oscillator: https://www.youtube.com/watch?v=w4dvw9PT97k (Visiting date: 15.07.2021)
6. Noise analysis in Multisim: https://knowledge.ni.com/KnowledgeArticleDetails?id=kA03 q000000YG05CAG&l=en-TR (Visiting date: 15.07.2021)
7. Pole Zero analysis with Multisim: https://zone.ni.com/reference/en-XX/help/372062L-01/ multisim/polezeroanalysis/ (Visiting date: 15.07.2021)
8. Worst case analysis with Multisim: https://zone.ni.com/reference/en-XX/help/375482B-01/ multisim/worstcaseanalysis/ (Visiting date: 15.07.2021)
9. Distortion analysis in Multisim: https://zone.ni.com/reference/en-XX/help/375482B-01/ multisim/distortionanalysis/ (Visiting date: 15.07.2021)

Chapter 4
Simulation of Digital Circuits with Multisim™

4.1 Introduction

Digital electronics is a branch of electronics which deals with digital signals to perform the various task to meet various requirement. The input signal applied to these circuits is of digital form, which is represented in 0's and 1's binary language format. These circuits are designed by using logical gates like AND, OR, NOT, NANAD, NOR, XOR gates which perform logical operations.

This chapter focus on the simulation of digital circuits with Multisim.

4.2 Example 1: Simulation of a Full Adder

We want to simulate a simple digital circuit in this example. A great deal of digital IC's is modeled in Multisim (Figs. 4.1 and 4.2).

© The Author(s), under exclusive license to Springer Nature Switzerland AG 2022   611
F. Asadi, *Essential Circuit Analysis using NI Multisim™ and MATLAB®*,
https://doi.org/10.1007/978-3-030-89850-2_4

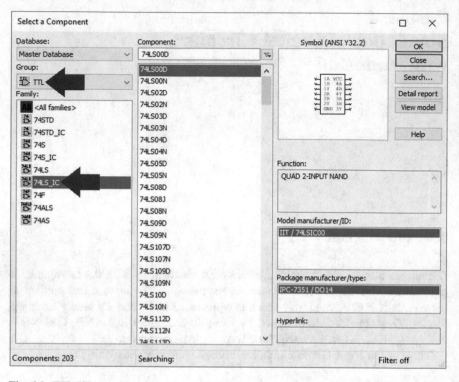

Fig. 4.1 TTL IC's

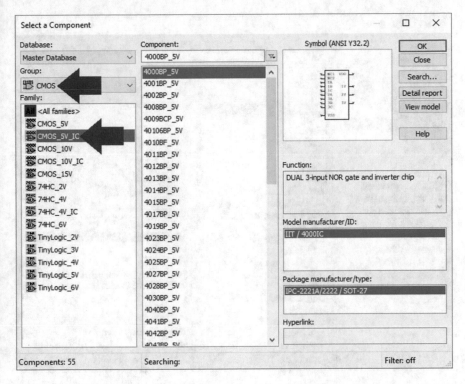

Fig. 4.2 CMOS IC's

We want to simulate a full adder circuit. The full adder truth table is shown in Table 4.1.

Table 4.1 Truth table for a full adder

Inputs			Outputs	
A	B	C (Carry in)	Sum	Carry
0	0	0	0	0
0	0	1	1	0
0	1	0	1	0
0	1	1	0	1
1	0	0	1	0
1	0	1	0	1
1	1	0	0	1
1	1	1	1	1

The equation of Sum and Carry outputs can be written as: $Sum = A \oplus B \oplus C$ and $Carry = AB + AC + BC$. You can draw the schematic for these functions easily with the aid of gates available in the TIL family (Fig. 4.3).

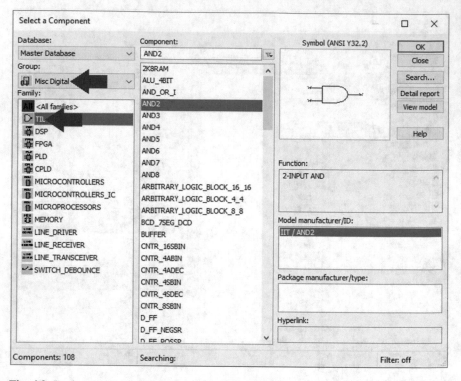

Fig. 4.3 Logic gates

The schematic of full adder is shown in Fig. 4.4.

Fig. 4.4 Full adder circuit

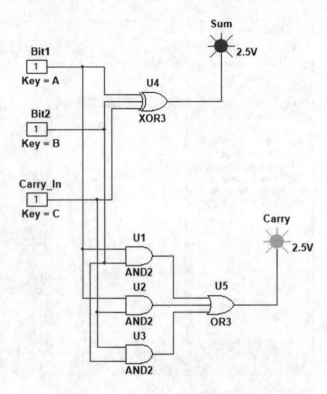

The inputs of the circuit are given by the interactive digital constant block (Fig. 4.5). Each interactive digital constant block is controlled with a keyboard key. Double click the block and select the desired control key from key to toggle drop down list (Fig. 4.6).

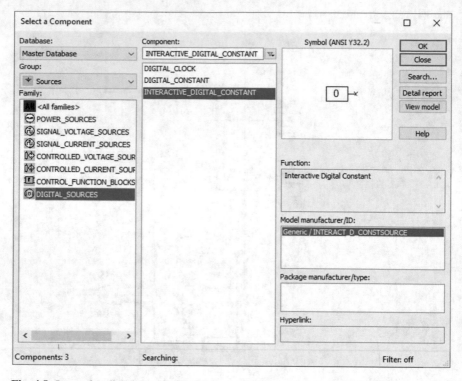

Fig. 4.5 Interactive digital constant block

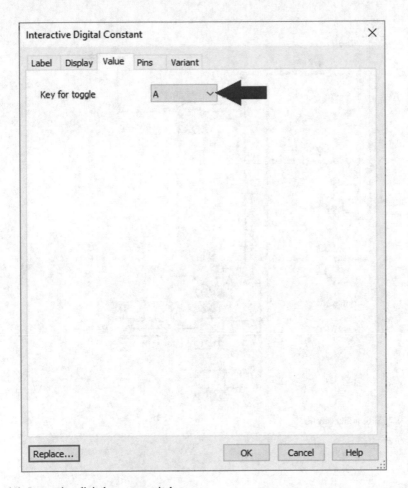

Fig. 4.6 Interactive digital constant window

You can use Single Pole Double Through (SPDT) switches to give the inputs as well (Fig. 4.7). Figure 4.7 used the probe_red and probe_green elements (Fig. 4.8) to show the status of output. The 2.5 V label which is shown in Fig. 4.7 behind the probe elements, shows the turn on threshold of probe. If voltage is equal or greater than 2.5 V, then the lamp turns on. The lamp is off for voltages bellow 2.5 V. You can change the probe threshold voltage by double clicking on it and enter the new value to the threshold voltage box.

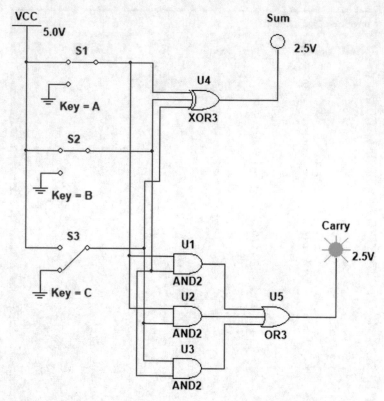

Fig. 4.7 Full adder circuit

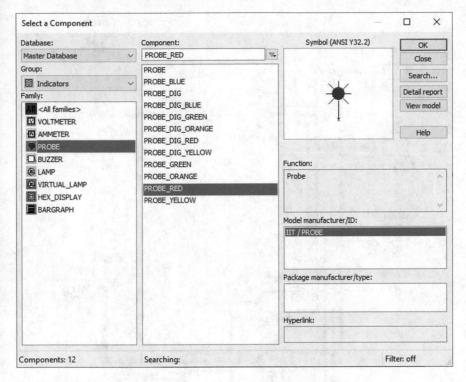

Fig. 4.8 Different types of probe

You can use digital probe (Fig. 4.9) to show the status of outputs as well (Fig. 4.10).

Fig. 4.9 Digital probe

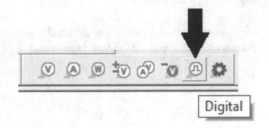

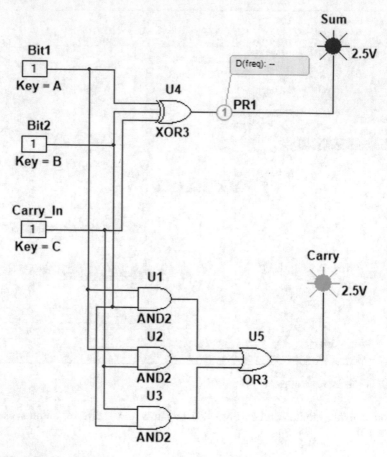

Fig. 4.10 Full adder circuit

You can use DC voltmeters to see the status of outputs as well (Fig. 4.11). 0 V shows low state and 5 V shows high state.

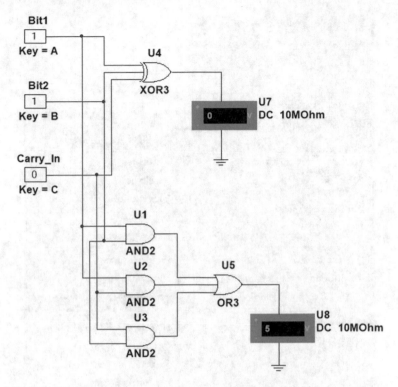

Fig. 4.11 Full adder circuit

4.3 Example 2: Logic Converter Block

Logic converter (Fig. 4.12) is one of the important tools in simulation of digital circuits. You can add a logic converter by using the simulate> instruments> logic converter as well (Fig. 4.13).

Fig. 4.12 Logic converter

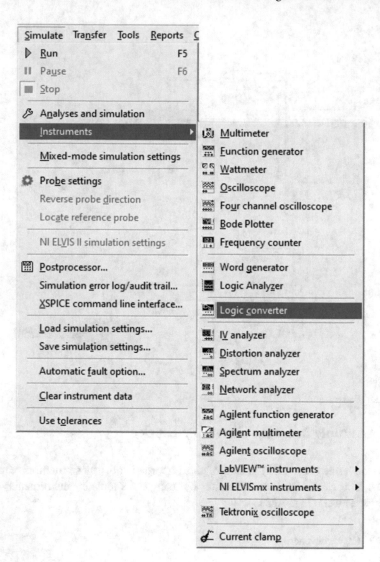

Fig. 4.13 Simulate> Instruments> Logic converter

Add a logic converter to the schematic (Fig. 4.14).

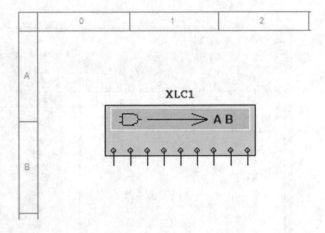

Fig. 4.14 Addition of logic converter to the schematic editor

Double click the logic converter (Fig. 4.15). You can have up to eight inputs (A, B, C, D, E, F, G, H), however you have only one output. If you have a table with more than one output, you need to enter the output columns one by one. You need to click the circle behind the input to activate it.

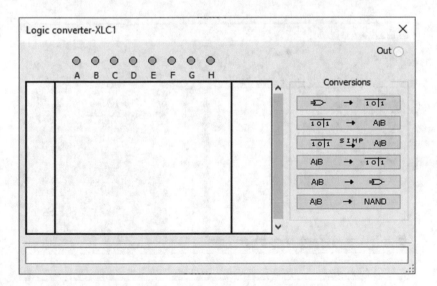

Fig. 4.15 Logic converter window

Let's simulate the truth Table 4.1. This table has three inputs. So, click the input A, B and C circles (Fig. 4.16). Table 4.1 has two outputs. We need to enter only one of outputs since logic converter has only one output. Click the question marks (Fig. 4.16) to have correct output (Sum column of Table 4.1) in front of each row (Fig. 4.17).

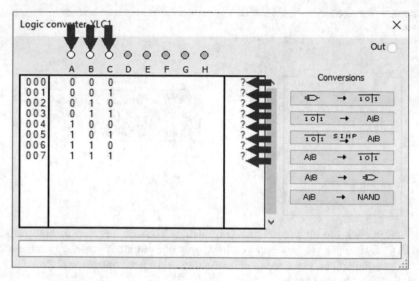

Fig. 4.16 Activation of A, B and C inputs

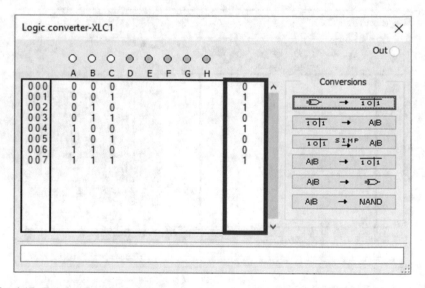

Fig. 4.17 Entering the output values

Click the truth table to Boolean function conversion button. This button extracts the Boolean function of the truth table. If you click the button with SIMP label, then it will produce the simplified Boolean function for you (Fig. 4.18).

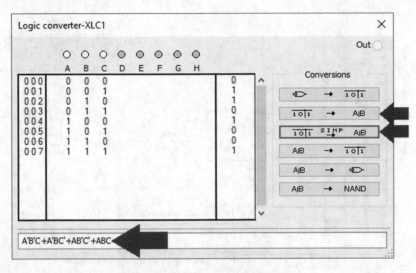

Fig. 4.18 Calculation of Boolean expression of the table

Click the Boolean function to logic gates conversion button (Fig. 4.19). Then click on the schematic editor to add the drawn logic diagram to it (Fig. 4.20). Note that you need to calculate the Boolean function of the table before clicking the Boolean function to logic gates conversion button, otherwise the error message shown in Fig. 4.21 appears.

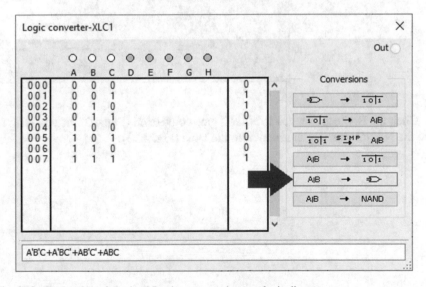

Fig. 4.19 Conversion of obtained Boolean expression to a logic diagram

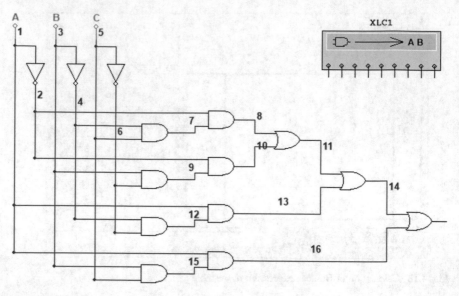

Fig. 4.20 Conversion from Boolean expression to logic diagram

Fig. 4.21 Error message

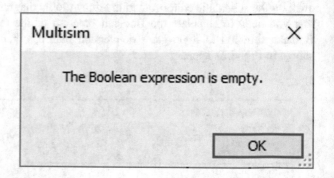

Click the Boolean function to NAND gate conversion (Fig. 4.22) if you want to see the NAND gate implementation of the table (Fig. 4.23).

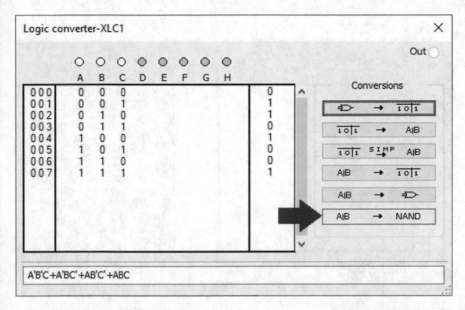

Fig. 4.22 Conversion of obtained Boolean expression to a logic diagram with NAND gates only

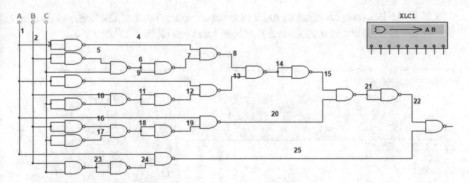

Fig. 4.23 Conversion from Boolean expression to NAND logic diagram

Assume that you have a Boolean expression and you want to see its truth table. To do this, enter the Boolean function to the logic converter (Fig. 4.24).

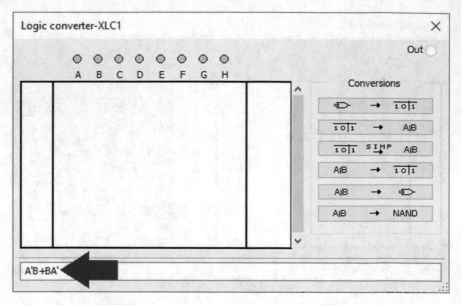

Fig. 4.24 Entering a Boolean expression

Click the Boolean function to truth table conversion button. The logic converter shows the truth table of the entered Boolean function (Fig. 4.25).

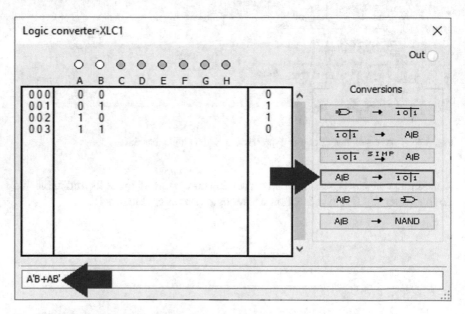

Fig. 4.25 Obtaining the truth table for entered expression

4.4 Example 3: Obtaining the Truth Table for a Digital Circuit

In this example we show how logic converter can be used to obtain the truth table of a given schematic. Assume that you want to obtain the truth table of the schematic shown in Fig. 4.26.

Fig. 4.26 Schematic of example 3

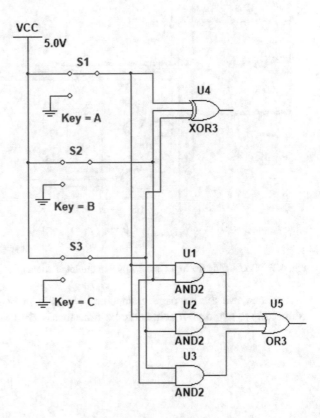

Connect the inputs of the circuit to the first three terminal of logic converter and connect the output of the circuit to the last terminal of the logic converter (Fig. 4.27).

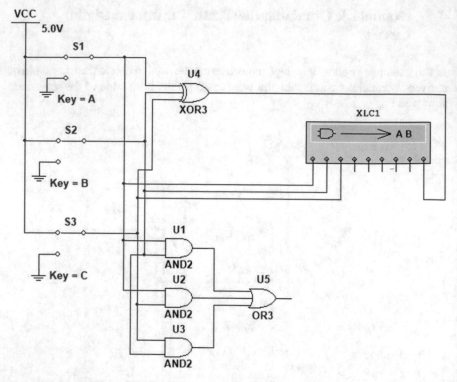

Fig. 4.27 Connecting the input and outputs to the logic converter

You can use the on page connector (available in place> connector> on page connector) to avoid your schematic become messy (Fig. 4.28).

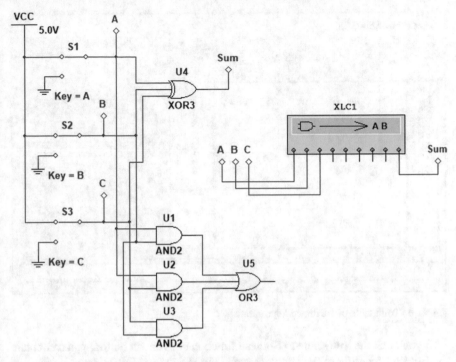

Fig. 4.28 Use of page connector to make the connection

Without running the circuit, double click the logic converter and click the logic gate to truth table conversion button (Fig. 4.29). Calculated table is shown in Fig. 4.30.

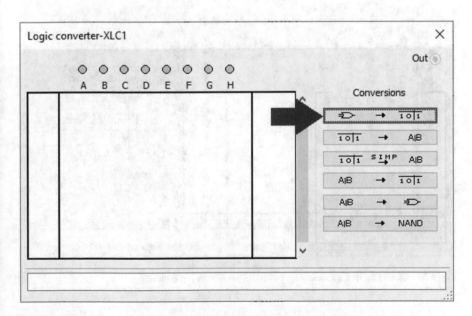

Fig. 4.29 Obtaining the truth table of the drawn logic circuit

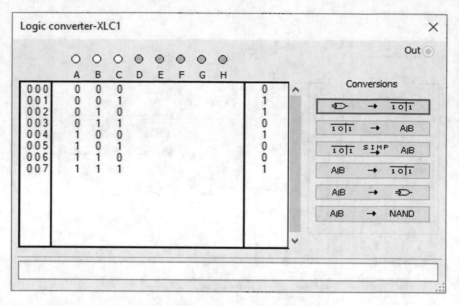

Fig. 4.30 Truth table of the drawn logic circuit

If you click the truth table to Boolean function conversion button, you can obtain the Boolean function of the circuit as well (Fig. 4.31).

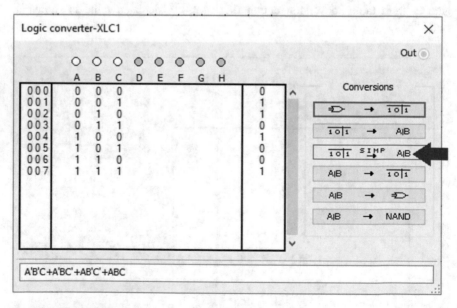

Fig. 4.31 Obtaining the Boolean expression of the drawn logic circuit

4.5   Example 4: Word Generator Block

Word generator (Figs. 4.32 or 4.33) permits you to produce digital test inputs for a digital circuit. In this example, we want to use the word generator to generate test signals for our full adder circuit.

Fig. 4.32   Word generator

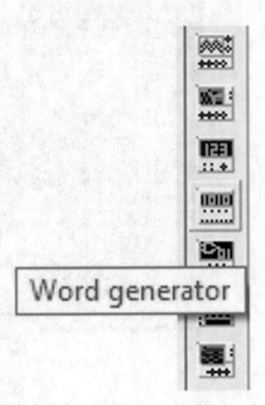

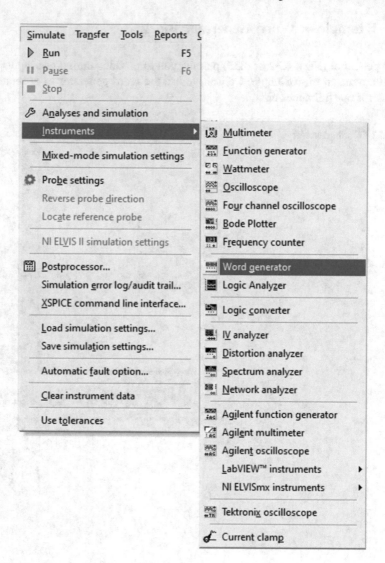

Fig. 4.33   Simulate> instruments> word generator

Add the word generator to the schematic and connect the inputs of the full adder to the word generator as shown in Fig. 4.34. In this schematic, digital probes are used to show the status of outputs.

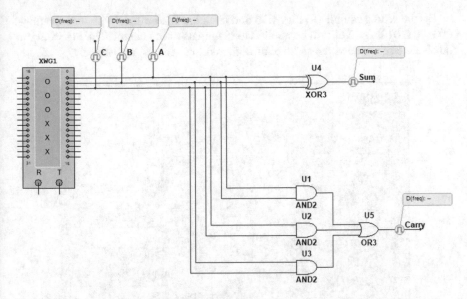

Fig. 4.34 Schematic of example 4

Double click the word generator and click the set button (Fig. 4.35).

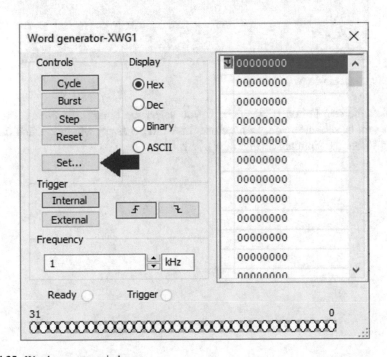

Fig. 4.35 Word generator window

Do the settings similar to Fig. 4.36 and click the OK button. With these settings, 000, 001, 010, ..., 111 will be applied to the inputs of the circuit. The high and low boxes determine the voltages level of high and low signals, respectively.

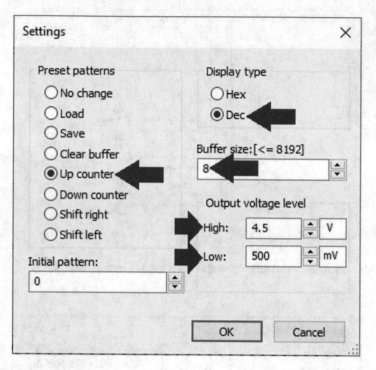

Fig. 4.36 Settings window

Click the step button (Fig. 4.37). If you select the binary, the binary value of inputs will be shown to you. Don't close the word generator window. Click the run button to run the simulation (Fig. 4.38).

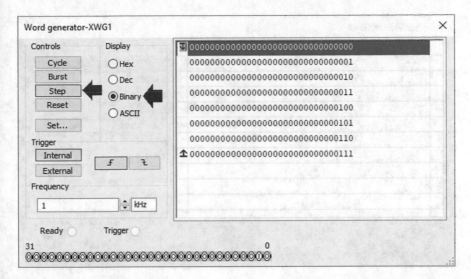

Fig. 4.37 000 is applied to the circuit

Fig. 4.38 Run button

After clicking the run button, the word generator applies 000 to the circuit and simulate it. Result of simulation is shown in Fig. 4.39. The play button of word generator goes to the next input and stops there (Fig. 4.40).

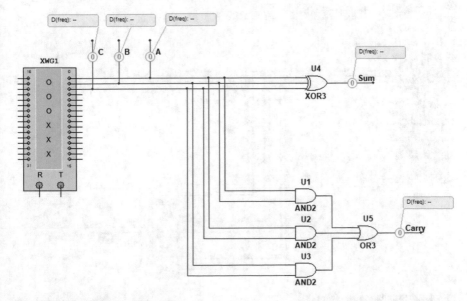

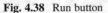

Fig. 4.39 Simulation result for 000

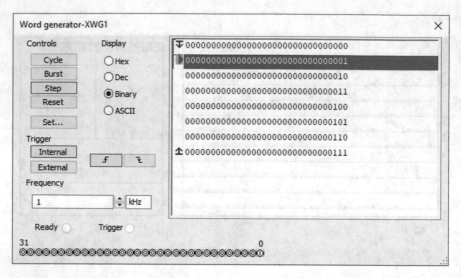

Fig. 4.40 001 is applied to the circuit

Press the run button again (Fig. 4.41). The word generator applies 001 to the circuit and simulate it. Result of simulation is shown in Fig. 4.42. The play button of word generator goes to the next input and stops there (Fig. 4.43).

Fig. 4.41 Run button

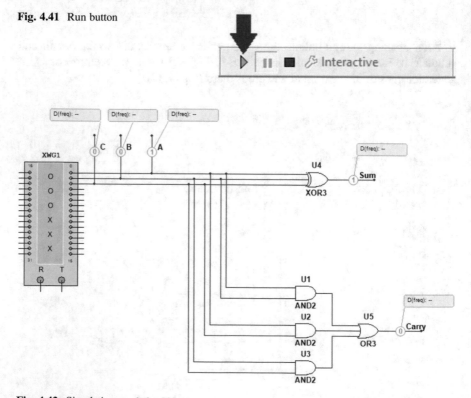

Fig. 4.42 Simulation result for 001

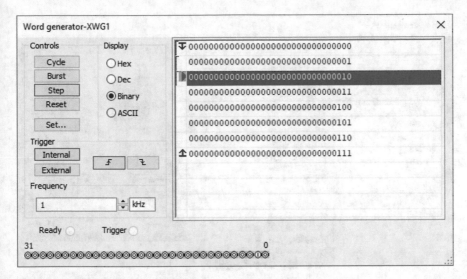

Fig. 4.43 010 is applied to the circuit

You can apply the reminded inputs to the circuit by clicking the run button.

4.6 Example 5: Seven Segment Display

Multisim has ready to use common anode/cathode seven segment displays (Fig. 4.44).

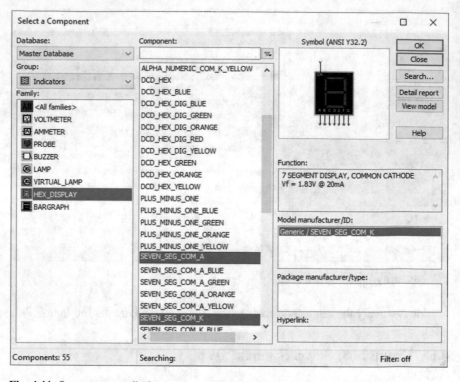

Fig. 4.44 Seven segment display

Multisim has a block called DCD-HEX (Fig. 4.45). This display accepts a four bit binary code and show its equivalent hexadecimal value. For instance, if you give 0101 to it, it displays 5 (Fig. 4.46) and if you give 1010, it shows A.

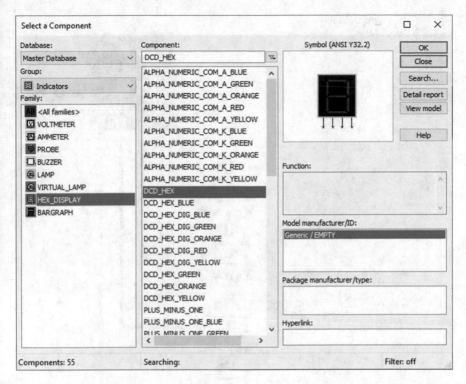

Fig. 4.45   DCD hex display

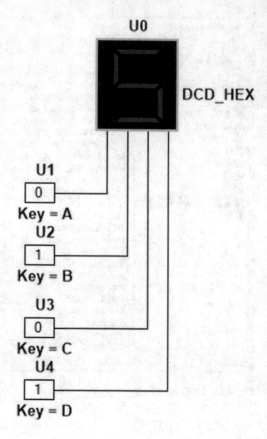

Fig. 4.46 Applying 0101 to the DCD hex display

The schematic shown in Fig. 4.47 is a binary counter. The DCD-HEX display permits you to see the output of the counter. Place of the IC is shown in Fig. 4.48.

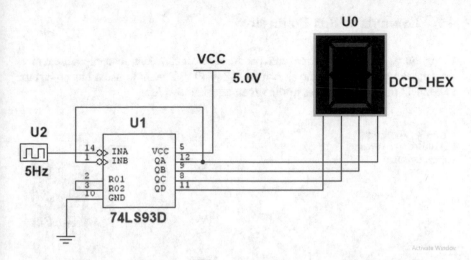

Fig. 4.47 Binary counter

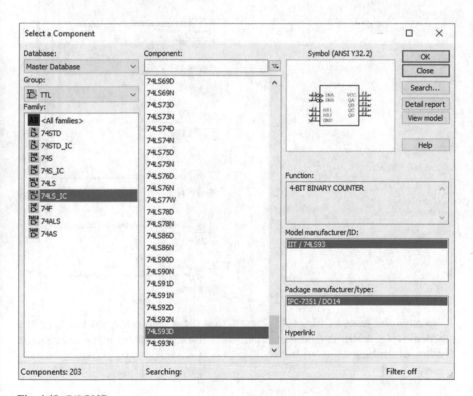

Fig. 4.48 74LS93D

4.7   Example 6: Bus Connector

Using buses simplifies wiring and produces cleaner, more readable schematics.
Assume the simple schematic shown in Fig. 4.49. We want to use a bus connector
to connect the digital sources to the seven segment display.

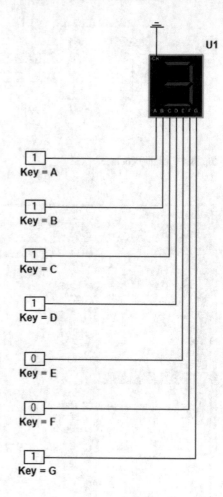

Fig. 4.49 Applying
1111001 to the seven
segment display

Remove the wires of Fig. 4.49. Press Ctrl+U and then draw two buses on the
schematic (Fig. 4.50). You need to double click the end point of the bus to finish the
drawing of the bus.

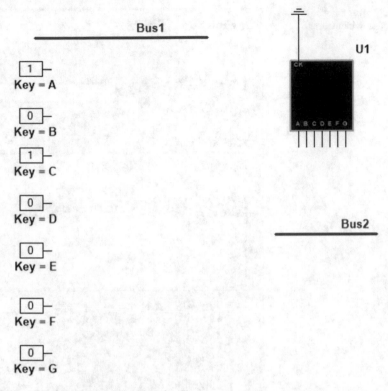

Fig. 4.50 Addition of bus to the schematic editor

Draw a wire from the upper digital source to the Bus1. When the wire comes close enough to the bus, it changes into a broken line. When the broken line is appeared, click on the bus to connect the wire to it (Fig. 4.51).

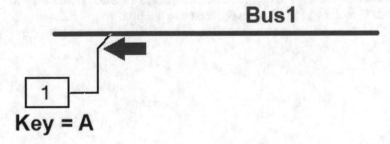

Fig. 4.51 Connecting a wire to the bus

The window shown in Fig. 4.52 appears after connecting the wire to the bus. Give the name A to the wire.

Fig. 4.52 Bus entry
connection window

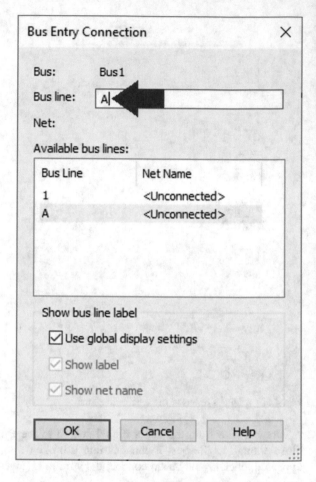

Repeat this procedure to connect the other wires to the Bus1. Give names B, C, D, E, F and G to the wires which are connected to the Bus1 (Fig. 4.53).

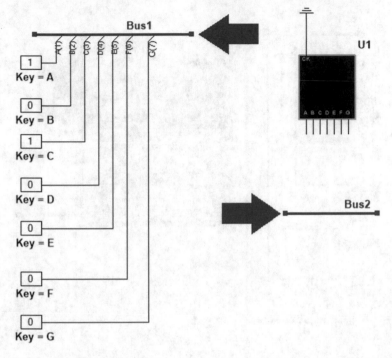

Fig. 4.53 Selection of bus1 and bus2

After connecting the wires to the Bus1, click on Bus1, hold down Shift key of the keyboard and click on Bus2. Then right click on Bus1 and select merge selected buses (Fig. 4.54).

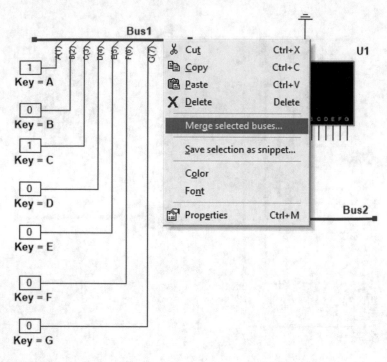

Fig. 4.54 Merging the bus1 and bus2

After clicking the merge selected buses, the window shown in Fig. 4.55 appears. Click OK in this window. After clicking the OK button, name of two buses changes to Bus1 (Fig. 4.56).

Fig. 4.55 Resolve bus name conflict window

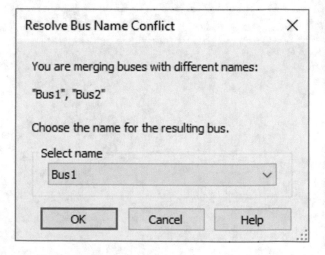

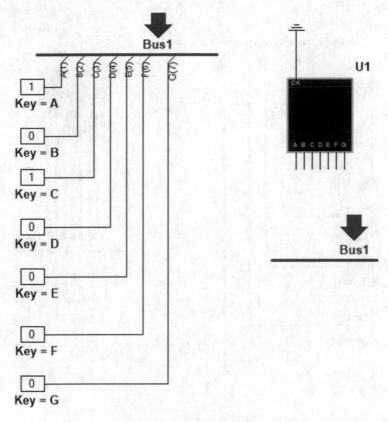

Fig. 4.56 Two buses are merged

Connect the seven segment to the bus (Fig. 4.57). Terminal A of seven segment must be connected to wire A of bus, terminal B of seven segment must be connected to wire B of bus, terminal C of seven segment must be connected to wire C of bus and so on.

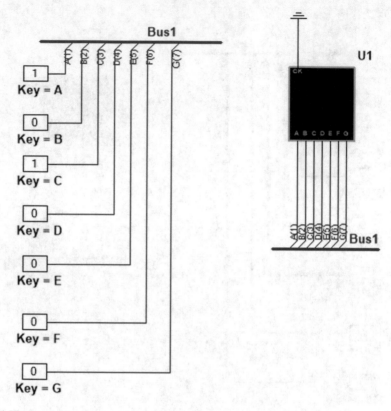

Fig. 4.57 Connecting the seven segment to the bus

Now run the simulation and test the circuit (Fig. 4.58).

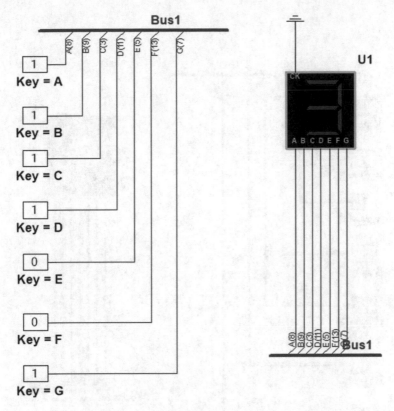

Fig. 4.58 Data is transferred from the input sources to the seven segment with the aid of bus

If you measure the current of seven segment display, you see a very large current (Fig. 4.59). Let's add resistors to limit the current.

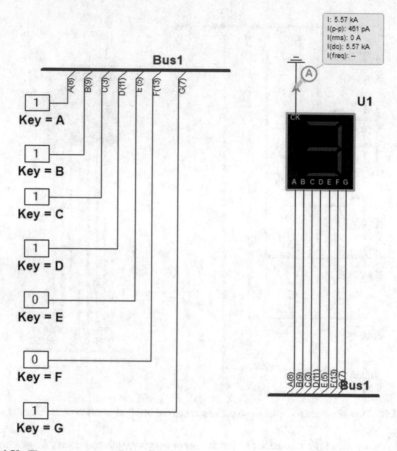

Fig. 4.59 The seven segment draws 5.57 kA

Add a 7 line isolated block (Fig. 4.60) to the schematic (Fig. 4.61). Now the current is in the mA range.

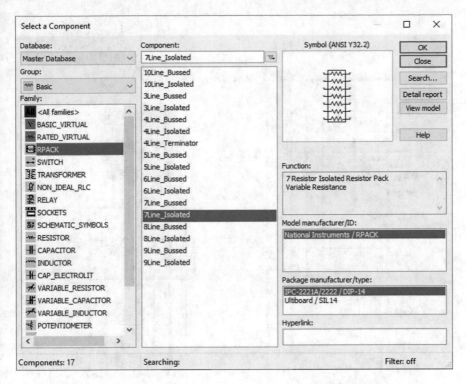

Fig. 4.60 7 line isolated block

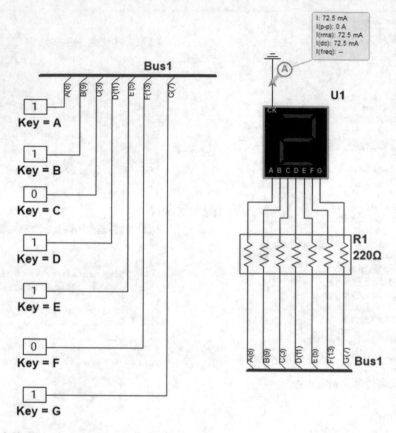

Fig. 4.61 The seven segment draws 72.5 mA

4.8   Example 7: Frequency Divider

In this example we will use a flip flop to do frequency division. A D flip flop can be used to divide the frequency of an input square wave by factor of two (Fig. 4.62).

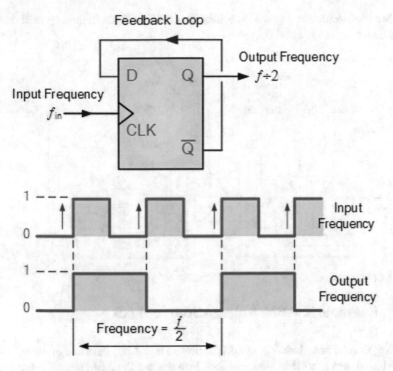

Fig. 4.62 Input frequency is divided by two

Assume the schematic shown in Fig. 4.63. Each 74LS74 contains two D flip flops. So, we can divide the input frequency by 2, 4, 8 and 16 using the cascade connection of these flip flops.

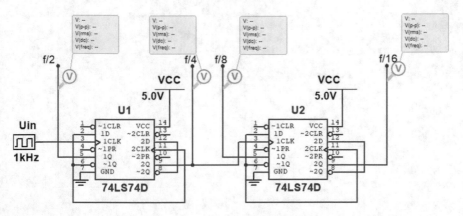

Fig. 4.63 Schematic of example 7

Run the simulation and read the frequencies. The input frequency is divided by 2, 4, 8 and 16 as expected (Fig. 4.64).

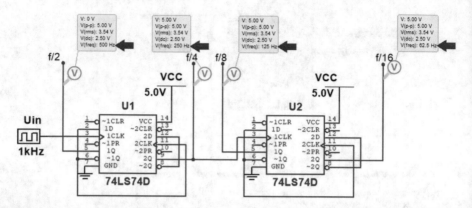

Fig. 4.64 Simulation result

4.9   Example 8: ABM Voltage Source Block

Analog Behavioral Modeling (ABM) sources (Fig. 4.65) generate an output waveform based on a user-defined expression. You can use the ABM source to convert the levels of a pulse. For instance, assume that you have a pulse with levels of low level of -1 V and high level of 1 V. We want to convert this pulse to a pulse with low level of 0 V and high level of 13 V. The schematic shown in Fig. 4.66 do this job. Settings of ABM voltage source is shown in Fig. 4.67.

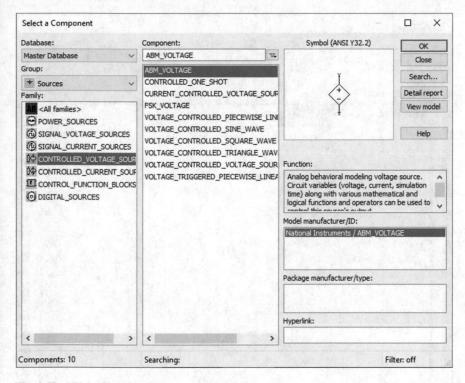

Fig. 4.65 ABM voltage block

Fig. 4.66 Schematic of
example 8

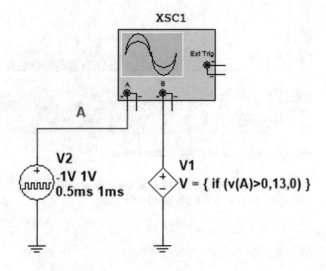

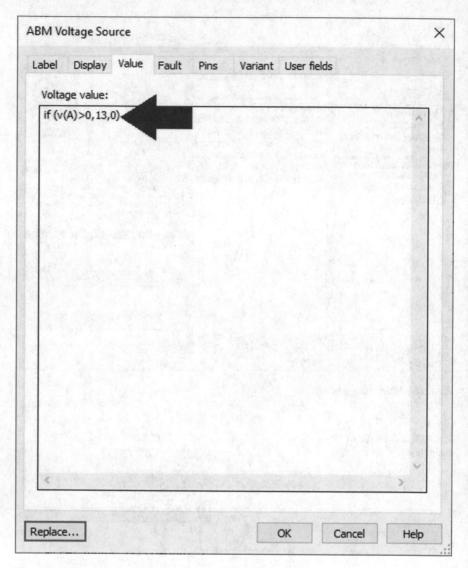

Fig. 4.67 Value tab of ABM voltage source

Run the simulation. Simulation result is shown in Fig. 4.68. Output is a pulse with low level of 0 V and high level of 13 V.

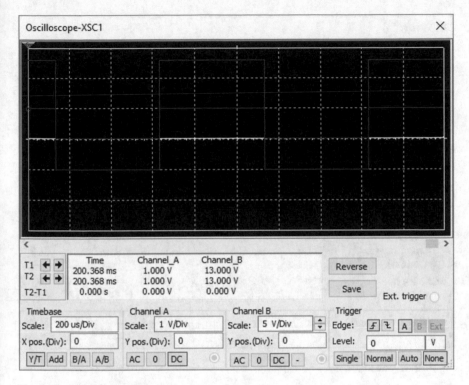

Fig. 4.68 Simulation result

4.10 More Examples of Digital Circuits

Multisim has some ready to use sample simulations related to digital circuits. Click the open samples button (Fig. 4.69) and then go to digital folder to see the samples (Fig. 4.70).

Fig. 4.69 Open sample button

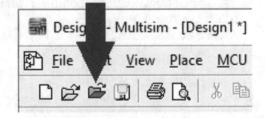

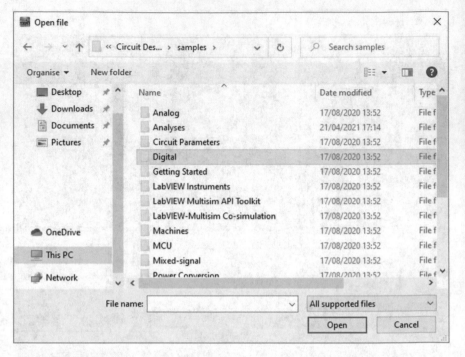

Fig. 4.70 Digital folder

4.11   Exercises

1. Implement the following Boolean function with Multisim.

   F = (A + B)(A'B + AC' + B'C)

2. **(a)** Simulate an 8×1 multiplexer in Multisim and test it.
   (b) Simulate a 4-bit shift register in Multisim and test it. Use the D flip flop.
3. Simulate a 4 bit Johnson counter in Multisim and test it.
4. Use Tools> Circuit wizards>555 timer wizard to produce a 1 Hz clock pulse
   (**Hint:** Use Astable operation).
5. Assume the state diagram shown in Fig. 4.71.

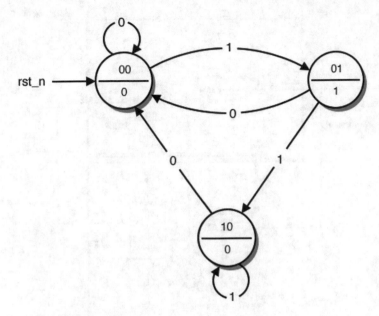

Fig. 4.71 State diagram of exercise 5

The state table of this state diagram is shown in Table 4.2.

Table 4.2 State table of Fig. 4.71

Current State			Next State		
A	B	Input I	A_{next}	B_{next}	Outputs Y
0	0	0	0	0	0
0	0	1	0	1	0
0	1	0	0	0	1
0	1	1	1	0	1
1	0	0	0	0	0
1	0	1	1	0	0
1	1	0	X	X	X
1	1	1	X	X	X

(a) Use hand analysis to ensure that the circuit in Fig. 4.72 implements the given state diagram.

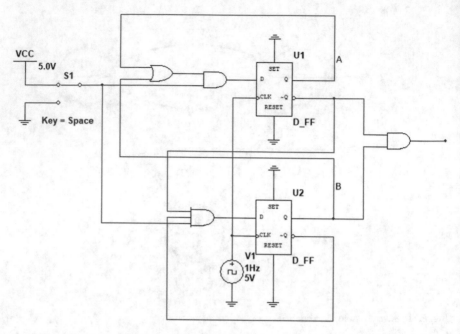

Fig. 4.72 D flip flop realization of state diagram of Fig. 4.71

(b) Simulate the circuit shown in Fig. 4.72 in Multisim and ensure that it works similar to Table 4.2.

(c) Draw the JK flip flop realization of the given state diagram and test it in Multisim.

Further Readings

1. M. Mano, M.D. Ciletti, *Digitla Design*, 6th edn. (Pearson, 2018)
2. V. Nelson, H.T. Nagle, H.T. Irvin, B.D. Carrol, *Digital Logic Circuit Analysis and Design* (Pearson, 1995)
3. T. Floyd, *Digital Fundaments*, 11th edn. (Pearson, 2014)
4. A.B. Marcovitz, *Introduction to Logic Design*, 3rd edn. (Mc-Graw Hill, 2009)
5. D. Harris, S. Harris, *Digital Design and Computer Architecture* (Morgan Kaufmann, 2012)
6. Multisim microcontroller functionality: https://www.ni.com/en-tr/support/documentation/supplemental/07/ni-multisim-microcontroller-functionality%2D%2Dfeature-chart.html, https://www.youtube.com/watch?v=wbJv8oyKNU8

Chapter 5
Simulation of Power Electronic Circuits with Multisim™

5.1 Introduction

Power electronics is the application of solid-state electronics to the control and conversion of electric power with high efficiency. Multsim can be used to simulate power electronics circuits as well. This chapter shows how Multisim can be used to simulate power electronics circuits.

5.2 Example 1: Switching Behavior of Thyristor

Silicon Controlled Rectifier (SCR) or thyristor is a four-layered, three-junction semiconductor switching device. It has three terminals: anode, cathode, and gate. Thyristor is also a unidirectional device like a diode, which means it flows current only in one direction.

Gate terminal used to trigger the SCR by providing small current to this terminal. SCR starts conducting when the gate receives a current trigger, and continuing to conduct until the voltage across the device is reversed biased, or until the anode currents falls below the holding current of thyristor.

This example shows the basic of basic of thyristors. Assume the circuit shown in Fig. 5.1.

© The Author(s), under exclusive license to Springer Nature Switzerland AG 2022
F. Asadi, *Essential Circuit Analysis using NI Multisim™ and MATLAB®*,
https://doi.org/10.1007/978-3-030-89850-2_5

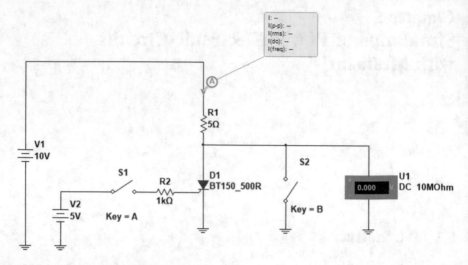

Fig. 5.1 Schematic of example 1

Run the simulation (Fig. 5.2). No trigger pulse is applied to the thyristor and the thyristor acts as open switch.

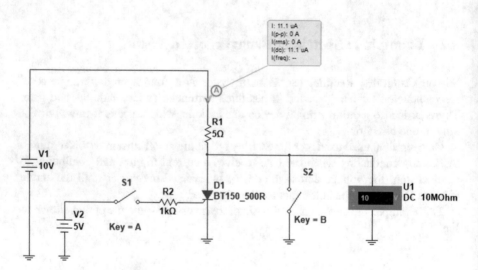

Fig. 5.2 Simulation result

Close the switch S1. The thyristor starts to conduct current. The voltage drop across the thyristor is 1.1 V (Fig. 5.3). Load current is $\frac{10-1.1}{5} = 1.78$ A.

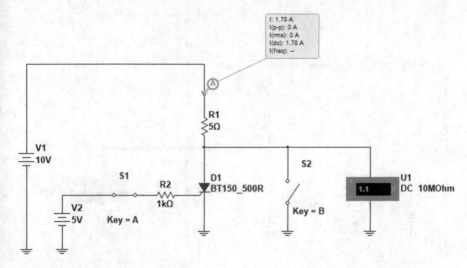

Fig. 5.3 Simulation result

Open the switch S1. Note that the thyristor continue to conduct current although the gate current is zero (Fig. 5.4).

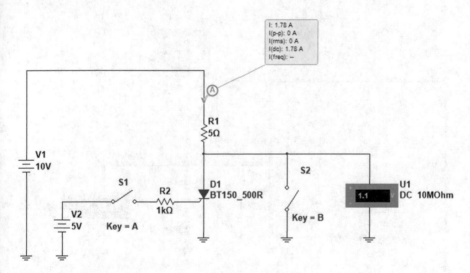

Fig. 5.4 Simulation result

Close the switch S2 (Fig. 5.5). After closing the S2, the load current pass from the mechanical switch S2, not the thyristor.

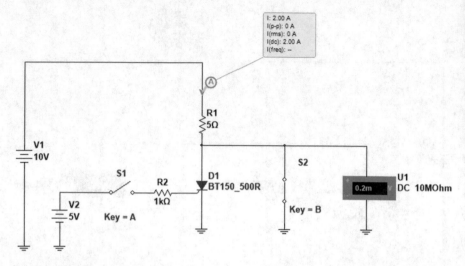

Fig. 5.5 Simulation result

Open the switch S2. As shown in Fig. 5.6, the thyristor is not conducting. It is in the blocking state until the next triggering pulse comes.

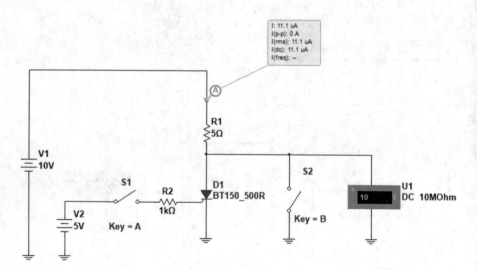

Fig. 5.6 Simulation result

5.3   Example 2: Single Phase Controlled Rectifiers

When you use a diode rectifier, the output voltage is not controllable, i.e. you can't increase or decrease it. When you need controllable output voltage, you need to use thyristor rectifiers. In this example, we will study a signal phase controlled rectifier. Assume the schematic shown in Fig. 5.7. Diode D3 is freewheeling diode. This schematic uses the simplified SCR block (Fig. 5.8).

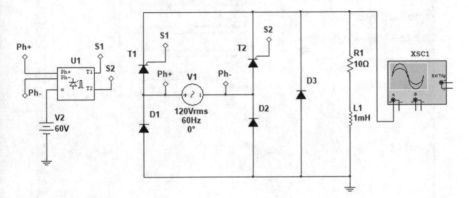

Fig. 5.7   Schematic of example 2

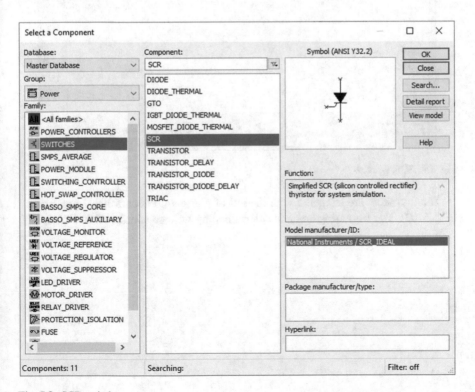

Fig. 5.8   SCR switch

Settings of the thyristors are shown in Fig. 5.9. The voltage drop of the thyristors are ignored.

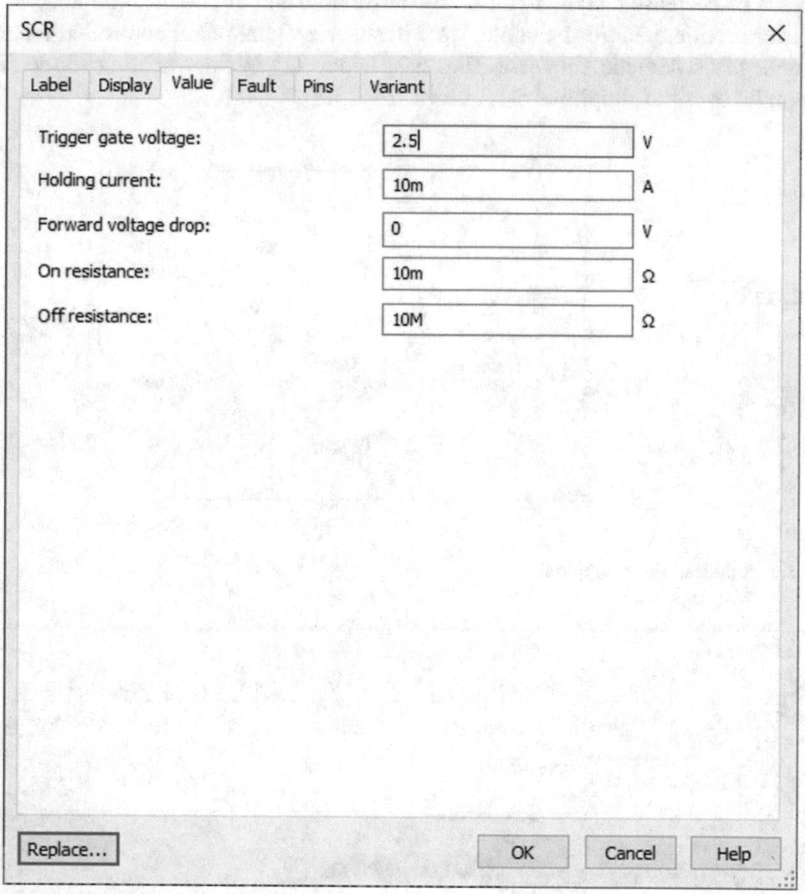

Fig. 5.9 SCR window

The thyristors are triggered with the aid of phase angle controller 2 pulse block (Fig. 5.10). The triggering angle is determined by voltage source V2 in Fig. 5.7.

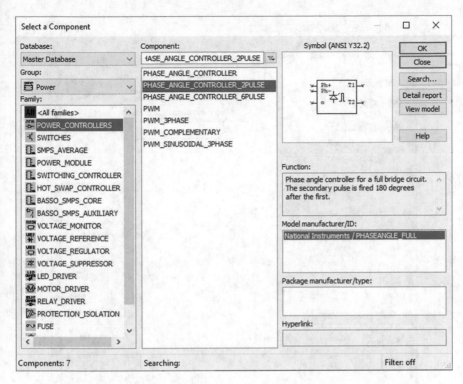

Fig. 5.10 Phase_angle_controller_2pulse block

Double click the phase angle controller 2 pulse block and enter the line frequency (Fig. 5.11).

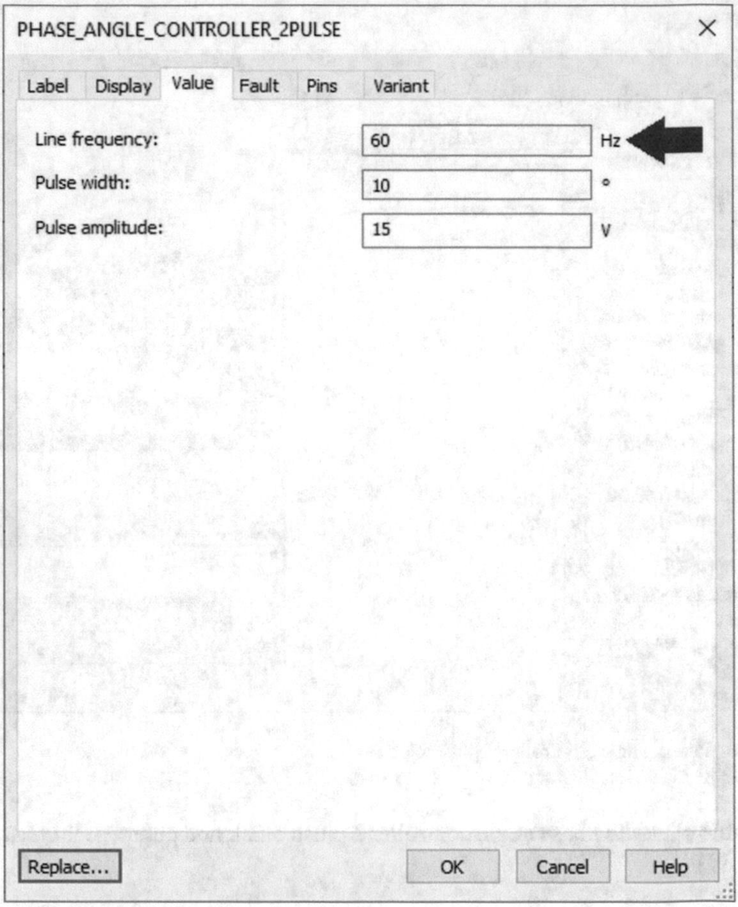

PHASE_ANGLE_CONTROLLER_2PULSE   ×

Label   Display   Value   Fault   Pins   Variant

Line frequency: 60   Hz ◀

Pulse width: 10   °

Pulse amplitude: 15   V

Replace...   OK   Cancel   Help

Fig. 5.11 Phase_angle_controller_2pulse block settings

Enter the desired triggering angle to source V2 and run the simulation. For instance, for triggering angle of 60°, voltage of V2 must be set to 60 V. Figures 5.12 and 5.13 show the output waveform for triggering angle of 0° and 60°, respectively. Note that the frequency of these waveforms are two times bigger then the frequency of AC source.

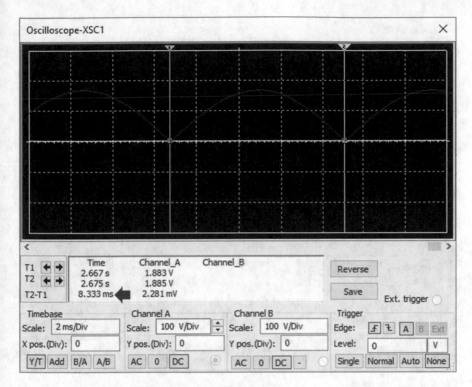

Fig. 5.12 Simulation result for triggering angle of 0 degrees

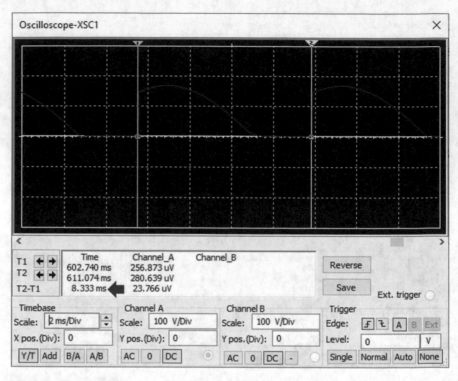

Fig. 5.13 Simulation result for triggering angle of 60 degrees

You can use probes to measure the average value of output voltage/current and load power (Fig. 5.14).

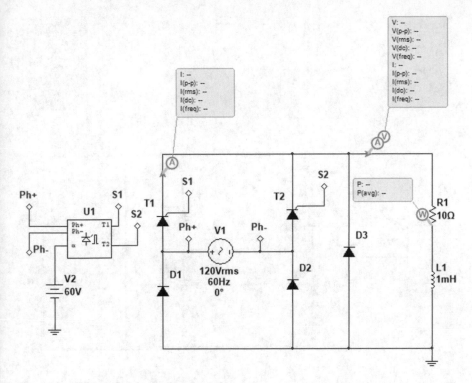

Fig. 5.14 Addition of probes to the schematic

Figures 5.15 and 5.16 show the values of voltage/current and load power for triggering angle of 0° and 60°, respectively. Note that as you increase the triggering angle, the average and RMS value of load voltage/current decreases.

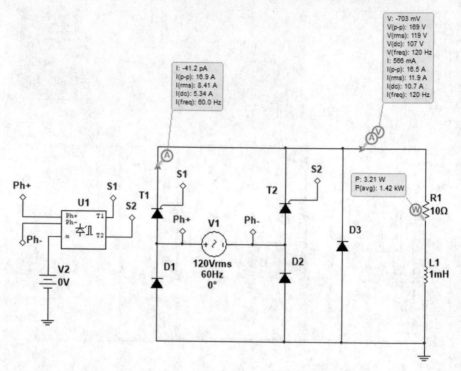

Fig. 5.15 Simulation result

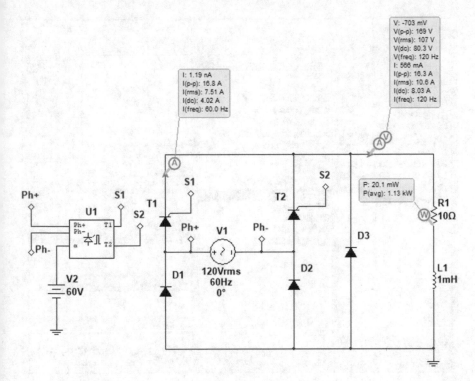

Fig. 5.16 Simulation result

5.4   Example 3: Triggering the Thyristors with Pulse Voltage Block

Using the phase angle controller 2 pulse block (Fig. 5.10) is not the only available way to trigger the thyristors. You can use the pulse voltage block (Fig. 5.17) to trigger the thyristors, as well. In this example we will trigger the thyristors of previous example with the aid of pulse voltage block.

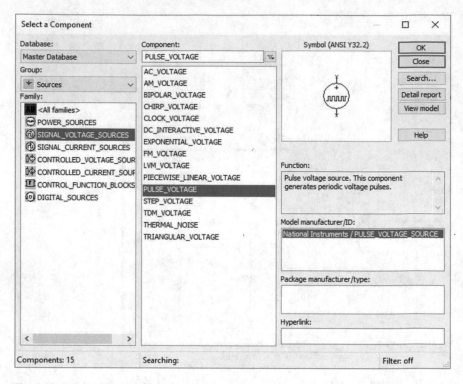

Fig. 5.17 Pulse voltage block

Assume the schematic shown in Fig. 5.18.

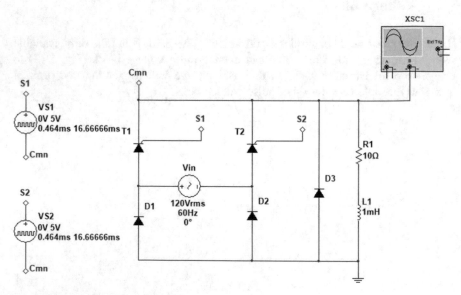

Fig. 5.18 Schematic of example 3

Settings of VS1 and VS2 blocks (for triggering angle of 60.479°) are shown in Figs. 5.19 and 5.20, respectively. Period of triggering pulse is $T = 1/60 = 16.666666\ldots$ ms (the decimal point part of this number is repeating). Ensure to enter the period with 5–6 digits after the decimal point otherwise your simulation will not work correctly after a while.

The delay time determines the triggering angle. For instance, for delay of 2.8 ms, the triggering angle will be 60.479° (Fig. 5.21). Note that the delay of VS2 must be equal to $T_{D1} + \frac{T}{2}$ where T_{D1} is the number entered into the delay time box of VS1 and T is the period of AC source. For instance, in Fig. 5.19, $T_{D1} = 2.8$ ms, so $T_{D2} = T_{D1} + \frac{T}{2} = 2.8$ ms $+ 8.33333$ ms $= 11.13333$ ms.

Pulse width box is filled with the same number for both of the blocks. The pulse width box determines the width of pulses. This box is similar to the pulse width box in Fig. 5.11. For 60 Hz system, pulse width of 10° equals to 0.464 ms (Fig. 5.22).

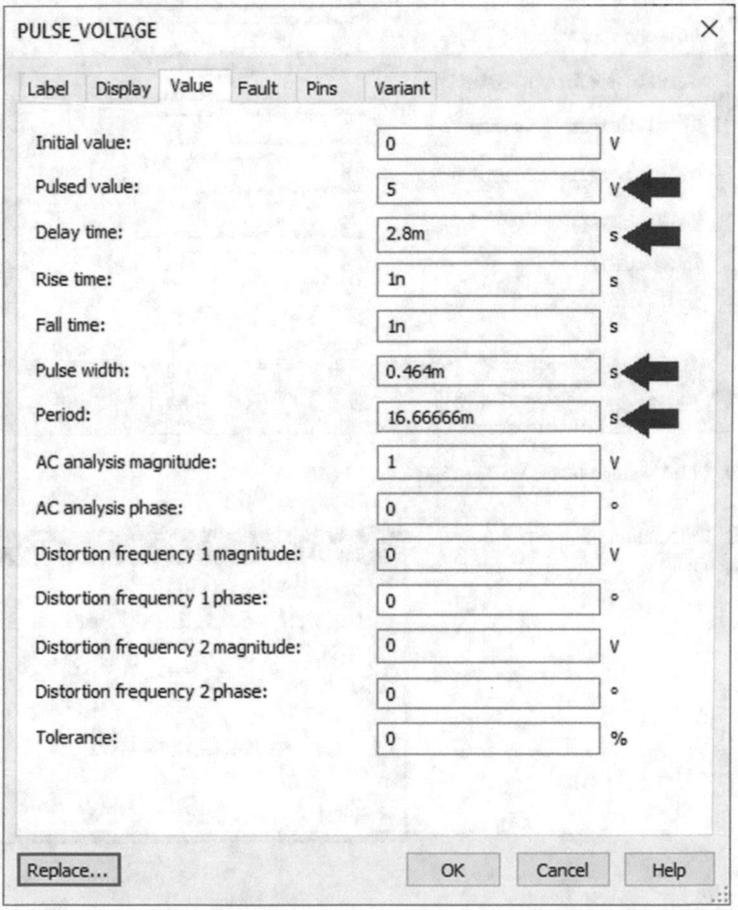

Fig. 5.19 Pulse voltage block VS1 settings

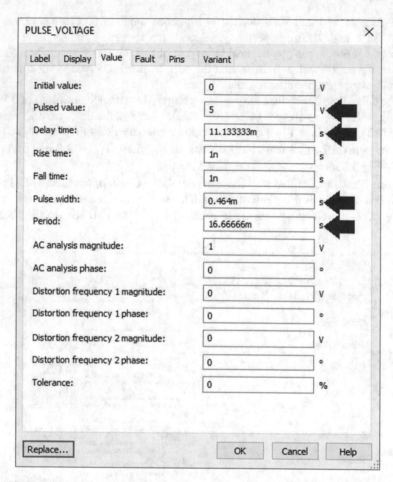

Fig. 5.20 Pulse voltage block VS2 settings

Fig. 5.21 Calculation of
triggering angle

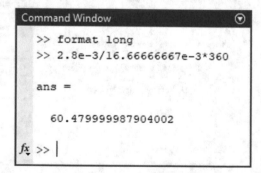

Fig. 5.22 Calculation of pulse width in degrees

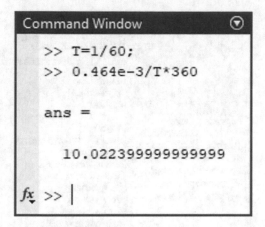

The following MATLAB code helps you to calculate the required value of delay time for VS1 and VS2.

```
Angle=60; %desired angle
f=60;
format long
Delay_time_VS1=Angle/360*1/f;
Delay_time_VS2=Angle/360*1/f+1/2/f;
disp ('Delay time for Vs1 (in ms) : ')
disp (Delay_time_VS1*1000)
disp ('Delay time for Vs2 (in ms) : ')
disp(Delay_time_VS2*1000)
format short
```

You can use the schematic shown in Fig. 5.23 instead of schematic shown in Fig. 5.18. This schematic uses only one pulse voltage block and uses a time delay block (Fig. 5.24) to apply the delayed version of trigger pulse of thyristor T1 to thyristor T2. The amount of delay equals to half of the AC source period (Fig. 5.25).

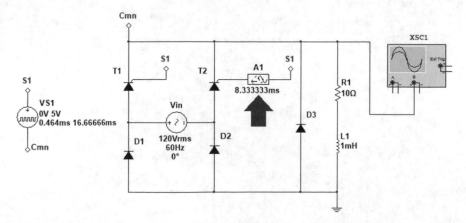

Fig. 5.23 Addition of a time delay block to gate of T2

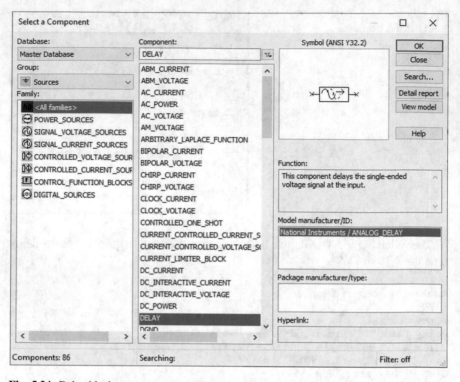

Fig. 5.24 Delay block

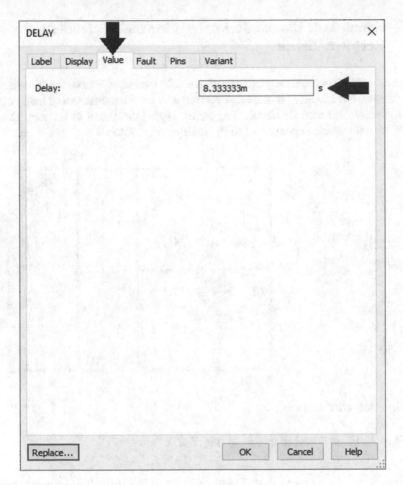

Fig. 5.25 Settings of delay block

The following MATLAB code helps you to calculate the required value of delay time for VS1.

```
Angle=60; %desired angle
f=60;
format long
Delay_time_VS1=Angle/360*1/f;
disp ('Delay time for Vs1 (in ms) :')
disp (Delay_time_VS1*1000)
format short
```

5.5 Example 4: Harmonic Analysis for the Controlled Rectifier Circuit

You can see the harmonic content of a signal with the aid of Fourier analysis section of Multsim. For instance, assume that you want to see the harmonics of load voltage for the controlled rectifier circuit. Triggering angle is assumed to be zero. Double click the wire which is connected to the resistor (Fig. 5.26).

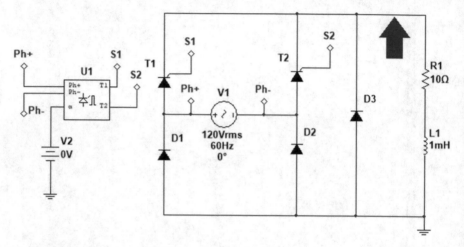

Fig. 5.26 Schematic of example 4

Give the name Vload to this node (Fig. 5.27).

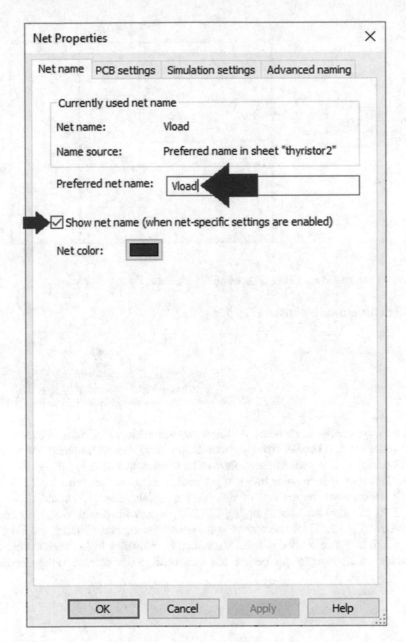

Fig. 5.27 Giving the name Vload to the upper terminal of resistor R1

Name of node will be added to the schematic (Fig. 5.28).

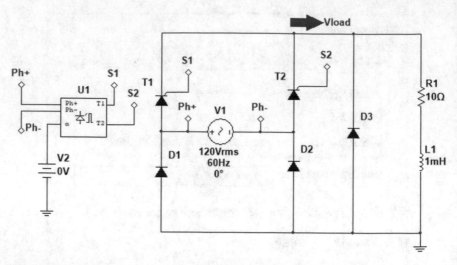

Fig. 5.28 The entered node name is shown on the schematic

Click the interactive button (Fig. 5.29).

Fig. 5.29 Interactive button

Go to the Fourier section and do the settings similar to Fig. 5.30. Frequency of
load voltage is 120 Hz. So, frequency resolution (fundamental frequency) box must
be filled with 120. We want to see values of harmonics up to fifth harmonic. Because
of this, number of harmonics box is filled with 5. Stop time for sampling (TSTOP)
box is used to set the amount of time during which sampling should occur. In
Fig. 5.30, the stop time for sampling (TSTOP) box is filled with 0.333, so about
40 period (Fig. 5.31) of the output voltage will be analyzed during the Fourier
analysis. It is recommended to specify a sampling frequency big enough to obtain a
minimum of 10 samples per period. In other words, value entered to the sampling

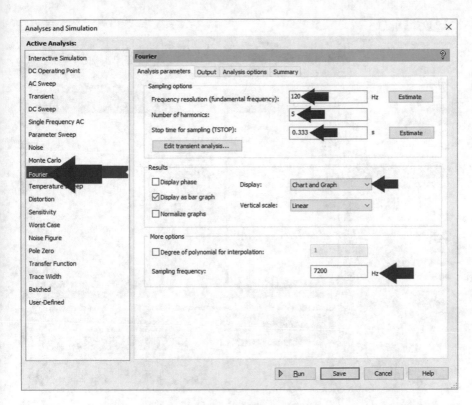

Fig. 5.30 Analysis parameters tab of Fourier analysis

Fig. 5.31 MATLAB calculations

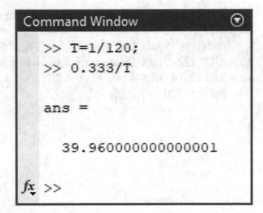

frequency box must be at least ten times bigger than the value entered into the frequency resolution (fundamental frequency) box.

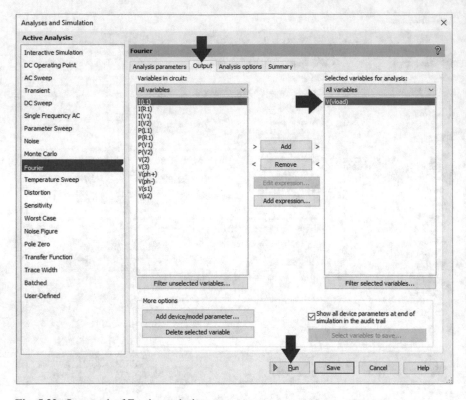

Fig. 5.32 Output tab of Fourier analysis

Go to the output tab and add V(load) to the right list. Then click the run button (Fig. 5.32).

Simulation result is shown in Fig. 5.33. According to this result, $v_o(t)=107.122+71.8862\sin(\omega t-104.39°)+14.3809\sin(2\omega t-118.78°)+6.16583\sin(3\omega t-133.17°)+3.42749\sin(4\omega t-147.56°)+2.18277\sin(5\omega t-161.95°)$ where $\omega = 2 \times \pi \times 120 = 753.98 \frac{Rad}{s}$.

1	Fourier analysis for V(vload):				
2	DC component:	107.122			
3	No. Harmonics:	5			
4	THD:	22.4883 %			
5	Grid size:	128			
6	Interpolation Degree:	1			
7					
8	Harmonic	Frequency	Magnitude	Phase	Norm. Mag
9	0	0	107.122	0	1.49015
10	1	120	71.8862	-104.39	1
11	2	240	14.3809	-118.78	0.200051
12	3	360	6.16583	-133.17	0.0857721
13	4	480	3.42749	-147.56	0.0476793
14	5	600	2.18277	-161.95	0.0303642
15					

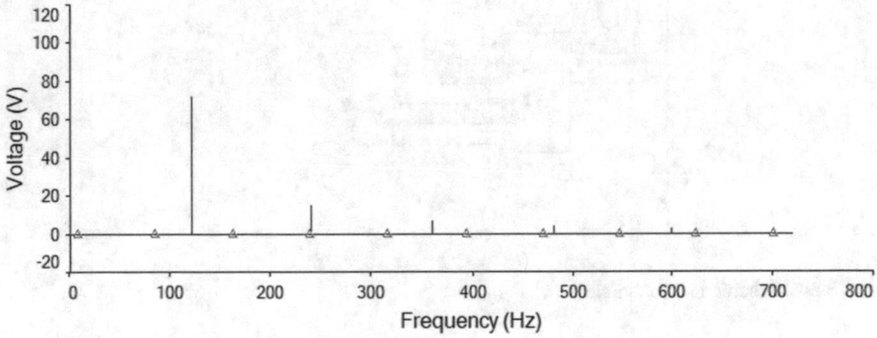

Fig. 5.33 Simulation result

5.6   Example 5: Three Phase Controlled Rectifiers

We will study a three phase controlled rectifier in this example. Assume the schematic shown in Fig. 5.34. Settings of the thyristors are shown in Fig. 5.35. The voltage drop of thyristors are assumed to be zero. The schematic of Fig. 5.34 uses the phase angle controller 6 phase block (Fig. 5.36) to trigger the thyristors. Triggering angle is set with the variable DC source which is connected to the phase angle controller 6 phase block.

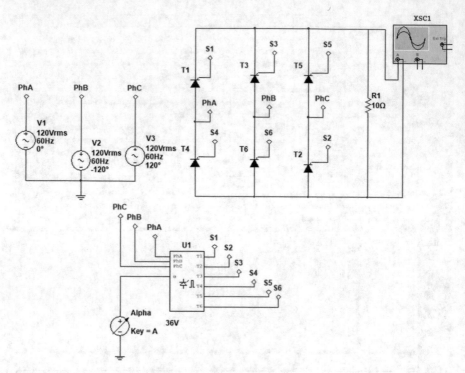

Fig. 5.34 Schematic of example 5

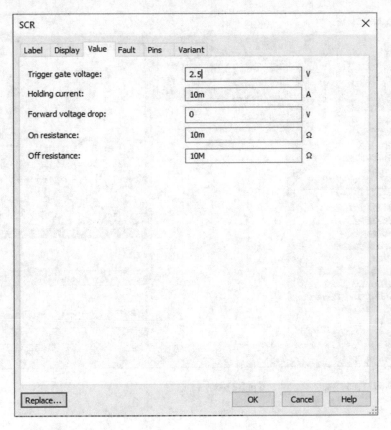

Fig. 5.35 SCR settings

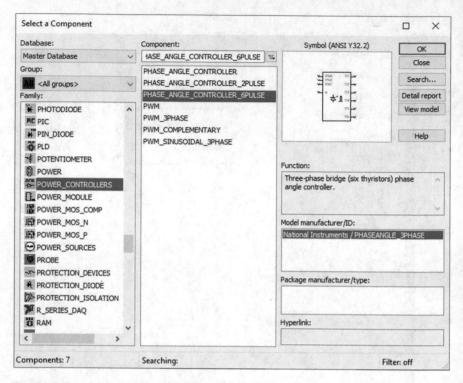

Fig. 5.36 Phase angle controller 6 pulse block

Figures 5.37, 5.38, 5.39, and 5.40 shows the load voltage for triggering angles of 0°, 15°, 30° and 45°. Note that the frequency of output voltage is six times bigger than the AC source frequency.

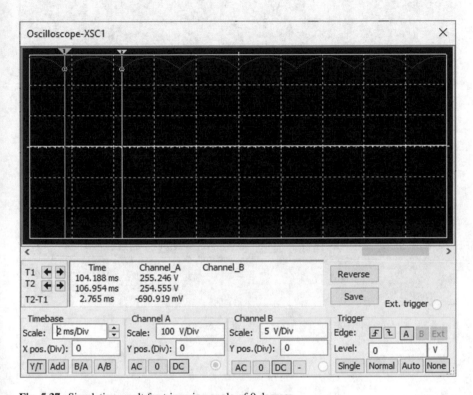

Fig. 5.37 Simulation result for triggering angle of 0 degrees

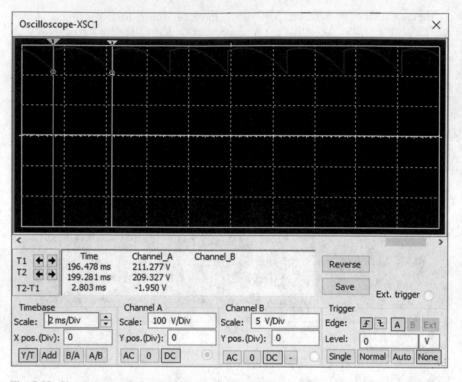

Fig. 5.38 Simulation result for triggering angle of 15 degrees

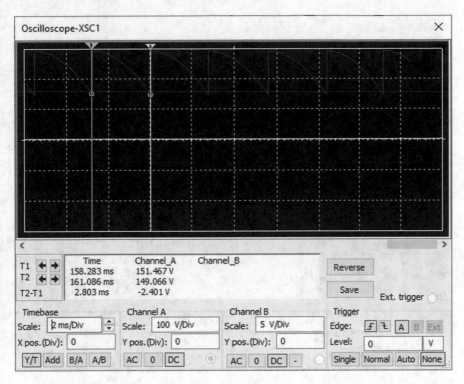

Fig. 5.39 Simulation result for triggering angle of 30 degrees

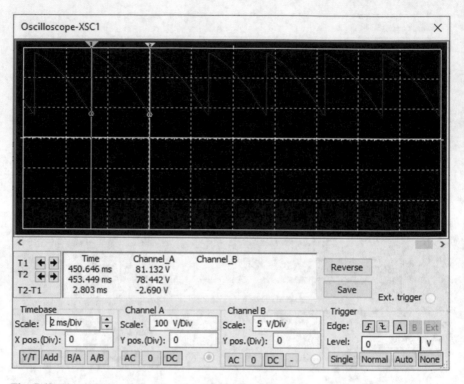

Fig. 5.40 Simulation result for triggering angle of 45 degrees

You can use the differential voltage probes to measure the load voltage (Fig. 5.41).

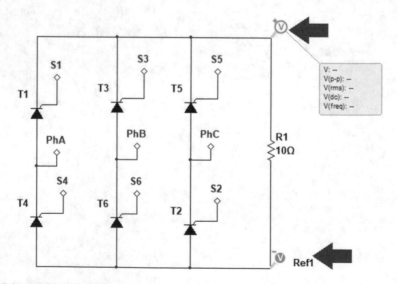

Fig. 5.41 Addition of differential probe to the circuit

RMS and dc values of load voltage is shown in Fig. 5.42.

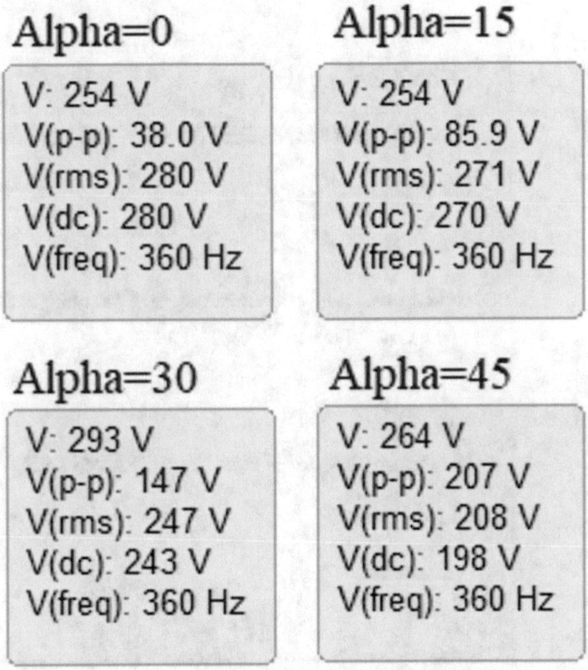

Fig. 5.42 Simulation results

Alpha=0

V: 254 V
V(p-p): 38.0 V
V(rms): 280 V
V(dc): 280 V
V(freq): 360 Hz

Alpha=15

V: 254 V
V(p-p): 85.9 V
V(rms): 271 V
V(dc): 270 V
V(freq): 360 Hz

Alpha=30

V: 293 V
V(p-p): 147 V
V(rms): 247 V
V(dc): 243 V
V(freq): 360 Hz

Alpha=45

V: 264 V
V(p-p): 207 V
V(rms): 208 V
V(dc): 198 V
V(freq): 360 Hz

Let's verify the Multisim results. Average value (DC component) of output voltage can be calculated using the $V_{dc} = \frac{3\sqrt{3}}{\pi} V_s \cos(\alpha)$ formula. V_s and α show the peak of input voltage and triggering angle, respectively. Result of calculations in Fig. 5.43 are quite close to Multisim result shown in Fig. 5.42.

```
Command Window                                                   ⊙

>> alpha=0;Vdc=3*sqrt(3)/pi*(120*sqrt(2))*cosd(alpha)

Vdc =

  280.6908

>> alpha=15;Vdc=3*sqrt(3)/pi*(120*sqrt(2))*cosd(alpha)

Vdc =

  271.1265

>> alpha=30;Vdc=3*sqrt(3)/pi*(120*sqrt(2))*cosd(alpha)

Vdc =

  243.0854

>> alpha=45;Vdc=3*sqrt(3)/pi*(120*sqrt(2))*cosd(alpha)

Vdc =

  198.4784

fx >>
```

Fig. 5.43 MATLAB calculations

Current probes can be used to measure the load current and thyristor currents. Simulation result for triggering angle of zero degrees is shown in Fig. 5.44. Note that the average value of thyristor T1 current is one third of load current. This is expected since each thyristor carries the load current for one third of the period.

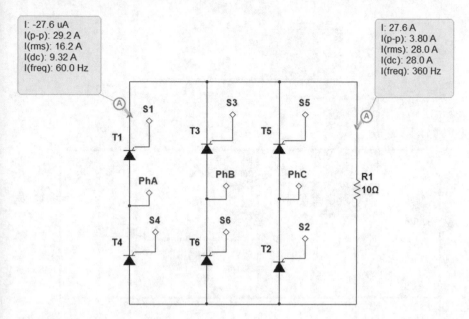

Fig. 5.44 Probe reading for triggering angle of 0 degrees

5.7   Example 6: Dimmer Circuit

A dimmer circuit controls the amount of power which is applied to a load. If the load is an incandescent light bulb, then the change of applied power changes the light intensity. Dimmers can be used to control the light intensity of incandescent light bulb, temperature of heaters and speed of small universal motors.

Schematic of a simple dimmer circuit is shown in Fig. 5.45. Place of used components are shown in Figs. 5.46, 5.47, and 5.48. You can use a resistor instead of the light bulb as well.

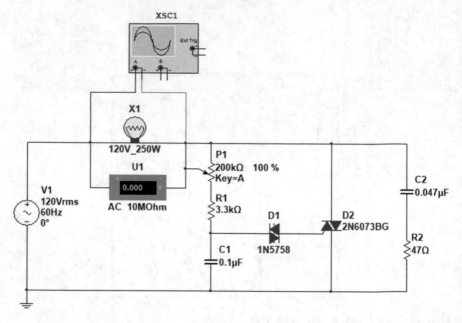

Fig. 5.45 Schematic of example 6

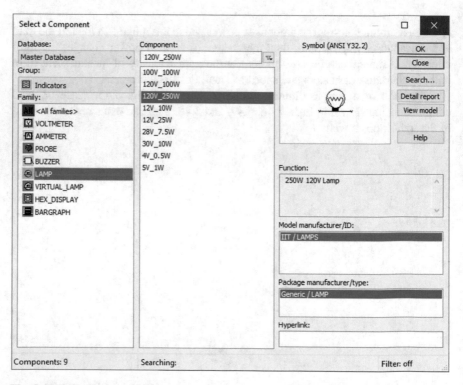

Fig. 5.46 Different types of lamps

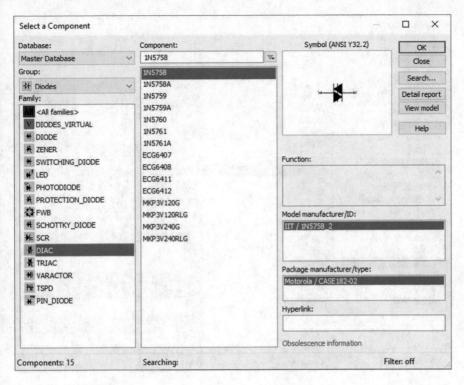

Fig. 5.47 1 N5758 diode

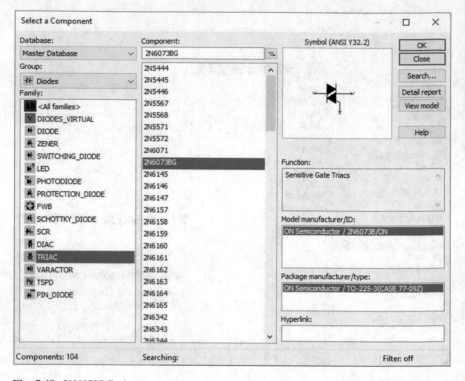

Fig. 5.48 2N6073BG triac

Run the simulation. Change the potentiometer P1 value and see its effect on the load voltage. Figures 5.49, 5.50, and 5.51 shows the load voltage waveform for different values of P1.

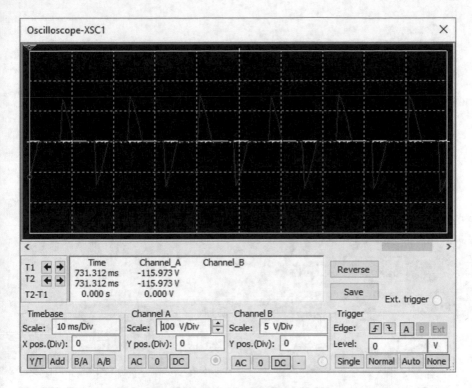

Fig. 5.49 Potentiometer is set to 100%. Voltmeter U1 shows 65.6 Vrms

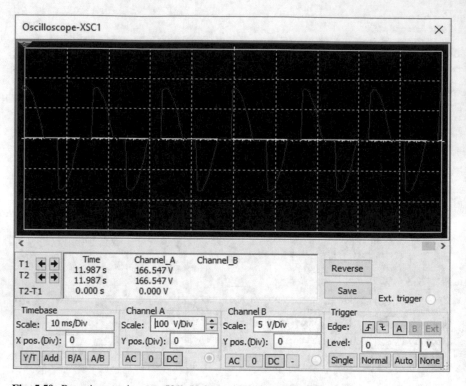

Fig. 5.50 Potentiometer is set to 50%. Voltmeter U1 shows 101.7 Vrms

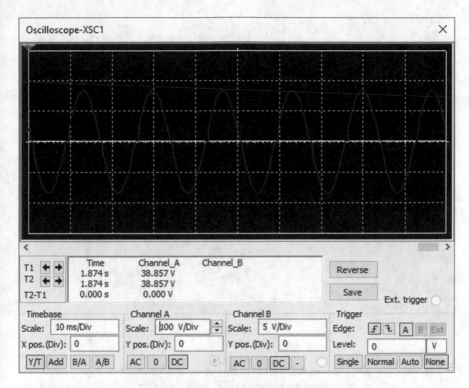

Fig. 5.51 Potentiometer is set to 0%. Voltmeter U1 shows 118.8 Vrms

5.8   Example 7: Switching of MOSFET

MOSFET's are very important switches in power electronics. We want to study the switching behavior of MOSFET's in this example. Assume the circuit shown in Fig. 5.52.

Fig. 5.52 Schematic of example 7

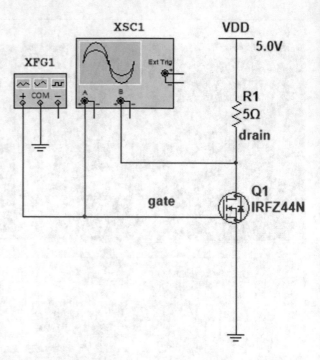

Double click the signal generator and do the settings similar to Fig. 5.53. Definition of amplitude and offset are shown in Fig. 5.54. Note that the offset box sets the average value (DC component) of the signal. The settings shown in Fig. 5.53 produce the waveform which is shown in Fig. 5.55.

Fig. 5.53 Function generator window

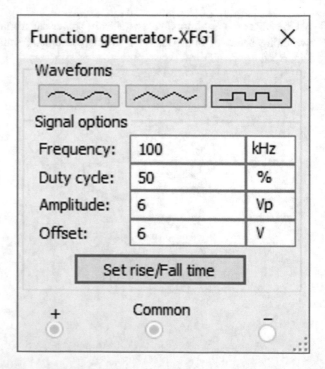

Fig. 5.54 Visualization of
offset and amplitude

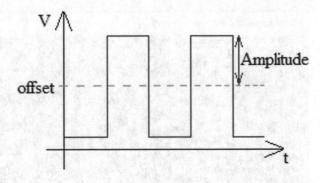

Fig. 5.55 Waveform
generated with settings of
Fig. 5.53

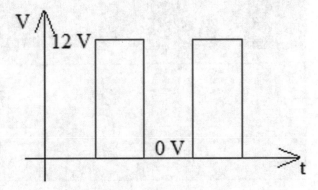

Run the simulation. The result shown in Fig. 5.56 is obtained. When the control
signal is high, the MOSFET is on and act as a closed switch. In this case the output
voltage is about zero. When the control signal is low, the MOSFET is off and act as
an open switch. In this case the output voltage is equal to VDD. The behavior of this
circuit is similar to NOT gate.

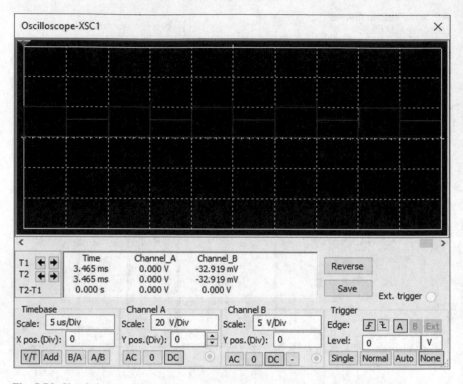

Fig. 5.56 Simulation result

You can use the Y pos. (Div) to see the waveforms better (Fig. 5.57).

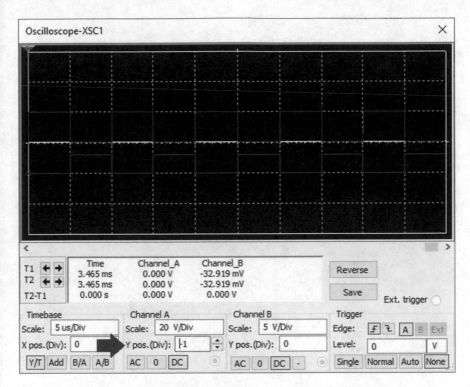

Fig. 5.57 Using the Y pos. (div) box to mode the waveform of channel 1

You can use the probes to measure the RMS and average of currents and voltages (Fig. 5.58).

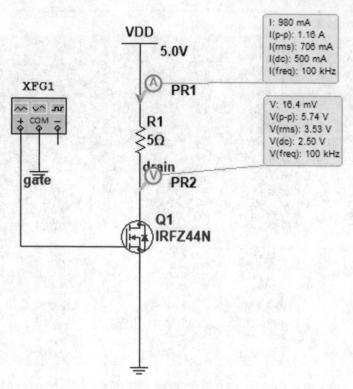

Fig. 5.58 Simulation result

Let's verify the results. The voltage waveform at drain is a square waveform with maximum of (about) 5 V. The average and RMS value for square waveform can be calculated using the $\frac{V_{peak}}{2}$ and $V_{peak}\sqrt{0.5}$, respectively. Similarly, the drain current is a square wave current with amplitude of $5/5 = 1A$. The calculation in Fig. 5.59 shows that the values given by the probes are correct.

Fig. 5.59 MATLAB
calculations

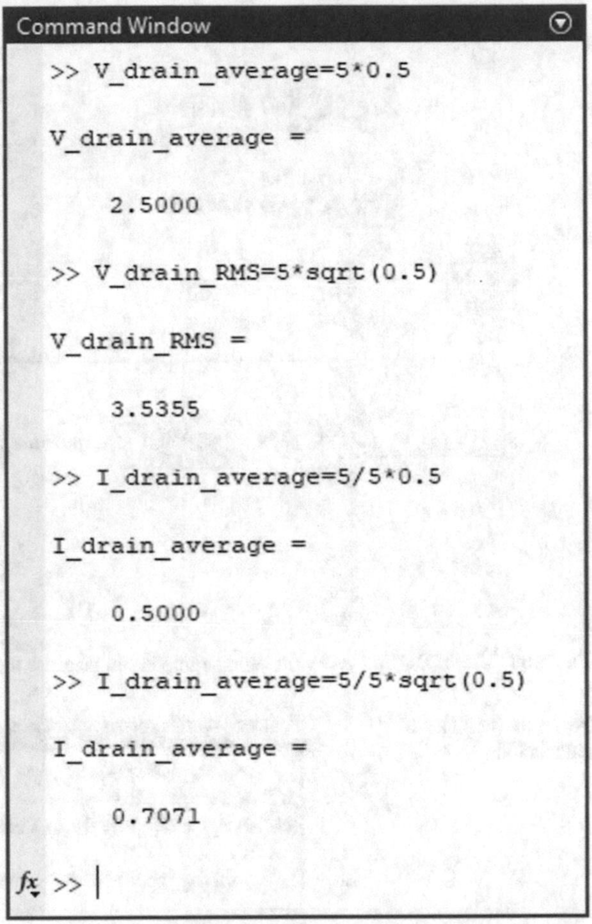

```
Command Window                                          ⊙

  >> V_drain_average=5*0.5

  V_drain_average =

        2.5000

  >> V_drain_RMS=5*sqrt(0.5)

  V_drain_RMS =

        3.5355

  >> I_drain_average=5/5*0.5

  I_drain_average =

        0.5000

  >> I_drain_average=5/5*sqrt(0.5)

  I_drain_average =

        0.7071

fx >> |
```

You can use voltmeter and ammeter to measure the average value and RMS of
AC component of the signals (Fig. 5.60). You can use the calculation shown in
Fig. 5.61 to obtain the RMS values of drain current and its voltage.

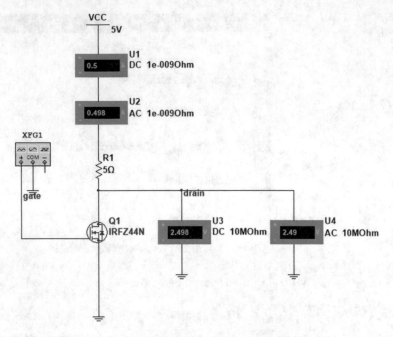

Fig. 5.60 Use of DC and AC voltmeters/ammeters to measure the RMS

Fig. 5.61 MATLAB
calculations

```
Command Window                                    ⊙

>> Idc_drain=0.5;
>> Iac_drain=0.5;
>> I_drain_RMS=sqrt(Idc_drain^2+Iac_drain^2)

I_drain_RMS =

    0.7071

>> Vdc_drain=2.498;
>> Vac_drain=2.49;
>> V_drain_RMS=sqrt(Vdc_drain^2+Vac_drain^2)

V_drain_RMS =

    3.5271

fx >> |
```

Let's use the transient analysis to see the dissipated power waveform. In order to do this, click the interactive button (Fig. 5.62).

Fig. 5.62 Interactive button

Select the transient and do the settings similar to Fig. 5.63.

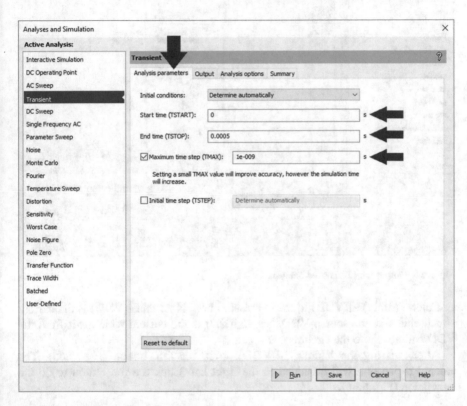

Fig. 5.63 Analysis tab of transient analysis

Go to the output tab and click the add expression (Fig. 5.64).

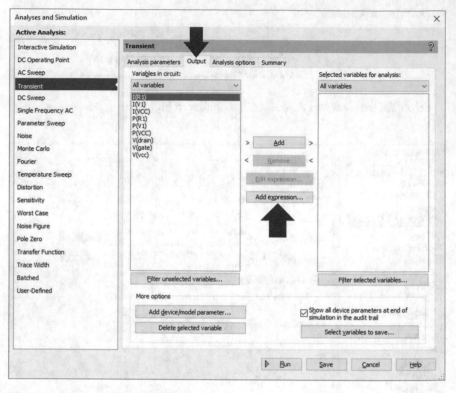

Fig. 5.64 Output tab of transient analysis

Enter V(drain)*-I(VDD) to the expression box. Note that I(VDD) is the current which enters to the source VDD. So, -I(VDD) is the current which exit from the VDD and goes into the circuit.

After entering the V(drain)*-I(VDD), click the OK button (Fig. 5.65). The entered expression will be added to the right list. Click the run button to do the simulation (Fig. 5.66).

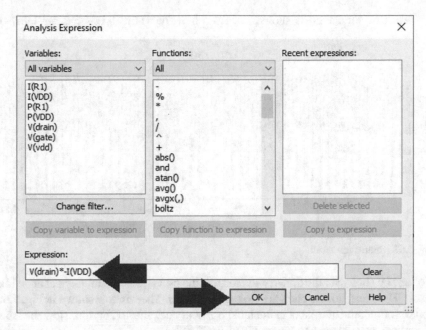

Fig. 5.65 Addition of V(drain)*-I(VDD) to the expression box

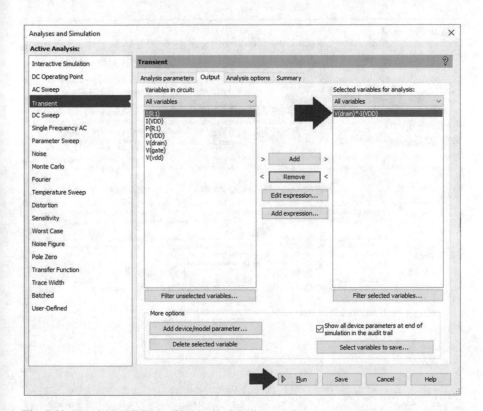

Fig. 5.66 V(drain)*-I(VDD) is added to the right list

Result of simulation is shown in Fig. 5.67. The vertical axis has the label of

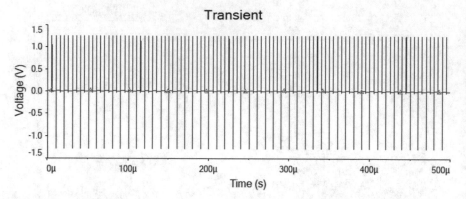

Fig. 5.67 Simulation result

Voltage (V) however the drawn waveform is the power waveform. Let's change this label. To do this, double click the Voltage (V) label. The window shown in Fig. 5.68 appears. Enter the new label in the label box and click the OK button. Now the label of vertical axis changes to Power (W) (Fig. 5.69).

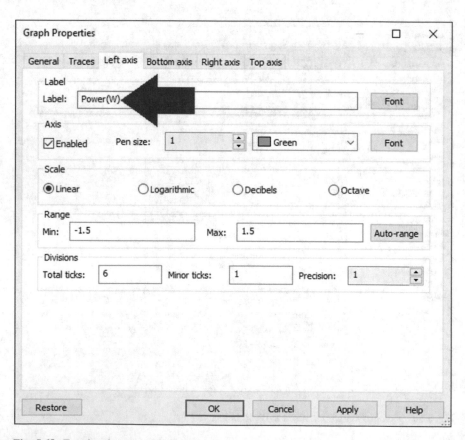

Fig. 5.68 Entering the y axis label

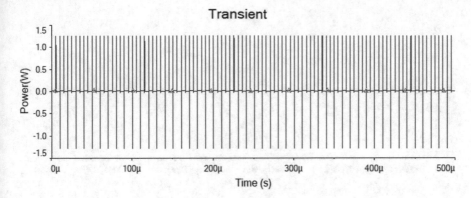

Fig. 5.69 Simulation result

Use the zoom in area tool (Fig. 5.70) to zoom in the obtained waveform. Power waveform for turn off and turn on instants are shown in Figs. 5.71 and 5.72, respectively.

Fig. 5.70 Zoom in area button

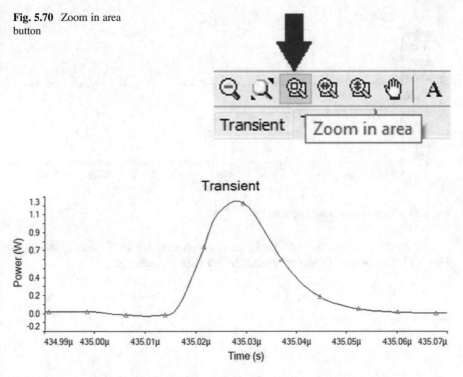

Fig. 5.71 Power waveform for turn off

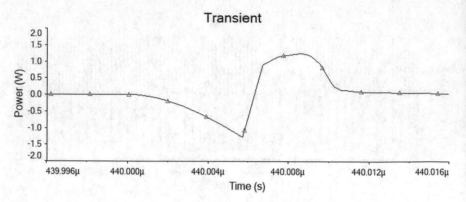

Fig. 5.72 Power waveform for turn on

Zoom in the waveform to see the conduction losses. The conduction loss for the time interval which the MOSFET is on is shown in Fig. 5.73.

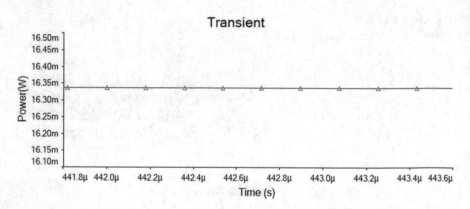

Fig. 5.73 Conduction loss for on MOSFET

The conduction loss for the time interval which the MOSFET is off is shown in Fig. 5.74. The power dissipated during the off state is negligible.

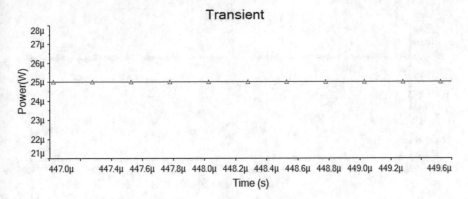

Fig. 5.74 Conduction loss for off MOSFET

You can measure the average dissipated power easily. To do this, draw the graph of the expression shown in Fig. 5.75. The graph of this expression is shown in Fig. 5.76.

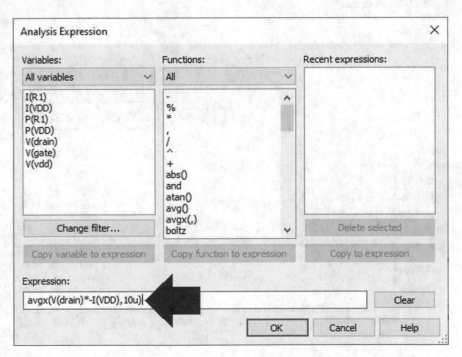

Fig. 5.75 Entering avgx(V(drain)*-I(VDD),10u) to the expression box

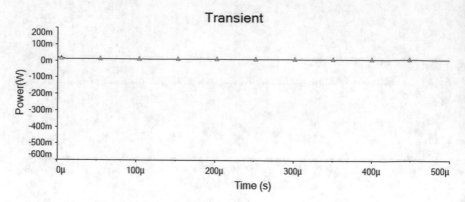

Fig. 5.76 Simulation result

Use the zoom in area (Fig. 5.77) to zoom in the waveform (Fig. 5.78). The average of dissipated power is about 10.4755 mW.

Fig. 5.77 Zoom in area button

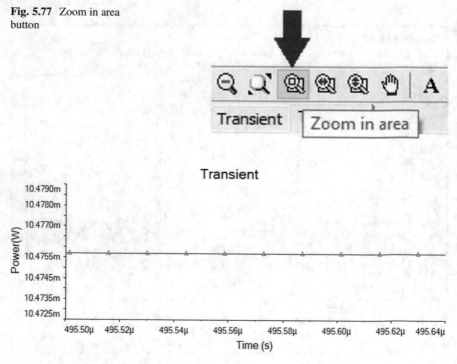

Fig. 5.78 Average dissipated power for VDD = 5 V

If you change the VDD to 25 V, the average of dissipated power increases to 241.5 mW (Fig. 5.79).

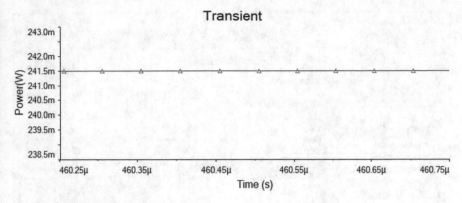

Fig. 5.79 Average dissipated power for VDD = 25 V

5.9   Example 8: Buck Converter

We want to simulate a buck converter in this example. Assume the schematic shown in Fig. 5.80.

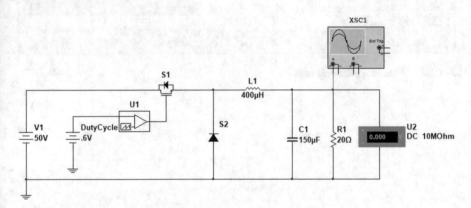

Fig. 5.80 Schematic of example 8

Place of the used MOSFET and diode is shown in Figs. 5.81 and 5.82. Settings of these blocks are shown in Figs. 5.83 and 5.84.

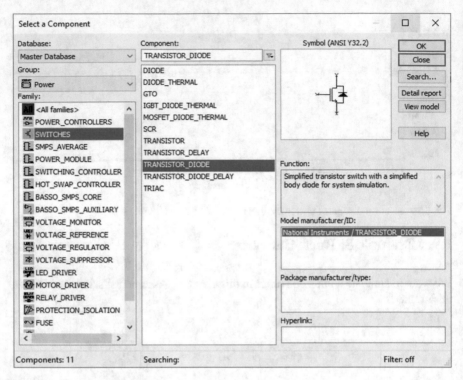

Fig. 5.81 Transistor_diode element

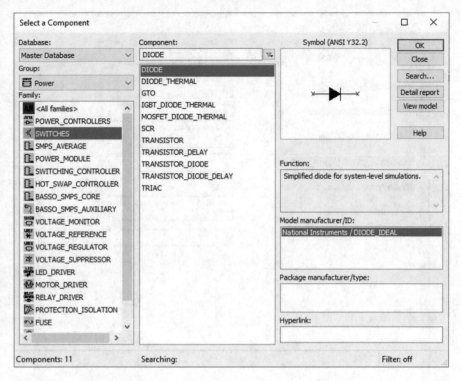

Fig. 5.82 Diode element

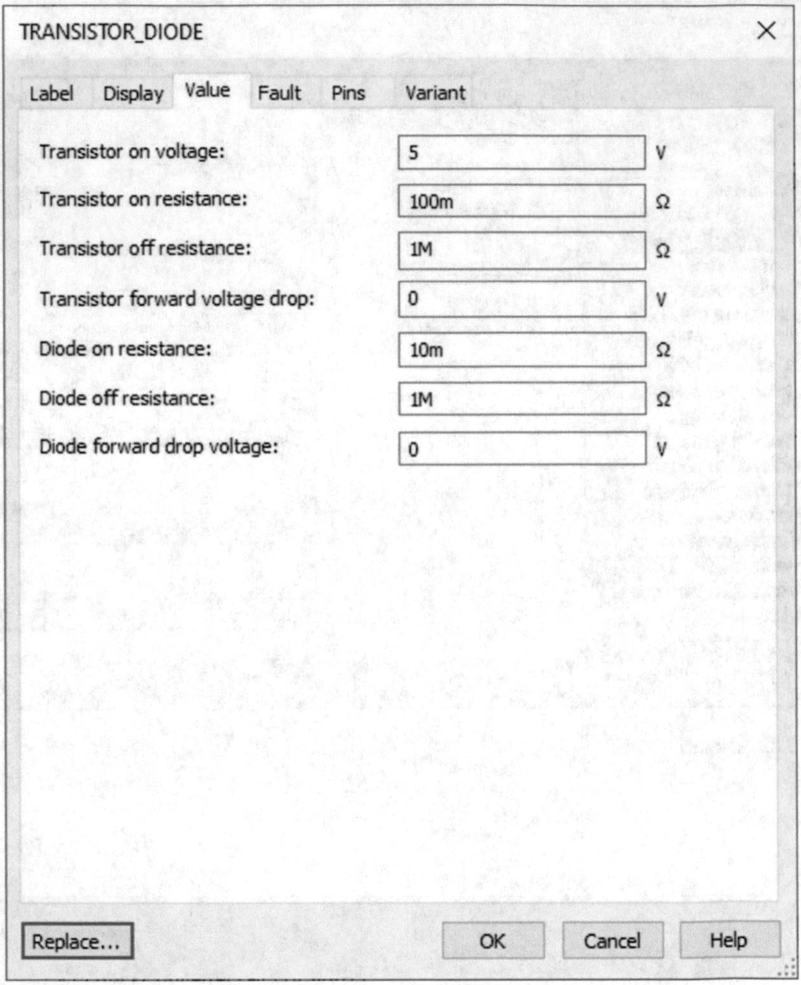

TRANSISTOR_DIODE   ✕

| Label | Display | Value | Fault | Pins | Variant |

Transistor on voltage:	5	V
Transistor on resistance:	100m	Ω
Transistor off resistance:	1M	Ω
Transistor forward voltage drop:	0	V
Diode on resistance:	10m	Ω
Diode off resistance:	1M	Ω
Diode forward drop voltage:	0	V

Replace...   OK   Cancel   Help

Fig. 5.83 Settings of transistor_diode

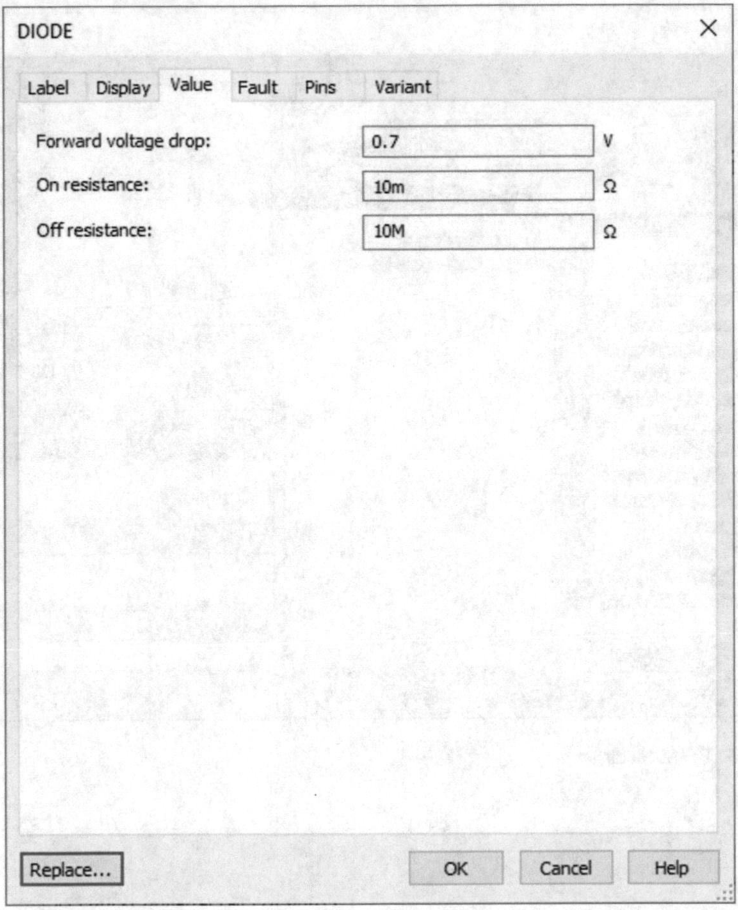

DIODE   ✕

Label   Display   Value   Fault   Pins   Variant

Forward voltage drop: | 0.7 | V

On resistance: | 10m | Ω

Off resistance: | 10M | Ω

| Replace... | | OK | | Cancel | | Help |

Fig. 5.84 Settings of diode

The MOSFET is controlled with PWM block (Fig. 5.85). The input for this block is the desired duty cycle. Settings of the PWM block is shown in Fig. 5.86.

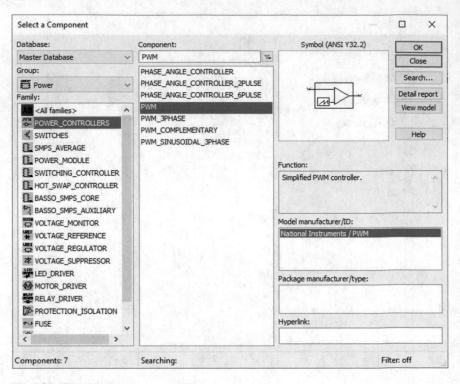

Fig. 5.85 PWM block

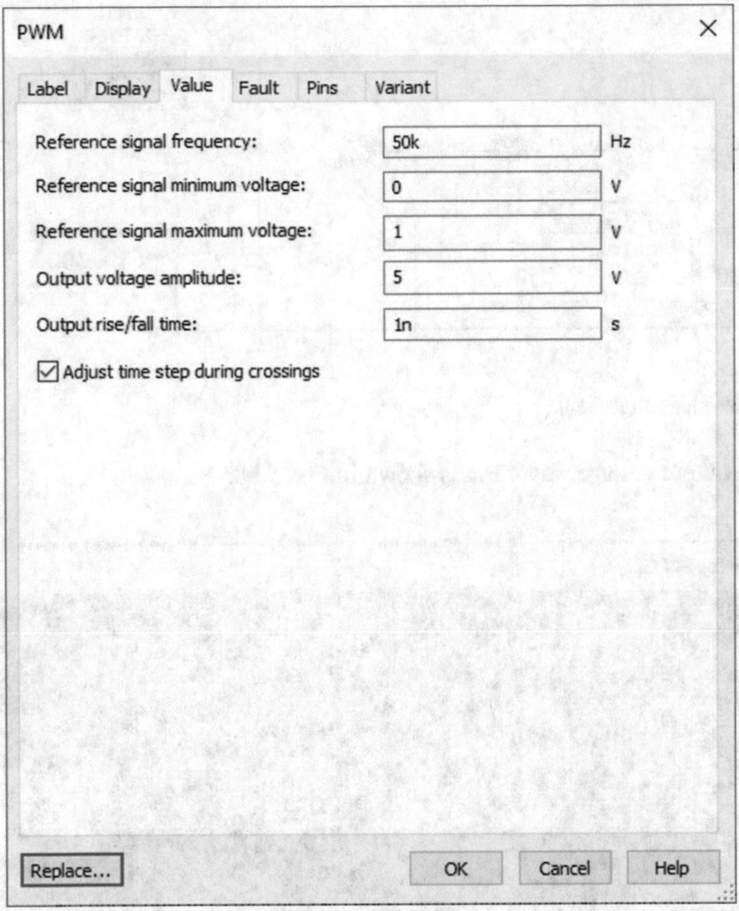

Fig. 5.86 PWM block settings

Run the simulation. The output voltage is about 30 V (Fig. 5.87).

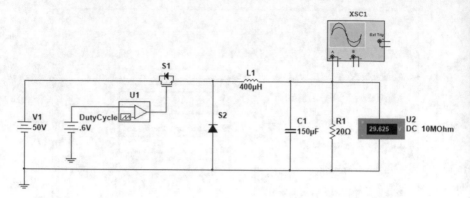

Fig. 5.87 Simulation result

The output voltage waveform is shown in Fig. 5.88.

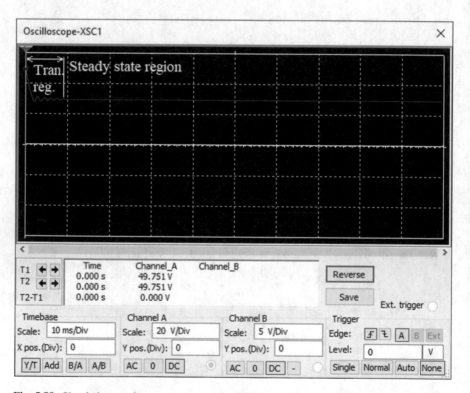

Fig. 5.88 Simulation result

You can use the AC mode of oscilloscope to measure the output voltage ripple. According to Fig. 5.89, the peak-peak of output voltage ripple is about 10 mV.

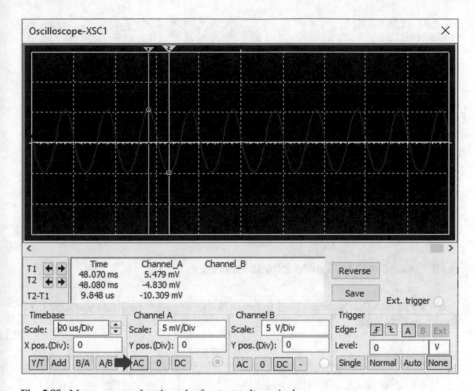

Fig. 5.89 Measurement of peak-peak of output voltage ripple

Let's measure the converter efficiency. We need to measure the input and output power. According to Fig. 5.90, the efficiency is about 99% (Fig. 5.91).

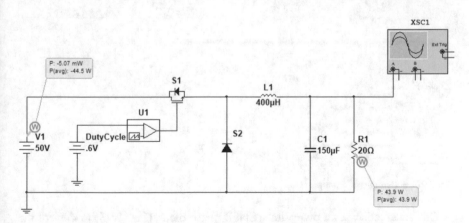

Fig. 5.90 Measurement of input and output power

Fig. 5.91 Calculation of converter efficiency

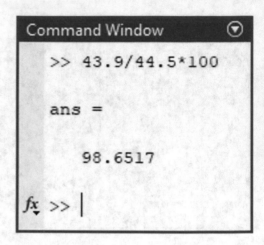

5.10 Example 9: Three Phase Inverter

We want to simulate a three phase inverter in this example. Assume the schematic shown in Fig. 5.92.

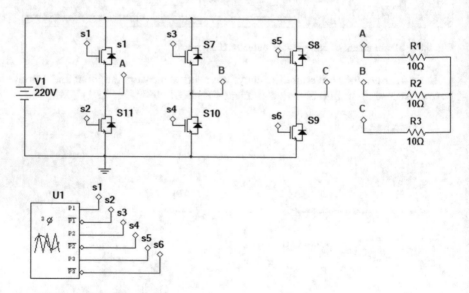

Fig. 5.92 Schematic of example 9

This schematic uses a pwm sinusoidal 3 phase block (Fig. 5.93) to control the switches. Settings of pwm sinusoidal 3 phase are shown in Fig. 5.94.

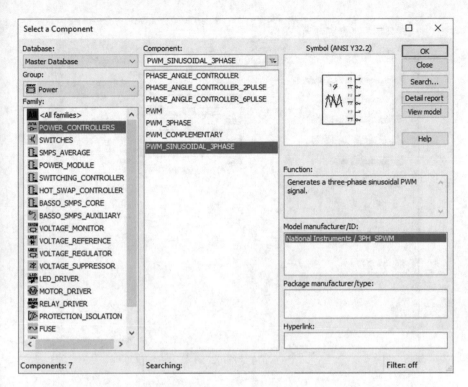

Fig. 5.93 PWM sinusoidal 3 phase block

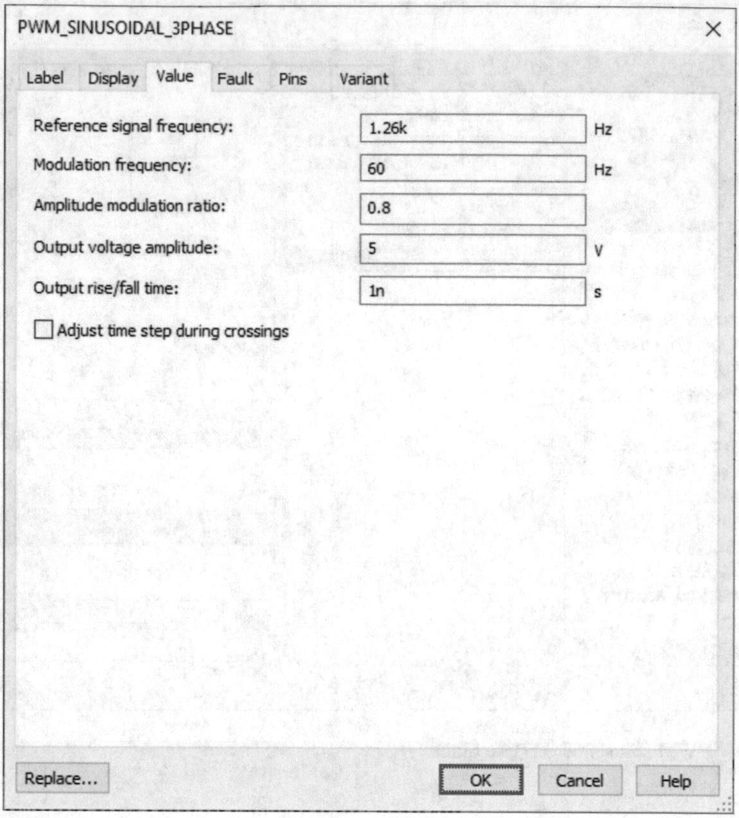

PWM_SINUSOIDAL_3PHASE	×

Label   Display   Value   Fault   Pins   Variant

Reference signal frequency:	1.26k	Hz
Modulation frequency:	60	Hz
Amplitude modulation ratio:	0.8	
Output voltage amplitude:	5	V
Output rise/fall time:	1n	s

☐ Adjust time step during crossings

Replace...	OK Cancel Help

Fig. 5.94 PWM sinusoidal 3 phase block settings

Run the simulation. You can use the probes to measure the line-line and phase voltages (Fig. 5.95).

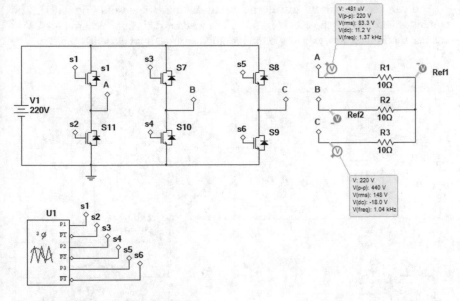

Fig. 5.95 Measurement of load voltage

The load current can be measured with current probes (Fig. 5.96).

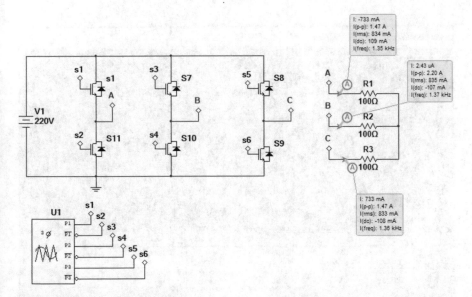

Fig. 5.96 Measurement of load currents

You can see the current waveforms using the current sensors (Fig. 5.97). The current waveforms are shown in Fig. 5.98. The load current contains harmonics. You can use the Fourier analysis to see the harmonics of the load current/voltage.

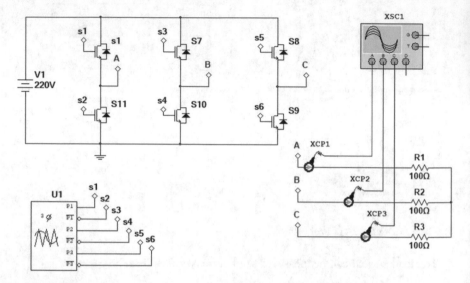

Fig. 5.97 Use three current probe to see the current waveforms with an oscilloscope

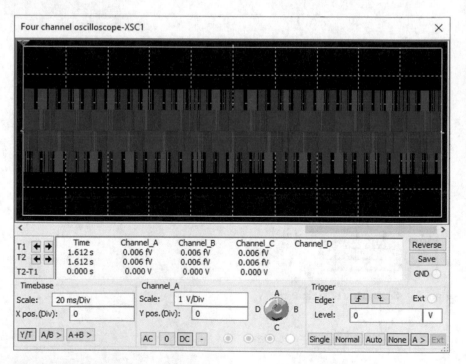

Fig. 5.98 Simulation result

Current harmonics decreases for RL loads. The impedance of the inductor increases with frequency and it decreases the high frequency current components. For instance, the current waveform of R = 10 Ω and L = 10 mH (Fig. 5.99) is shown in Fig. 5.100. This current waveform has less harmonics in comparison to Fig. 5.98.

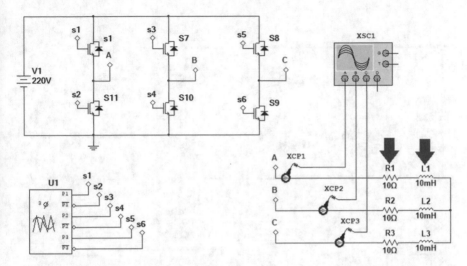

Fig. 5.99 Inverter supplies an inductive load

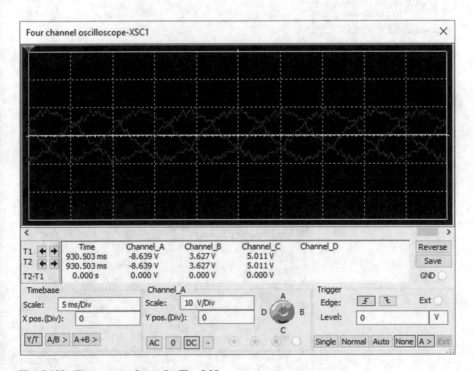

Fig. 5.100 Current waveforms for Fig. 5.99

5.11   Example 10: Calculation of THD for Inverters

In this example we will use the distortion analyzer block to measure the Total Harmonic Distortion (THD) of inverters. Assume the schematic shown in Fig. 5.101.

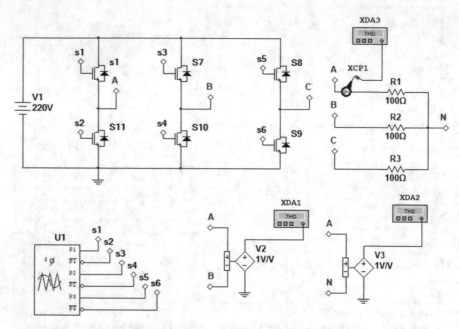

Fig. 5.101 Schematic of example 10

Run the simulation and double click the distortion analyzers to see the THD's. Simulation results are shown in Figs. 5.102, 5.103, and 5.104. According to Fig. 5.102, the line-line voltage THD is 21.495%, the line-neutral voltage THD is 22.257% and the current THD is 22.257%.

Fig. 5.102 Simulation result

Distortion analyzer-XDA1	✕
Total harmonic distortion (THD)	
21.495 %	

Start	Fundamental freq.	60	Hz
Stop	Resolution freq.	5 Hz	
		5 Hz	

| Controls | | | Display | | |
| THD | SINAD | Set... | % | dB | In ◉ |

Fig. 5.103 Simulation result

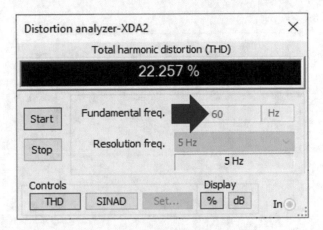

Fig. 5.104 Simulation result

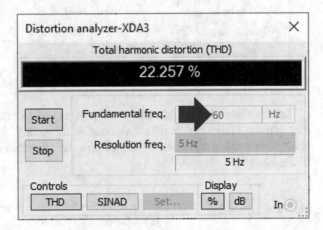

Change the load to RL type and run the simulation (Fig. 5.105).

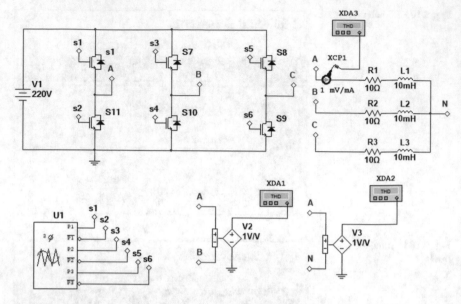

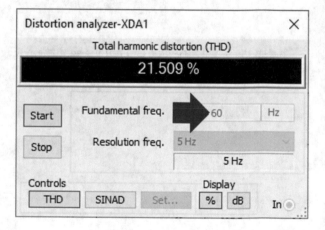

Fig. 5.105 Measurement of current THD for inverter with inductive load

Results of simulation for RL load is shown in Figs. 5.106, 5.107, and 5.108. Note that current THD (Fig. 5.108) decreased considerably. The voltage THD didn't change since the modulation technique and modulation index is the same for both of the loads.

Fig. 5.106 Simulation result

Fig. 5.107 Simulation result

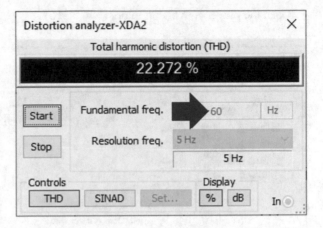

Fig. 5.108 Simulation result

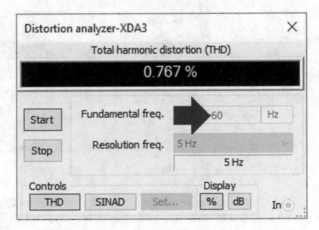

5.12 More Examples Related to Power Electronics and Electrical Drives

You can see more examples about the power electronics circuits in the samples folder of the Multisim. Click the open samples button (Fig. 5.109).

Fig. 5.109 Open samples button

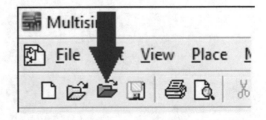

Open the power conversion folder (Fig. 5.110).

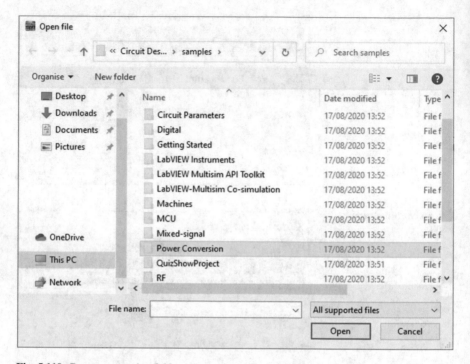

Fig. 5.110 Power conversion folder

The circuits related to power electronics are in these folders. For instance, to see samples related to rectifiers, you need to open the AC-DC folder (Fig. 5.111).

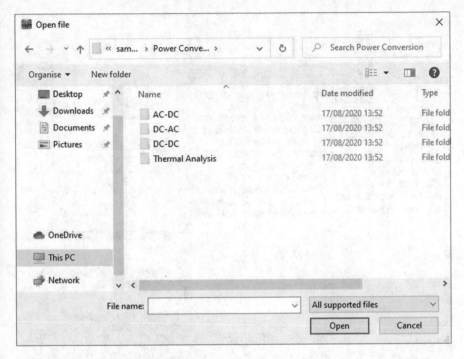

Fig. 5.111 Folders inside the power conversion folder

Multisim has model of different electrical machines (Fig. 5.112). This models permits you to be able to simulate the electrical drives.

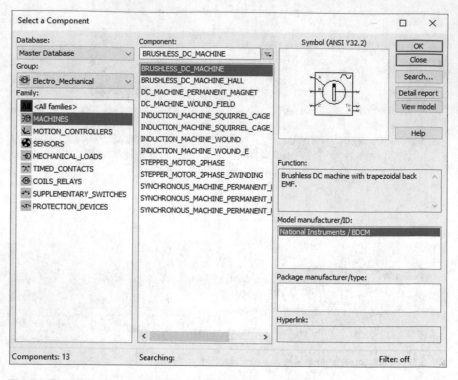

Fig. 5.112 Different types of machines available in the Multisim

Sample simulation related to electrical drives can be found in the machines folder (Fig. 5.113).

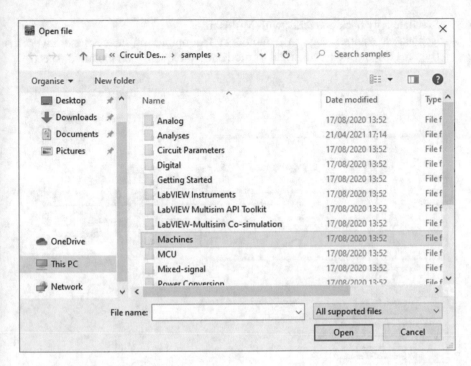

Fig. 5.113 Machines folder

5.13 Exercises

1. **(a)** Measure the power factor of Fig. 5.28 for triggering angle of 30°.
   (b) Measure the power factor of Fig. 5.34 for triggering angle of 45°.
2. Simulate the boost converter shown in Fig. 5.114 with Multisim and measure the output voltage, output voltage ripple, output power and efficiency. Signal S1 has frequency of 200 kHz and duty cycle of 0.5. Voltage drop of diode is assumed to be 0.7 V.

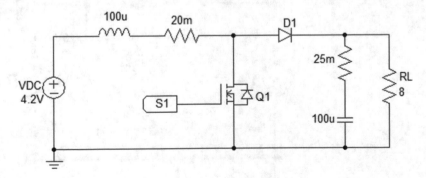

Fig. 5.114 Schematic of exercise 2

3. Simulate a fly back converter with Multisim.
4. Connect a squirrel cage AC motor to the output of three phase inverter (Fig. 5.115) and study the effect of modulation frequency on the motor speed.

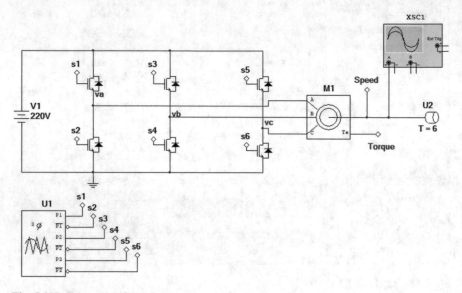

Fig. 5.115 Schematic of exercise 4

5. A single phase voltage controller is shown in Fig. 5.116. SCR S1 is triggered at $2k\pi + \alpha$ and SCR S2 is triggered at $(2k + 1)\pi + \alpha$ angles (k = 0, 1, 2, ...). The load voltage/current waveforms are shown in Fig. 5.117.

Fig. 5.116 Schematic of exercise 5

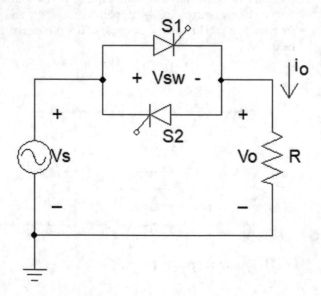

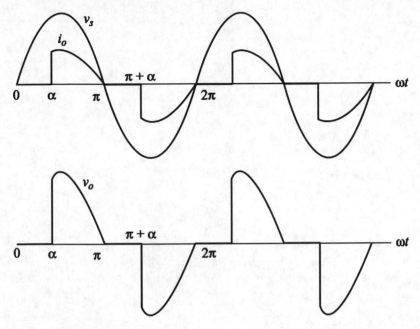

Fig. 5.117 Waveforms of circuit in Fig. 5.116

Use Multisim to simulate the circuit for $\alpha = 30°$, $60°$ and $90°$.

Further Readings

1. D.W. Hart, *Power Electronics* (McGraw-Hill, 2010)
2. N. Mohan, T.M. Undeland, W.P. Robbins, *Power Electronics: Converters, Applications, and Design*, 3rd edn. (John Wiley and Sons, 2007)
3. R.W. Erikson, D. Maksimovic, *Fundamentals of Power Electronics*, 3rd edn. (Springer, 2020)
4. F.L. Luo, H. Ye, *Advanced DC-DC Converters* (CRC Press, 2016)
5. F. Asadi, K. Eguchi, *Dynamics and Control of DC-DC Converters* (Morgan & Claypool, 2018)
6. F. Asadi, K. Eguchi, *Simulation of Power Electronics Converters Using PLECS* (Academic Press, 2019)
7. Square wave inverter: https://www.youtube.com/watch?v=MNp5V8IZgNA (Visiting date: 15.07.2021)
8. H bridge inverter: https://www.instructables.com/H-Bridge-Inverter-Simulation-Using-NI-Multisim-and/ (Visiting date: 15.07.2021)

Appendix: Review of Some of the Important Theoretical Concepts

This appendix reviews some of the important theoretical concepts used in the book.

Instantaneous Power

The instantaneous power of a device ($p(t)$) is defined as:

$$p(t) = v(t) \times i(t) \tag{A.1}$$

where $v(t)$ is the voltage across the device and $i(t)$ is the current through the device. The instantaneous power is generally a time-varying quantity. If the passive sign convention illustrated in Fig. A.1 is observed, the device is absorbing power if $p(t)$ is positive at a specified value of time t. The device is supplying power if $p(t)$ is negative.

Fig. A.1 Passive sign convention: $p(t) > 0$ indicates power is being absorbed

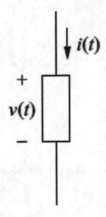

© The Editor(s) (if applicable) and The Author(s), under exclusive license to Springer Nature Switzerland AG 2022
F. Asadi, *Essential Circuit Analysis using NI Multisim™ and MATLAB®*,
https://doi.org/10.1007/978-3-030-89850-2

For instance, consider the simple circuit shown in Fig. A.2. In this circuit, $v_{in}(t) = 311 \sin (377t)$ and $R = 50\,\Omega$.

Fig. A.2 A simple resistive circuit

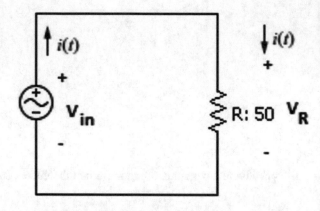

According to Ohm's law, $i(t) = \frac{v_{in}(t)}{R} = 6.22 \sin (377t)$ and instantaneous power for resistor R is:

$$p_R(t) = 311 \sin (377t) \times 6.22 \sin (377t) = 2345 \sin^2 (377t). \qquad (A.2)$$

Obtained result is positive for all the times, i.e., $\forall t,\ \sin^2(377t) > 0$. This is expected since resistor dissipates power.

The instantaneous power of AC source can be calculated with the aid of Fig. A.3.

Fig. A.3 Calculation of instantaneous power of input AC source

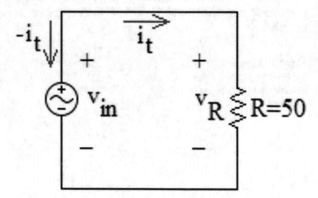

The instantaneous power of AC source is:

$$p_{V_{in}}(t) = 311 \sin (377t) \times -6.22 \sin (377t) = -2345 \sin^2(377t) \qquad (A.3)$$

Obtained result is negative for all the time. We expect this result since the AC source supplies the power into the load. For instance, at $t = 12$ ms, $p_{V_{in}}(t) = -2.263$ kW and $p_R(t) = +2.263$ kW. This means that at $t = 12$ ms, AC source supplies 2.263 kW and resistor absorbs 2.263 kW. Figure A.4 shows the instantaneous power waveforms on the same graph.

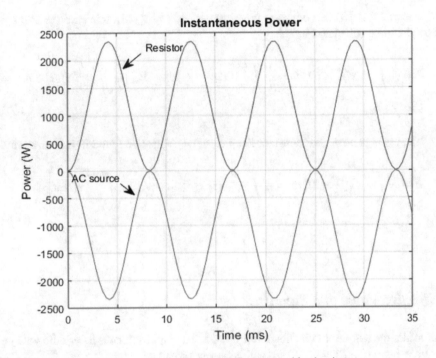

Fig. A.4 Graph of instantaneous power for input AC source and load resistor

Average Power

Function $f(t)$ is periodic if,

$$\exists T > 0, \forall t \, f(t + T) = f(t) \qquad (A.4)$$

T is called the period. For instance, $f(t) = \sin (t)$ is periodic since $f(t + 2\pi) = f(t)$.

If device voltage ($v(t)$) and current ($i(t)$) are periodic, i.e., $v(t) = v(t + T)$ and $i(t) = i(t + T)$, then the instantaneous power will be a periodic since

$$p(t) = v(t) \times i(t)$$

$$p(t+T) = v(t+T) \times i(t+T) = v(t) \times i(t) = p(t) \tag{A.5}$$

The average power for such a periodic waveform is defined as:

$$P = \frac{1}{T}\int_{t_0}^{t_0+T} p(t)dt = \frac{1}{T}\int_{t_0}^{t_0+T} v(t) \times i(t)dt \tag{A.6}$$

Assume that $v(t)$ is a constant function, i.e., $v(t) = V_{dc}$. In this case the average power can be calculated by the

$$P = \frac{1}{T}\int_{t_0}^{t_0+T} v(t) \times i(t)dt = \frac{1}{T}\int_{t_0}^{t_0+T} V_{dc} \times i(t)dt = V_{dc}\left[\frac{1}{T}\int_{t_0}^{t_0+T} i(t)dt\right]$$

$$= V_{dc}I_{avg} \tag{A.7}$$

The average power for constant $i(t)$, i.e., $i(t) = I_{dc}$ can be found in the same way.

$$P = \frac{1}{T}\int_{t_0}^{t_0+T} v(t) \times i(t)dt = \frac{1}{T}\int_{t_0}^{t_0+T} v(t) \times I_{dc}dt = I_{dc}\left[\frac{1}{T}\int_{t_0}^{t_0+T} v(t)dt\right]$$

$$= I_{dc}V_{avg} \tag{A.8}$$

Effective Value of a Signal

Consider the simple circuit shown in Fig. A.5. The input source is a periodic voltage source, i.e., $v(t + T) = v(t)$. The load is purely resistive.

Fig. A.5 A resistor is connected to a periodic voltage source

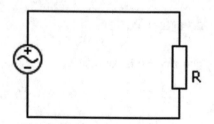

The average power consumed by the resistor is:

$$P = \frac{1}{T}\int_0^T p(t)dt = \frac{1}{T}\int_0^T v(t) \times i(t)dt = \frac{1}{T}\int_0^T \frac{v(t)^2}{R}dt$$

$$= \frac{1}{R}\left[\frac{1}{T}\int_0^T v(t)^2 dt\right] \tag{A.9}$$

Now consider the circuit shown in Fig. A.6. The input source is a constant DC voltage source, i.e., $v(t) = V_{dc}$.

Fig. A.6 The same resistor (as the one in Fig. A.5) is connected to a DC source

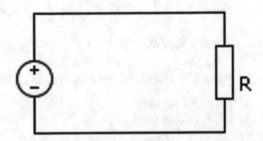

In this case the power consumed by the resistor is $\frac{V_{dc}^2}{R}$. Power consumption of both circuits are the same when $V_{dc} = \sqrt{\frac{1}{T}\int_0^T v(t)^2 dt}$. Since,

$$\frac{1}{R}\left[\frac{1}{T}\int_0^T v(t)^2 dt\right] = \frac{V_{dc}^2}{R} \Rightarrow V_{dc} = \sqrt{\frac{1}{T}\int_0^T v(t)^2 dt} \tag{A.10}$$

The $\sqrt{\frac{1}{T}\int_0^T v(t)^2 dt}$ is called Root Mean Square (RMS) or effective value of signal $v(t)$. So, RMS value of periodic signal $v(t)$ is a DC value which produce the same amount of heat in the resistive load as the periodic signal $v(t)$.

The RMS can be defined for the current waveforms as well.

$$I_{rms} = \sqrt{\frac{1}{T}\int_0^T i(t)^2 dt} \tag{A.11}$$

Example A.1
Determine the RMS value of the periodic pulse waveform shown in Fig. A.7.

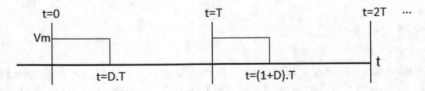

Fig. A.7 Waveform of Example A.1

Solution

$$v(t) = \begin{cases} V_m & 0 < t < DT \\ 0 & DT < t < T \end{cases}$$

$$V_{rms} = \sqrt{\frac{1}{T}\int_0^T v(t)^2 dt} = \sqrt{\frac{1}{T}\left(\int_0^{DT} V_m^2 dt + \int_{DT}^T 0 dt\right)} = \sqrt{\frac{1}{T}(V_m^2 DT)}$$

$$= V_m\sqrt{D}$$

Example A.2 Determine the RMS values of the following waveforms $\left(\omega = \frac{2\pi}{T}\right)$.

(a) $v(t) = V_m \sin(\omega t)$.

(b) $v(t) = |V_m \sin(\omega t)|$.

$$\text{(c) } v(t) = \begin{cases} V_m \sin(\omega t) & 0 < t < \dfrac{T}{2} \\ 0 & \dfrac{T}{2} < t < T \end{cases}.$$

Solution

(a)

$$V_{rms} = \sqrt{\frac{1}{T}\int_0^T (V_m \sin(\omega t))^2 dt} = \sqrt{\frac{1}{T} \times V_m^2 \int_0^T \sin^2(\omega t) dt}$$

$$= \sqrt{\frac{V_m^2}{T}\int_0^T \frac{1 - \cos(2\omega t)}{2} dt} = \sqrt{\frac{V_m^2}{T}\int_0^T \frac{1}{2} dt - \int_0^T \frac{\cos(2\omega t)}{2} dt}$$

$$= \sqrt{\frac{V_m^2}{T} \times \left(\frac{T}{2} - \frac{\sin(2\omega t)}{4\omega}\right) \Big|_0^T}$$

$$= \sqrt{\frac{V_m^2}{T} \times \frac{T}{2} - 0}$$

$$= \sqrt{\frac{V_m^2}{2}}$$

$$= \frac{V_m}{\sqrt{2}}$$

(b) RMS value of $v(t) = |V_m \sin(\omega t)|$ is the same as $v(t) = V_m \sin(\omega t)$. Since $(|V_m \sin(\omega t)|)^2 = (V_m \sin(\omega t))^2$. So, RMS value of $v(t) = |V_m \sin(\omega t)|$ is $\frac{V_m}{\sqrt{2}}$. Graph of $v(t) = |V_m \sin(\omega t)|$ is shown in Fig. A.8. Such a waveform is called Full Wave Rectified in power electronics.

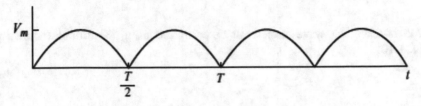

Fig. A.8 Full wave rectified sinusoidal waveform

(c) Graph of $v(t) = \begin{cases} V_m \sin(\omega t) & 0 < t < \dfrac{T}{2} \\ 0 & \dfrac{T}{2} < t < T \end{cases}$ is shown in Fig. A.9. Such a waveform is called Half Wave Rectified in power electronics.

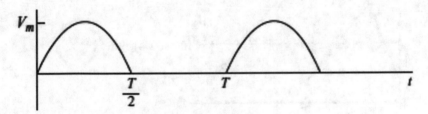

Fig. A.9 Half wave rectified sinusoidal waveform

$$V_{rms} = \sqrt{\frac{1}{T}\left(\int_0^{\frac{T}{2}}(V_m \sin(\omega t))^2 dt + \int_{\frac{T}{2}}^{T} 0 dt\right)} = \sqrt{\frac{1}{T} \times V_m^2 \int_0^{\frac{T}{2}} \sin^2(\omega t) dt}$$

$$= \sqrt{\frac{V_m^2}{T}\int_0^{\frac{T}{2}}\frac{1 - \cos(2\omega t)}{2}dt} = \sqrt{\frac{V_m^2}{T}\int_0^{\frac{T}{2}}\frac{1}{2}dt - \int_0^{\frac{T}{2}}\frac{\cos(2\omega t)}{2}dt}$$

$$= \sqrt{\frac{V_m^2}{T} \times \left(\frac{t}{2} - \frac{\sin(2\omega t)}{4\omega}\right)\Big|_0^{\frac{T}{2}}}$$

$$= \sqrt{\frac{V_m^2}{T} \times \frac{T}{4} - 0}$$

$$= \sqrt{\frac{V_m^2}{4}}$$

$$= \frac{V_m}{2}$$

RMS of triangular wave shapes can be calculated using the formulas shown in Fig. A.10.

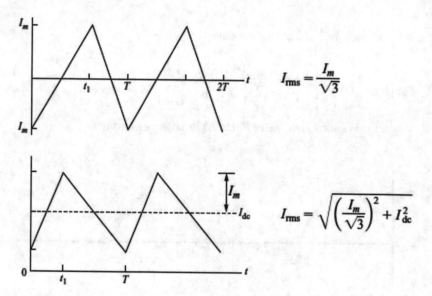

$$I_{rms} = \frac{I_m}{\sqrt{3}}$$

$$I_{rms} = \sqrt{\left(\frac{I_m}{\sqrt{3}}\right)^2 + I_{dc}^2}$$

Fig. A.10 RMS value of triangular waveforms

Effective Value of Sum of Two Periodic Signals

Consider two periodic waveforms, i.e., $v_1(t + T) = v_1(t)$, $v_2(t + T) = v_2(t)$. The RMS value of sum of two waveforms ($v(t) = v_1(t) + v_2(t)$) is:

$$
\begin{aligned}
V_{rms}^2 &= \frac{1}{T} \int_0^T (v_1 + v_2)^2 dt = \frac{1}{T} \int_0^T \left(v_1^2 + 2v_1 v_2 + v_2^2\right) dt \\
&= \frac{1}{T} \int_0^T v_1^2 dt + \frac{1}{T} \int_0^T 2v_1 v_2 dt + \frac{1}{T} \int_0^T v_2^2 dt
\end{aligned}
\tag{A.12}
$$

Sometime the $\frac{1}{T}\int_0^T v_1(t)v_2(t)dt$ term is zero. The $\frac{1}{T}\int_0^T v_1(t)v_2(t)dt$ is the iner product of $v_1(t)$ and $v_2(t)$. When $\frac{1}{T}\int_0^T v_1(t)v_2(t)dt = 0$, the signals $v_1(t)$ and $v_2(t)$ are called orthogonal. Table A.1 shows some of the important orthogonal functions.

Table A.1 Some of the important orthogonal functions ($\omega = \frac{2\pi}{T}$, $n \neq m$ and k is a constant)

No.	$v_1(t)$	$v_2(t)$
1	$\sin(n \times \omega \times t + \varphi_1)$	$\sin(m \times \omega \times t + \varphi_2)$
2	$\sin(n \times \omega \times t + \varphi_1)$	$\cos(m \times \omega \times t + \varphi_2)$
3	$\cos(n \times \omega \times t + \varphi_1)$	$\cos(m \times \omega \times t + \varphi_2)$
4	$\sin(n \times \omega \times t + \varphi_1)$	k
5	$\cos(m \times \omega \times t + \varphi_1)$	k

For instance according to the second row of the table, $\sin(n \times \omega \times t + \varphi_1)$ and $\cos(m \times \omega \times t + \varphi_2)$ (when $n \neq m$) are orthogonal since $\frac{1}{T}\int_0^T \sin(n\omega t + \varphi_1) \times \cos(m\omega t + \varphi_2)dt = 0$.

For orthogonal functions,

$$V_{rms}^2 = \frac{1}{T}\int_0^T (v_1 + v_2)^2 dt = \frac{1}{T}\int_0^T (v_1^2 + 2v_1v_2 + v_2^2)dt$$

$$V_{rms}^2 = \frac{1}{T}\int_0^T v_1^2 dt + \frac{1}{T}\int_0^T 2v_1v_2 dt + \frac{1}{T}\int_0^T v_2^2 dt$$

$$V_{rms}^2 = \frac{1}{T}\int_0^T v_1^2 dt + \frac{1}{T}\int_0^T v_2^2 dt$$

$$V_{rms} = \sqrt{V_{1,rms}^2 + V_{2,rms}^2} \qquad (A.13)$$

RMS value of sum of more than two orthogonal functions (each two terms are assumed to be orthogonal) can be calculated in the same way:

$$\left(v(t) = \sum_{n=1}^{N} v_n(t) \forall k, l\, 1 \leq k \leq N, o1 \leq l \leq N, ok \neq l, o\, \frac{1}{T}\int_0^T v_k(t)v_l(t)dt = 0 \right) \Rightarrow$$

$$V_{rms} = \sqrt{V_{1,rms}^2 + V_{2,rms}^2 + V_{3,rms}^2 + \ldots} = \sqrt{\sum_{n=1}^{N} V_{n,rms}^2} \qquad (A.14)$$

Example A.3 Determine the RMS value of $v(t) = 4 + 8\sin(\omega_1 t + 10°) + 5\sin(\omega_2 t + 50°)$ under the following conditions.

(a) $\omega_2 = 2\omega_1$
(b) $\omega_2 = \omega_1$

Solution:

(a) When $\omega_2 = 2\omega_1$, the $v(t) = 4 + 8 \sin(\omega_1 t + 10°) + 5 \sin(2\omega_1 t + 50°)$. According to Table A.1, all the functions are orthogonal to each other, so

$$V_{rms} = \sqrt{V_{1,rms}^2 + V_{2,rms}^2 + V_{3,rms}^2} = \sqrt{4^2 + \left(\frac{8}{\sqrt{2}}\right)^2 + \left(\frac{5}{\sqrt{2}}\right)^2} = 7.78 \text{ V}$$

(b) When $\omega_2 = \omega_1$, the $v(t) = 4 + 8 \sin(\omega_1 t + 10°) + 5 \sin(\omega_1 t + 50°)$. $8 \sin(\omega_1 t + 10°)$ and $5 \sin(\omega_1 t + 50°)$ are not orhtogonal to each other. So, we can't use the previous formullas. Note that $a \times \sin(\omega t) + b \times \cos(\omega t) = \sqrt{a^2 + b^2} \sin\left(\omega t + \tan^{-1}\left(\frac{b}{a}\right)\right)$. So,

$$v(t) = 4 + 8 \sin(\omega_1 t + 10°) + 5 \sin(\omega_1 t + 50°)$$

$$= 4 + 12.3 \sin(\omega_1 t + 25.2°)$$

The two terms of last equation are orthogonal to each other (See Table A.1). So, the RMS is

$$V_{rms} = \sqrt{4^2 + \left(\frac{12.3}{\sqrt{2}}\right)^2} = 9.57 \text{ V}$$

Example A.4 In this example we show how RMS values can be calculated with the aid of MATLAB®. Assume

$v(t) = 311 \sin(2\pi \times 60t) + 100 \sin(2\pi \times 2 \times 60t) + 20 \sin(2\pi \times 3 \times 60t)$ is given. The RMS can be calculated easily:

$$V_{rms} = \sqrt{\left(\frac{311}{\sqrt{2}}\right)^2 + \left(\frac{100}{\sqrt{2}}\right)^2 + \left(\frac{20}{\sqrt{2}}\right)^2} = 231.43 \text{ V}$$

The commands shown in Fig. A.11 calculates the RMS value of given signal. The first two lines sample a period of given signal. The sampling time is $\frac{1}{6000} = 166.7\mu s$. The rms command is used to calculate the RMS value of sampled signal.

```
Command Window                                                              ⊙
  >> t=0:1/6000:1/60;
  >> v=311*sin(2*pi*60*t)+100*sin(2*pi*2*60*t)+20*sin(2*pi*3*60*t);
  >> rms(v)

  ans =

     230.2829

fx >> |
```

Fig. A.11 Calculation of RMS value of $v(t) = 311 \sin(2\pi \times 60t) + 100 \sin(2\pi \times 2 \times 60t) + 20 \sin(2\pi \times 3 \times 60t)$ with $\frac{1}{6000}$ steps

The result is 230.283 which is a little bit lower than the expected value of 231.43. if you decrease the sampling time from $166.7\mu s$ to $16.67\mu s$ you get a more accurate result (Fig. A.12).

```
Command Window                                                              ⊙
  >> t=0:1/60000:1/60;
  >> v=311*sin(2*pi*60*t)+100*sin(2*pi*2*60*t)+20*sin(2*pi*3*60*t);
  >> rms(v)

  ans =

     231.3158

fx >> |
```

Fig. A.12 Calculation of RMS value of $v(t) = 311 \sin(2\pi \times 60t) + 100 \sin(2\pi \times 2 \times 60t) + 20 \sin(2\pi \times 3 \times 60t)$ with $\frac{1}{60000}$ steps

Measurement of RMS of Signals

The cheap multimeters are not suitable devices to measure the RMS value of signals inside a power electronics converters. The cheap multimeters are able to measure the RMS value of pure sinusoidal signals, i.e. the one shown in Fig. A.13.

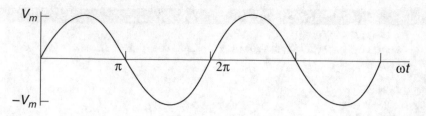

Fig. A.13 Pure sinusoidal waveform

Figure A.14 shows one of the methods that cheap multimeter uses measure the RMS of a signal. V_X is the signal under measurement. Assume that V_X is a pure sinusoidal waveform, i.e. a signal such as the one shown in Fig. A.13. Then the capacitor is charged up to Vm Volts (voltage drop of diode is neglected) where Vm is the peak value of voltage under measurement. So, Analog-to-Digital converter reads the maximum of input signal. The read value is simply multiplied by $\frac{1}{\sqrt{2}}$, and the result, i.e., $\frac{Vm}{\sqrt{2}}$, is the RMS value of input signal. This method only works for pure sinusoidal signals and doesn't produce correct result if the input signal is not pure sinusoidal.

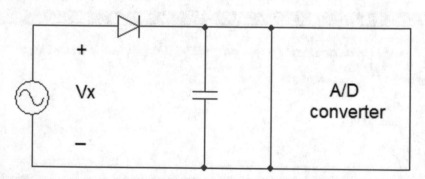

Fig. A.14 A simple circuit for detection of input AC signal peak value

The expensive multimeters samples the input waveform and uses a processor to calculate the RMS value. So, the wave shape of input signal doesn't affect the measurements. Such a multimeter has "TRUE RMS" label on it. So, ensure that your multimeter is TRUE RMS type if you want to measure the RMS of a signal of a power electronics converters. Digital oscilloscopes can be used to measure the RMS of signals as well.

Apparent Power and Power Factor

Apparent power (S) is the product of RMS of voltage and RMS of current magnitudes.

$$S = V_{rms} \times I_{rms} \tag{A.15}$$

The power factor of a load is defined as the ratio of average power to apparent power:

$$\text{pf} = \frac{P}{S} = \frac{P}{V_{rms}I_{rms}} \tag{A.16}$$

The above equation can be used to analyse both the linear circuits and non-linear circuits. In the linear circuit case $PF = cos\ (\alpha)$ where α shows the phase angle between the voltage and current sinusoids.

Power Computations for Linear Circuits

The steady state voltages and currents of a linear circuit which has sinusoidal AC sources are sinusoidal. Assume an element with the following voltage and current,

$$v(t) = V_m \cos(\omega t + \theta)$$
$$i(t) = I_m \cos(\omega t + \varphi) \tag{A.17}$$

Then the instantaneous power is:

$$p(t) = v(t)i(t) = [V_m \cos(\omega t + \theta)][I_m \cos(\omega t + \varphi)] \tag{A.18}$$

According to basic trigonometric identities:

$$(\cos A)(\cos B) = \frac{1}{2}[\cos(A + B) + \cos(A - B)] \tag{A.19}$$

So, instantaneous power can be written as:

$$p(t) = \left(\frac{V_m I_m}{2}\right)[\cos(2\omega t + \theta + \varphi) + \cos(\theta - \varphi)] \tag{A.20}$$

The average power can be calculated easily:

$$p(t) = \frac{1}{T} \int_0^T p(t)dt = \left(\frac{V_m I_m}{2}\right) \int_0^T [\cos(2\omega t + \theta + \varphi) + \cos(\theta - \varphi)]dt$$

$$= \left(\frac{V_m I_m}{2}\right) \cos(\theta - \varphi)$$

$$= V_{rms} I_{rms} \cos(\theta - \varphi)$$

(A.21)

So, the power factor of circuit is $\frac{V_{rms} I_{rms} \cos(\theta - \varphi)}{V_{rms} I_{rms}} = \cos(\theta - \varphi)$. The average power (measured with units of Watts, W) is the part of power which is consumed by the resistors in the circuit. In the steady state, no net power is absorbed by an inductor or a capacitor. The term reactive power (measured with units of Volt-Amper Reactive, VAR) is commonly used in conjunction with voltages and currents for inductors and capacitors. Reactive power is characterized by energy storage during one-half of the cycle and energy retrieval during the other half. Reactive power (Q) is calculated as:

$$Q = V_{rms} I_{rms} \sin(\theta - \varphi)$$

(A.22)

By convention, inductors absorb positive reactive power and capacitors absorb negative reactive power.

Complex Power (measured with units of Volt Amper, VA) is defined as $(j = \sqrt{-1})$:

$$S = P + jQ$$

(A.23)

Apparent power is the magnitude of complex power:

$$S = |S| = \sqrt{P^2 + Q^2}$$

(A.24)

Example A.5 In the following circuit (Fig. A.15), $v_1(t) = 311 \sin(2 \times \pi \times 50 \times t)$, $L = 0.1\,H$ ve $R = 40\,\Omega$. Determine the apparent power, average (active) power, reactive power and power factor.

Fig. A.15 Circuit of Example A.5

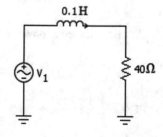

Solution:

$$Z = R + j \times L \times \omega = 40 + 31.415j$$

$$\varphi = \tan^{-1}\left(\frac{L\omega}{R}\right) = \tan^{-1}\left(\frac{31.415}{40}\right) = 38.14° = 0.666 \text{ Rad}$$

$$V = \frac{311}{\sqrt{2}} < 0° = 219.92 \angle 0°$$

$$I = \frac{V}{Z} = \frac{219.92e^{j0}}{40 + 31.42j} = 3.4 - 2.67j = 4.323e^{-0.666j}$$

$$S = |V \times I| = 950.824 \text{ VA}$$

$$P = V \times I \times \cos(\varphi) = 747.63 \text{ W}$$

$$Q = V \times I \times \sin(\varphi) = 587.46 \text{ VAR}$$

$$PF = \cos(\varphi) = 0.786$$

Fourier Series

A periodic and non-sinusoidal signal $f(t)$ that satisfy certain conditions (Dirichlet conditions) can be written as the sum of sinusoids. The Fourier series of $f(t) = f(t + T)$ can be written as ($\omega_0 = \frac{2\pi}{T}$):

$$f(t) = a_0 + \sum_{n=1}^{\infty} [a_n \cos(n\omega_0 t) + b_n \sin(n\omega_0 t)] \tag{A.25}$$

where a_0, a_n and b_n are,

$$a_0 = \frac{1}{T} \int_{-\frac{T}{2}}^{\frac{T}{2}} f(t)dt$$

$$a_n = \frac{2}{T} \int_{-\frac{T}{2}}^{\frac{T}{2}} f(t)\cos{(n\omega_0 t)}dt \tag{A.26}$$

$$b_n = \frac{2}{T} \int_{-\frac{T}{2}}^{\frac{T}{2}} f(t)\sin{(n\omega_0 t)}dt$$

The $a_0 = \frac{1}{T}\int_{-\frac{T}{2}}^{\frac{T}{2}} f(t)dt$ is called the average value of $f(t)$. The above equations can be written in the following forms as well (remember that $a \times \sin{(\omega t)} + b \times \cos{(\omega t)} = \sqrt{a^2 + b^2}\sin{\left(\omega t + \tan^{-1}\left(\frac{b}{a}\right)\right)}$).

(A) Sum of sines

$$f(t) = a_0 + \sum_{n=1}^{\infty} C_n \sin{(n\omega_0 t + \theta_n)}$$

$$C_n = \sqrt{a_n^2 + b_n^2} \text{ and } \theta_n = \tan^{-1}\left(\frac{a_n}{b_n}\right) \tag{A.27}$$

(B) Sum of cosines

$$f(t) = a_0 + \sum_{n=1}^{\infty} C_n \cos{(n\omega_0 t + \theta_n)}$$

$$C_n = \sqrt{a_n^2 + b_n^2} \text{ and } \theta_n = \tan^{-1}\left(-\frac{b_n}{a_n}\right) \tag{A.28}$$

The following equation can be used to determine the RMS value of a signal using its Fourier series coefficients.

$$F_{rms} = \sqrt{\sum_{n=0}^{\infty} F_{n,rms}^2} = \sqrt{a_0^2 + \sum_{n=1}^{\infty} \left(\frac{C_n}{\sqrt{2}}\right)^2}$$

$$= \sqrt{a_0^2 + \sum_{n=1}^{\infty} \left(\frac{a_n^2 + b_n^2}{2}\right)} \tag{A.29}$$

Fourier Series of Important Wave Shapes

Fourier series of important wave shapes are shown in Figs. A.16, A.17, A.18, A.19, and A.20.

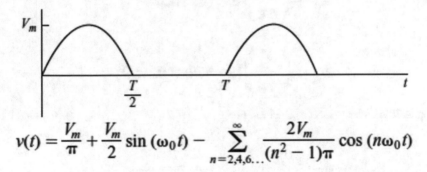

$$v(t) = \frac{V_m}{\pi} + \frac{V_m}{2} \sin(\omega_0 t) - \sum_{n=2,4,6\ldots}^{\infty} \frac{2V_m}{(n^2 - 1)\pi} \cos(n\omega_0 t)$$

Fig. A.16 Fourier series of a half wave rectified waveform

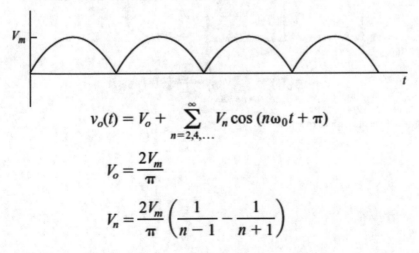

$$v_o(t) = V_o + \sum_{n=2,4,\ldots}^{\infty} V_n \cos(n\omega_0 t + \pi)$$

$$V_o = \frac{2V_m}{\pi}$$

$$V_n = \frac{2V_m}{\pi} \left(\frac{1}{n-1} - \frac{1}{n+1}\right)$$

Fig. A.17 Fourier series of a full wave rectified waveform

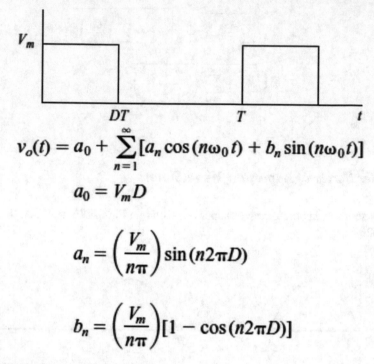

$$v_o(t) = a_0 + \sum_{n=1}^{\infty} [a_n \cos(n\omega_0 t) + b_n \sin(n\omega_0 t)]$$

$$a_0 = V_m D$$

$$a_n = \left(\frac{V_m}{n\pi}\right) \sin(n2\pi D)$$

$$b_n = \left(\frac{V_m}{n\pi}\right) [1 - \cos(n2\pi D)]$$

Fig. A.18 Fourier series of a pulsed waveform

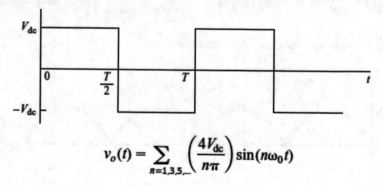

$$v_o(t) = \sum_{n=1,3,5,\dots} \left(\frac{4V_{dc}}{n\pi}\right) \sin(n\omega_0 t)$$

Fig. A.19 Fourier series of a square wave

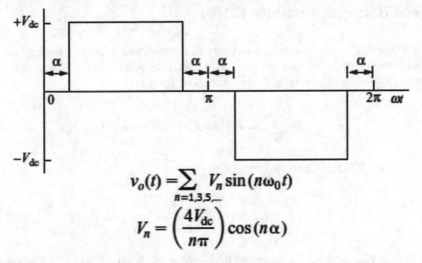

$$v_o(t) = \sum_{n=1,3,5,\dots} V_n \sin(n\omega_0 t)$$

$$V_n = \left(\frac{4V_{dc}}{n\pi}\right) \cos(n\alpha)$$

Fig. A.20 Fourier series of a modified square wave

Calculation of Average Power Using the Fourier Series

Assume that the Fourier series of voltage and current of an element is given as follows

$$v(t) = V_0 + \sum_{n=1}^{\infty} V_n \cos(n\omega_0 t + \theta_n)$$

$$i(t) = I_0 + \sum_{n=1}^{\infty} I_n \cos(n\omega_0 t + \varphi_n)$$

(A.30)

Then the average power (i.e., $\frac{1}{T}\int_0^T v(t)i(t)dt$) can be calculated as:

$$P = \sum_{n=0}^{\infty} P_n = V_0 I_0 + \sum_{n=1}^{\infty} V_{n,rms} I_{n,rms} \cos(\theta_n - \varphi_n)$$

(A.31)

or

$$P = V_0 I_0 + \sum_{n=1}^{\infty} \frac{V_{n,max} I_{n,max}}{2} \cos(\theta_n - \varphi_n)$$

(A.32)

For instance the average power for $v(t) = 10 + 20\cos(2\pi \times 60t) + 30\cos(4\pi \times 60t + 30°)$ and $i(t) = 2 + 2.65 \cos(2\pi \times 60t - 48.5°) + 2.43 \cos(4\pi \times 60t - 36.2°)$ is 52.2 W.

Total Harmonic Distortion (THD)

THD quantify the non-sinusoidal property of a waveform. THD is often applied in situations where the dc term is zero. Assume that the Fourier series of the signal is given ($f(t)$ can be either voltage or current waveform):

$$f(t) = \sum_{n=1}^{\infty} [a_n \cos (n\omega_0 t) + b_n \sin (n\omega_0 t)] \qquad (A.33)$$

Then the THD of signal is defined as:

$$THD = \sqrt{\frac{F_{rms}^2 - F_{1,rms}^2}{F_{1,rms}^2}} \qquad (A.34)$$

Where F_{rms} and $F_{1, rms}$ show the RMS value of signal $f(t)$ and RMS value of fundamental harmonic of $f(t)$ (note that $F_{rms} = \sqrt{\sum_{n=1}^{\infty} \frac{(a_n^2 + b_n^2)}{2}}$ and $F_{1,rms} = \sqrt{\frac{(a_1^2 + b_1^2)}{2}}$). For instance for a current waveform of i-$(t) = 4 \sin (\omega_0 t) + 1.5 \sin (3\omega_0 t) + 0.64 \sin (5\omega_0 t)$, the THD is $\sqrt{\frac{3.0545^2 - 2.829^2}{2.829^2}} = 0.408$. It is quite common to express the THD in percentage, so the THD for aforementioned current waveform is 40.8%.

Example A.6 Determine the THD of $v(t) = 311 \sin (2\pi \times 60t) + 100 \sin (2\pi \times 2 \times 60t) + 20 \sin (2\pi \times 3 \times 60t)$.

Solution

RMS of given waveforms is $V_{rms} = \sqrt{\left(\frac{311}{\sqrt{2}}\right)^2 + \left(\frac{100}{\sqrt{2}}\right)^2 + \left(\frac{20}{\sqrt{2}}\right)^2} = 231.43 \ V$. Peak value of fundamental harmonic is 311 V. So, the RMS value of fundamental harmonic is $V_{1,rms} = \frac{311}{\sqrt{2}} = 219.91 \ V$. Finally, the THD is:

$$THD = \sqrt{\frac{V_{rms}^2 - V_{1,rms}^2}{V_{1,rms}^2}} = \sqrt{\frac{231.43^2 - 219.91^2}{219.91^2}} = 0.33 \text{ or } 33\%$$

Example A.7 Determine the THD for the given voltage waveform

$$v(t) = \begin{cases} -100 & -1ms < t < 0 \\ +100 & 0 < t < 1ms \end{cases}$$

Solution

The graph of one period of given waveform is shown in Fig. A.21.
Fourier series of $v(t)$ is:

$$v(t) = \sum_{n=1,3,5}^{\infty} \frac{2 \times 200}{n\pi} \sin(n\omega_0 t) \tag{A.35}$$

RMS value of $v(t)$ is:

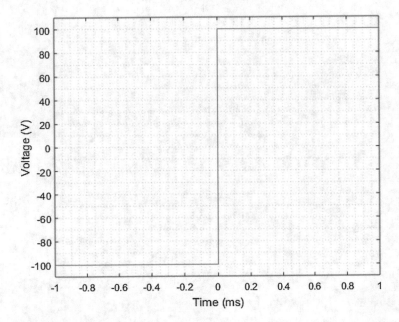

Fig. A.21 Graph of given $v(t)$

$$V_{rms} = \sqrt{\frac{1}{T} \int_0^T v(t)^2 dt} = \sqrt{\frac{1}{2m} \int_{-1m}^0 (-100)^2 dt + \frac{1}{2m} \int_0^{1m} (100)^2 dt}$$

$$= 100\ V \tag{A.36}$$

Fundamental harmonic (first harmonic) of $v(t)$ has the peak value of $V_1 = \frac{2 \times 200}{\pi} = 127.324\ V$. So the RMS value of fundamental harmonic is $V_{1,rms} = \frac{V_1}{\sqrt{2}} = \frac{127.324}{\sqrt{2}} = 90.0325\ V$. Finally, the THD is:

$$THD = \sqrt{\frac{V_{rms}^2 - V_{1,rms}^2}{V_{1,rms}^2}} = \sqrt{\frac{100^2 - 90.0325^2}{90.0325^2}} = 0.4834 \text{ or } 48.34\% \quad (A.37)$$

Index

© The Editor(s) (if applicable) and The Author(s), under exclusive license to
Springer Nature Switzerland AG 2022
F. Asadi, *Essential Circuit Analysis using NI Multisim™ and MATLAB®*,
https://doi.org/10.1007/978-3-030-89850-2